68000 Assembly Language
Programming and Interfacing

68000 Assembly Language Programming and Interfacing:

a unique approach
for the beginner

AMBROSE "BO" BARRY, P.E.

 REGENTS/PRENTICE HALL Englewood Cliffs, New Jersey 07632

Library of Congress Cataloging-in-Publication Data

Barry, Ambrose.
   68000 assembly language, programming and interfacing : a unique
approach for the beginner / Ambrose ''Bo'' Barry.
      p.   cm.
   Includes index.
   ISBN 0-13-805680-3
   1. Motorola 68000 (Microprocessor)—Programming.   2. Computer
interfaces. I. Title.
QA76.8.M6895B37   1992   91-31309
005.265—dc20   CIP

Editorial/production supervision and
   interior design: **Lillian Goode**
Cover design: **Wanda Lubelska**
Manufacturing buyers: **Ilene Levy/Ed O'Dougherty**
Acquisitions Editor: **Holly Hodder**
Editorial Assistant: **Cathy Frank**

Dedicated to my lovely wife, Hilda
and to my students who have persevered
and encouraged for the past three years.
                                    'Bo' Barry

 © 1992 by **REGENTS/PRENTICE HALL**
A Simon & Schuster Company
Englewood Cliffs, New Jersey 07632

Printed in the United States of America

10   9   8   7   6   5   4   3   2   1

ISBN 0-13-805680-3

Prentice-Hall International (UK) Limited, *London*
Prentice-Hall of Australia Pty. Limited, *Sydney*
Prentice-Hall Canada Inc., *Toronto*
Prentice-Hall Hispanoamericans, S.A., *Mexico*
Prentice-Hall of India Private Limited, *New Delhi*
Prentice-Hall of Japan, Inc., *Tokyo*
Simon & Schuster Asia Pte. Ltd., *Singapore*
Editors Prentice-Hall do Brasil, Ltds., *Rio de Janeiro*

contents

preface

About This Book

This book is about the **Motorola 68000** microprocessor, with emphasis on its programming and interfacing. It presents the material, thru an informal style, by introducing a series of examples, each of increasing complexity. The purpose of the book is to provide a background in both the software programming and hardware interfacing of a typical and powerful microprocessor.

The unique aspect of this book is that it is NOT another "Programming Manual", with nothing but pages upon pages of explanations of the operations of each and every instruction, with scant or meaningless examples.

Intended Audience Of This Book

This book is intended as an **introductory** book for anyone interested in computer programming and/or interfacing. It is ideally suited for a curriculum in two year engineering technology programs. Its goal is to introduce the material in such a way as to replace the Motorola 6800 which is STILL being taught to beginners. The book is NOT a complete treatise on the 68000 and is not a computer DESIGN text.

A 16/32 bit microprocessor was chosen as the entry level rather than the conventional 8 bit devices which are still deeply entrenched in industrial applications. The 68000, despite its complexity and power, is a 'clean' device to learn, and provides a good starting point for future developments in this rapidly changing area. For applications where an 8 or 4 bit device is more aptly suited, the 'backward' learning process is much easier after coverage of the 68000. While increasingly sophisticated industrial developments are best suited by a 32 or 16 bit μP, MANY and probably MOST applications will ALWAYS be best suited for an 8 (or 4) bit μP. This fact is often the driving force on what type of microprocessor is taught in the college environment. However, mention of only 8 bit experience in one's resume is looked at BY MANY as an indication of obsolescence.

The study of microprocessor architecture, assembly language programming, and interfacing techniques is an integral part of electrical and computer engineering technology curricula. This field of study does not have to be limited to the electrical/computer-inclined, however. The microprocessor is a powerful and fascinating device and can be conquered by anyone with an inquisitive mind (or an industrial need). Creativity and common sense assist the learning process as well. Little or no computer or digital logic background is needed.

How To Use This Book

This book is a combination textbook/laboratory manual, providing material for both classroom instruction and individual laboratory experimentation. The first 5 chapters are intended for either a review, an initial brief survey, or later reference. Work actually starts with Chapter 6, with introduction of the first example. Instructions and techniques are introduced in chapters 6 thru 9 as they are needed in order to complete the examples. A complete alphabetical listing of the 68000's instructions is included in the last chapter, along with details on actual execution of a portion of the examples on the three hardware and software sytems mentioned below.

Required Support Material

You should use the book with one of the following:
1. The PseudoCorp **PseudoMax** and **PseudoSam** cross assembler and simulator programs. (A student version is available from Prentice Hall).
2. The Motorola 68000 Educational Computer Board **(ECB).**
3. Micro Board Designs **MAX 68000** microcomputer system.

In addition, a MS-DOS computer is needed in order to utilize the PseudoCorp software and to connect to the two hardware systems. Alternatively, another type of personal computer could be used with a compatible cross assembler/simulator and a terminal program to connect to the two hardware systems.

Acknowledgments

My special thanks to the Engineering Technology students at The University of North Carolina, Charlotte that were involved with the examples used in this book. Their enthusiasm was greatly enhanced by the promise to mention them by name in the examples. The enthusiasm (and perseverance) of the students that were used as guinea pigs during development of the text is also appreciated.

Ambrose ''Bo'' Barry

ABOUT THE AUTHOR

Ambrose 'Bo' Barry first became involved with microprocessors in 1975, creating and teaching Chapman College's first microprocessor course (MC6800).

Before moving to North Carolina in 1986 to teach at the University of North Carolina, Charlotte, he taught calculus courses for the Florida Institute of Technology; math, electronics, and computer courses for Chapman College (Holloman AFB, NM); and math, electronics, and computer courses as an Assistant Professor at New Mexico State University. Consulting projects involved designs using the 6800, the Zilog Z-80, and the Intel 8052AH.

Bo received his Masters Degree in Electrical Engineering from Auburn University in 1971 and his Bachelors Degree in Electrical Engineering from Arizona State University in 1966. He is a licensed Professional Engineer and has both FCC Commercial (First Class) and Amateur Radio (Extra Class) licenses, having been licensed as W4GHV since 1954. He is a member of the ARRL and ASEE.

Bo is past editor/author of two computer newsletters: SanPic—for Sanyo computer users; and QZX—for amateur radio/Timex computer enthusiasts.

He can be contacted at 8118 Kensington Lane, Harrisburg, North Carolina 28075. Telephone (704) 547-4185 (work), or (704) 455-6153 (home). He will gladly provide MS-DOS disks containing the following shareware upon receipt of a disk (3.5/5.25″—any density), disk mailer, and return postage: (1) **Automenu** menuing program; (2) **Flowcharter** flowcharting program; (3) **EDRAW** schematic drawing program; (4) **QEDIT** editor program; (5) A terminal program; and (6) A disk (for instructors) with the lengthy and incomplete text examples.

PseudoCorp Software Development Tools

PseudoSam Cross-Assemblers

Our line of macro cross-assemblers are easy to use and full featured, including conditional assembly and unlimited include files. Available for most processors, these assemblers provide the sophistication and performance expected of a much higher priced package.

PseudoMax Simulators

Don't wait until the hardware is finished to debug your software. Our simulators can test your program logic on your PC before the target hardware is built.

PseudoSid Disassemblers

A minor glitch has shown up in the firmware, and you can't find the original source program. Our line of disassemblers can help you re-create the original assembly language source.

DeveloperPack

Buy our developer package and the next time your boss says "Get to work!", you'll be ready for anything. Each DeveloperPack contains a cross-assembler, simulator, and dis-assembler for a particular microprocessor.

Broad Range of Support

Intel 8748	RCA 1802,05	Intel 8051,31	Intel 8096
Motorola 6800	Motorola 6801	Motorola 68HC11	Motorola 6805
Hitachi 6301	Motorola 6809	MOS Tech. 6502	WDC 65C02
Rockwell 65C02	Intel 8080,85	Zilog Z80	NSC 800
Hitachi HD64180	Motorola 68000,8	Motorola 68010	Intel 80c196
Zilog Super 8	Zilog z8	Intel 8041	

and more in development.

To Be Released in 1992

C Cross-Compilers
Advanced Simulators
Linking Assemblers
Pascal Cross-Compilers

- -

PseudoCorp Software

For more information, send your name and address to:
PseudoCorp
Professional Development Products Group
716 Thimble Shoals Blvd., Suite E
Newport News, VA 23606

chapter 1

introduction to digital computers

1.1 APPLICATIONS FOR COMPUTERS

To get the setting established, we should try to visualize the various classifications for computers. Not only is this often a difficult task, the distinction between the classes is rapidly becoming more gray. We can classify them as:

1. A mainframe computer, a minicomputer, or a microcomputer
2. A 4 bit, 8 bit, 16 bit, 32 bit, or 64 bit computer
3. A business computer, a scientific computer, a home computer, or an industrial computer

A beginner would probably classify them into the mainframe, mini, or micro categories simply by observation of the size of the computer cabinet or by inspection of the purchase invoices. A professional, of course, is not worried about these two characteristics, but would classify them according to the number of instructions per second that can be achieved, the amount of memory that can be addressed, and other complex characteristics.

Rather than explore the mainframe and minicomputer fields, we will concentrate on a microcomputer. However, keep in mind that rumors indicate that a fast version of the **MICROcomputer (μC)** we are about to study can outperform some of the popular MINI-computers. As to the number of bits in our μC, let's say that it is a 16/32- bit computer, with an explanation in due time. Finally, our main concern here will be with industrial computers, those used to monitor and control industrial processes. Programming techniques covered here will, however, be useful for other potential applications of the Motorola 68000, 68020, 68030, and 68040. Members of this family of integrated circuits (ICs) are used in personal computers such as the Apple Macintosh, Atari ST, Sinclair QL, Commodore Amiga, and NeXT computers.

Considering only the class of computers known as *microcomputer systems,* a likely question would be: What are they used for? Here is a list found recently in a sales brochure, which represents only a minuscule portion of the total applications of a μC: CRT terminals, smart credit cards, telephones, blood analyzers, panel meters, radar detectors, brake systems, electric meters, water meters, digital multimeters, PBX systems, VCRs, talking toys, life-support monitors, automotive fan controls, automotive flashers and chimes, postage meters, radios, blenders, sewing machines, robots, microwave ovens, cameras, washers, dryers, video games, telephone systems, laser printers, pagers, entertainment centers, sonobouys, weather balloons, cellular telephones, bar code sensors, personal computers, ceiling fans, fish finders, hand-held transceivers, modems, motors, engines, dishwashers, glucose monitors, printers, gasoline pumps, and cable TV converters.

In most of these applications, the system was designed or *built from the ground up,* starting with the **microprocessor (µP)** integrated circuit itself. System *design* requires extensive training and is not undertaken by a beginner. However, the ability to *program* a µP in a µC system and to *interface* a readily available predesigned µC development board to an industrial application is well within the capability of an industrious beginner. Proper selection of the input/output devices, such as lights, switches, and relays, and connection to the development board, are relatively straightforward.

1.2 SELECTION OF A COMPUTER TO SUIT AN APPLICATION

Many factors should be, but are not always, considered during the selection process of the best computer to accomplish a specific application.

Cost

If very large quantities of a computer system are to be sold, and profits are to be maximized, the smallest computer that would minimally accomplish the task will be selected. Four-bit microprocessors, invented in 1971, are still used today for MANY applications.

Flexibility

Many a computer neophyte has been shocked to find that a handful of ordinary integrated circuits requiring *NO PROGRAMMING*, but simply wired together in a specific manner, can handle repetitive, decision-making applications. In most cases this IC approach will run faster than a µC system, handling more input/output (I/O) situations per second. A good circuit designer can put one together much faster than a programmer can create the software in these cases. But what happens when the application changes, resulting in a new input/output situation? The custom-designed circuit is now useless, perhaps. On the other hand, if a µC system had been chosen, it could in most cases readily be **reprogrammed** to handle the new situation. So, flexibility, spelled "re-programmability," is one of the µC's main advantages.

The use of a microprocessor to solve a task shifts the design burden from hardware to software. In most cases, students prefer "building something," to "writing something." The hardware/software analogy is apparent here. To get the setting established, we realize that both hardware and software design take a bit of creativity. In hardware tasks we have to decide how to hook up the relatively "dumb" ICs to perform a task. If it does not work, the problem is usually an error in the wiring itself, but may be a result of faulty design. In software tasks we have to decide how to program a much more "intelligent" IC to perform the task. If it does not work, the problem is usually an error in the software instead of in the relatively simple wiring of the external circuitry we have added to our µC system. The shift in emphasis is distinct, and the goal of this book is to include sufficient "hardware breaks" to stimulate and hold interest.

Capability

Closely coupled with *FLEXIBILITY* is the overall *capability* of a selected µC system. Situations requiring more than simple mathematical and logical operations, and more important, those requiring manipulation of large amounts of data, will probably require the use of a µC system instead of a custom-designed hardware approach.

Once it is decided to use a µC system, the choice of a 4/8/16/32-bit computer comes next. For low-volume applications, the cost of the µC chip chosen is insignificant in comparison with the expenditures for software programming. The cost differences between an 8-bit and a 16- or 32-bit device are becoming smaller each day. The 8-bit manufacturers are putting a complete system on a single chip, causing the 8-bit device to remain popular.

When massive amounts of data manipulation are required, usually meaning that a lot of computer memory is needed, a 16- or 32-bit computer is the usual choice. Without special circuitry, a typical 8-bit device can access only 65,536 memory locations; 16- and 32-bit devices can access many more locations.

The historical trend from 4, to 8, to 16/32, to 32 and to 64 bits will at some point, of course, approach a point of diminishing returns. Specific tasks will need or produce data of various sizes, dictated by the application, not by the size of the available μP. As mentioned before, a 4-bit μP is still the ideal choice for some applications. Eight-bit systems will be the natural choice for some. Color graphics applications may optimally need 24-bit capability, allowing definition of over 16 million colors. Applications such as artificial intelligence will, of course, need the capability to address vast amounts of memory, with perhaps a 32-bit system best suiting the task. A bold prediction is that a plateau will occur at the 32-bit level, with enhanced features being introduced to 32-bit systems as time progresses.

Speed

The hardest capability to pin down accurately is the speed of the system—not just the speed of the μC chip itself, but the overall speed of the operation, which is related to the number of input/output devices connected, the size of the *numbers* to be input or output, and the criticality of timing between an input or an output. In other words:

1. How many switches and sensors are to be monitored, and how many devices are to be turned on/off?
2. How big is the *number* we will be manipulating? An unsigned integer number between 0 and 65,535 can be represented by a 16-bit binary number. How computers handle such numbers, including positive and negative, is discussed in Chapter 4.
3. How much time can we tolerate between the stimulus (input) and the reflex (output)?

Almost any size μC can typically respond in a matter of a few microseconds if programmed properly. As programming examples are developed, procedures to calculate the elapsed time for the operations will be covered. One technique used in the speed comparisons of various computer chips and systems is a measure of the millions of instructions per second (MIPS) that it can execute. A 68000 running at 8 MHz performs up to 0.6 MIPS; a 68020 running at 17 MHz performs up to 2.4 MIPS; a 68030 is capable of up to 30 MIPS; and the latest, the 68040, can perform as fast as 40 MIPS.

When MIPS comparisons are discussed, two different concepts in processing are brought up. The 68000 is what is called a *Complex Instruction Set Computer* (CISC). It has powerful instructions useful for carrying out procedures used by high-level languages. While the basic instruction set and addressing modes are straightforward to the assembly language programmer, the 68000 has several THOUSAND instructions that it knows how to perform. However, each instruction requires several clock cycles to execute.

Another concept in processing uses what is called a *Reduced Instruction Set Computer* (RISC). They have a very small instruction set, perhaps as few as 40, but can execute each one in a SINGLE clock cycle. Again the elapsed time of an instruction is the number of required clock cycles divided by the processor's crystal frequency. One clock cycle at 1 MHz takes 1 microsecond (μs), and so on. The RISC IC does not contain the conventional *micro code* of a CISC. In other words, it is a super-fast collection of *ordinary* ICs (latches, gates, etc.) hardwired to perform a small number of operations.

As you probably guessed, when using a RISC the burden is placed on the programmer, requiring a more extensive software effort. The choice between a CISC and RISC is not clear and depends on several variables. The application, the amount of software time required, and the execution speed of the resultant program must all be taken into consideration. The decision will be further muddled by introduction of processors that have characteristics of both a CISC and a RISC! Motorola has introduced a RISC IC, the MC88100, said to have a speed of 10 MIPS.

Expansion

When one takes future expansion capability into serious consideration, a 16 or 32-bit device will probably be chosen.

1.3 SELECTION OF A PROGRAMMING LANGUAGE TO SUIT AN APPLICATION

To use a computer to solve a problem, the algorithms, which you have probably written in English, will have to be "translated" into a language understandable by a computer. As you probably know, computers really know ONLY 0's and 1's, which comprise the instructions and data that make up a computer program. **An *algorithm* is a step-by-step description of the procedures used to define the steps needed to create a computer program.**

Many, many programming languages are available. Following are but a few of them.

1. *BASIC:* **B**eginner's **A**ll-purpose **S**ymbolic **I**nstruction **C**ode. BASIC is a relatively simple language to learn since it uses cryptic English words to describe its actions. Once considered strictly for beginners, today's BASIC is powerful, and when compiled into a machine language program, its primary drawback of slow speed is eliminated.

2. *COBOL:* **CO**mmon **B**usiness **O**riented **L**anguage. COBOL is used almost exclusively in the business and finance world to handle tasks such as payroll and accounting applications.

3. *FORTRAN:* **FOR**mula **TRAN**slation. FORTRAN is used in the scientific community to do applications requiring a lot of number crunching.

4. *Pascal:* Named after a French mathematician, Blaise **Pascal.**

   Next we hit the beginning of a dividing line between what are called **high-level** and **low-level** languages.

5. *C or C language* (a language created after B language). C is a highly popular (at present) language in the gray area between high- and low-level languages. Considered a high level by most, it retains the ability to do efficient low-level (assembly-level) input/output functions. Portions of the C program can consist of assembly language routines, allowing a combination of both.

6. *Assembly language:* This highly cryptic language is computer dependent but provides the highest execution speed and allows direct control of the input and output devices. Programs are written using abbreviations of instructions called **mnemonics.** They do not completely describe the instruction, they simply jog your memory, and a reference manual will be needed if the "jog" is not successful. The mnemonics, chosen by the μP chip manufacturer, are unique to that μP. A **JUMP** on the 68000 is a **JMP,** while it is a **JP** on the Z-80 μP. To create computer programs in assembly language, an **assembler program** is used. Another approach is to use another type of computer, such as a MS-DOS computer, and a **cross-assembler** *progam* to create assembly language programs for a μP such as the 68000.

7. *Machine language:* This inefficient language involves insertion of the instructions into the computer in their native form (1's and 0's) without the benefit of an assembler program. A primitive technique called *hand assembly,* still used in some intro-

ductory classes, is the use of hexadecimal keypads to insert the machine language instructions, which have been looked up in a reference manual one by one.

Since *assembly language programming* is the thrust of this book, a look at the advantages and disadvantages of this level of programming is in order. The assembler (or cross-assembler) takes the **source program** (written in mnemonics) and creates the required **object code** along with a **listing file** showing the mnemonics, the memory locations used, and the hexadecimal equivalent of each instruction. The cross-assembler used with this book runs on a MS-DOS personal computer and creates the object and listing files for the 68000. The object code is transferred to the 68000 system for execution. For brevity, the cross-assembler program will be referred to as an assembler.

Although some portability exists among the high-level languages, assembly language programs are highly nonportable. A 68000 program will not run on a Z-80, for instance, and even a program written on different types of computers using the same μP will not run without modifications on another. This is due to the differing memory and input/output organization.

High-level languages usually perform, in each line or statement, a recognizable function. The compiled code that is ultimately transferred to the μP and executed will have several or many machine instructions for each statement. Compilers for the high-level languages have characteristically been inefficient, creating code that was much larger in size than could be created if done at the assembly language level. As a result, the programs executed slower. On the positive side, high-level language programming is usually easier, allowing faster creation of programs for many applications.

Fortunately (and one of the reasons for this book), some applications are best suited for assembly language programming. When fast or unique input/output conditions are needed, such as a process control application, assembly language provides a more efficient solution.

An important trend that may decrease the popularity and usefulness of assembly language involves the use of a **C cross-compiler**. C language cross-compilers are becoming available that allow creation of C programs on a personal computer, for example, then selection of the target computer, such as the 68000. It then compiles the source program, written in C, creating the needed 68000 object code. It works identically to other types of compilers, but it is anticipated that more and more efficient cross-compilers will become available, making this approach more feasible for some applications. It is doubted that assembly language programming will die out, as has machine language programming, but more opportunities will exist for those familiar with both C and assembly language.

Ultimately, EVERY computer program is simply a set of binary numbers, referred to as a machine language or object program. The object code created in any fashion can be programmed into an erasable programmable read-only memory (EPROM) IC, installed in the dedicated μC system, and then run in a stand-alone control application.

In summary, assembly language programs are not very portable. The programs are not only ''chip'' dependent but often use specific capabilities of an individual computer brand. Assembly language programs perform their tasks much faster than those written in higher-level languages. Their main strengths lie in input and output operations, not in complex arithmetic operations (unless specialized math processing ICs are added). On the other hand, high-level languages are more portable, occupy a lot more memory than a corresponding assembly language program, and although they may perform at a satisfactory speed, the program in assembly language fashion would run much faster. Perhaps the main factor that is used in the selection process is: How many hours will it take to write the program? Wages have higher importance nowadays, since computer memory expansion seems to be unlimited, and their speed capabilities are continually increasing.

The saving grace for assembly language is that in most cases when a high-level language is preferred (or selected), a knowledge of assembly language proves invaluable during the debug phases, and when unique input/output operations are involved, or where specific routines need to be speeded up.

1.4 SOFTWARE VERSUS HARDWARE, AND VICE VERSA

The beginner is often perplexed between what is called **software** and **hardware**. Skipping the philosophical definitions often encountered, we easily identify **hardware** as the collection of integrated circuits, resistors, capacitors, and so on, that are used to perform the computing operation. Simply stated, **software** is the program used in the computer to perform an operation or function.

Being a bit more accurate defining **hardware:** It includes much more than just integrated circuits, resistors, and so on. Included is the conceptual design of the system or component needed for a specific task, the circuit design, the printed circuit board layout, component considerations, timing and external noise considerations, and more.

Being a bit more accurate defining **software:** It is more than just the computer program being used with an assortment of hardware. To create a program:

1. The problem has to be defined in terms of tasks to be preformed.
2. A logical solution using algorithms and the most appropriate computer language has to be developed.
3. The source program has to be written.
4. The program is then run, debugged, and improved until satisfaction is achieved.

Software can often be programmed into a read-only-memory (ROM) IC and then is sometimes called *firmware*. Some tasks that are done in software can be done with hardware: for example, blinking a light. A software program turns it on, delays for a period of time, and then turns it off, delays a period of time, and so on. This function can be replaced by an IC, allowing normal use of the computer during the delay routine. Not all functions can be replaced with hardware, however. To **design** hardware takes detailed training. To **interface** to existing hardware is much more straightforward.

To become proficient in software programming, we have three or four tasks.

1. We must learn each of the instructions available for whatever programming language we plan to use. We need to know their format (syntax) and their function (semantics) in detail. For example, **MOVE.L D0,D1** is an instruction whose **semantics** is to move the longword contents of Data Register D0 to Data Register D1. The **syntax** of the instruction is as shown, with the .L tacked on to denote a longword (a 32-bit number), with a space before D0 and with a comma between D0 and D1.
2. We should develop thought processes that help us create effective algorithms for solution of the programming task at hand. These algorithms are basically independent of the programming language used. We need to create a readable flowchart depicting these algorithms. A flawed algorithm will prevent us from achieving the correct solution to our problem.
3. We should develop as many *programming techniques* as possible. Structured programming is one such approach. We try to structure our program in such a fashion that it is readable by others and tends to flow from the start to the end without a variety of loops and branching. Longer programs or repetitive processes are broken into small subprograms or subroutines, which are called from our main program loop. A novice programmer will often succeed with a rambling program but will be unable to decipher it for modifications or explain its operation a few months after creating it.
4. If we are programming at the assembly level, and to a lesser extent if programming in a higher-level language, we need knowledge of the specific configuration of our computer hardware. Part of this is called the **architecture,** which describes the internal structure of the μP being used. Programming in BASIC, for example, does not require knowledge of this architecture or even of the specific μP being used. The goals of this book then are to:

1. Describe the 68000 instruction set, its architecture, and its memory addressing modes
2. Describe algorithm and flowchart development
3. Describe as many programming techniques as possible

4. Develop software programs utilizing goals 1, 2, and 3

5. Describe interfacing techniques to allow construction of the hardware needed together with a software program to satisfy a task

1.5 THE MOTOROLA FAMILY OF COMPUTERS

Motorola's first introduction into the computer arena was an 8-bit μP, the MC6800 in 1974. It was expremely popular, preferred by many since it bore a resemblance to minicomputers produced by Digital Equipment Corporation. Subsequently, an offshoot, the 6502, was developed by MOS Technology Corporation, and it became extremely popular, probably due to its lower cost. In the 8-bit field, the 6800 and 6502 were considered to be easier to program and faster than most of their competition. As for "visibility," most viewed only what was available in personal computers, which began to blossom in the late 1970s. However, some μPs, such as the 6800, were manufactured in much larger quantities to serve in industrial control applications, such as automobiles. Several other μPs, related to the 6800, were the 6802, 6808, and 6809.

As the demand for higher MIPS rates grew, the μPs were redesigned to increase their basic clock speed. When one manufacturer had a short-lived lead on another in terms of their μP's clock speed, an impressive speed comparison (or benchmark) article always appeared. Rarely were benchmarks written for two different μPs running at the same speed, leaving the conclusions up to the reader.

The next push was to expand the 8-bit μPs to 16 bits and higher. Being able to read and write 16 bits at a time brought a distinct increase in computing speed. Initially, the number of bits that could be accommodated was a function of the largest chip package available. In the Motorola case, the hardware side said to the software side: "We are giving you a maximum of 64 pins to work with." They promptly took two of them away for the + power connection and two for the − ground, leaving the software experts with 60 pins. Perhaps the software side was already a bit ahead of their hardware counterparts or perhaps had the foresight to think into the future. They developed the 68000 in 1979, which was basically a 32-bit μP on the INSIDE but limited to be a 16-bit computer on the OUTSIDE. It could hold data 32 bits in length, but only 16 bits could be accommodated by the pins on the IC package. It could hold memory addresses 32 bits in length, but only 24 bits could be accommodated by the package. Details are provided in subsequent chapters. In summary, the 68000 was advertised as a 16/32-bit computer.

As for compatibility with its 6800 predecessor, the 68000 required a complete hardware redesign and was not software compatible with the 6800. Since 16- or 32-bit peripheral ICs were not available at the time, the 68000 designers wisely included the capability of easily interfacing the 6800's 8-bit family of peripheral chips. They are still the practical choice today for many applications. Sixteen-bit I/O chips are also available today.

The 68000 continually goes through upgrades, with the 68020, 68030, and 68040 being introduced to date along with rumors of the 68050. Since they all remain upwardly compatible from a software standpoint, the 68000 remains to be the most appropriate choice for an introduction to microprocessor programming and applications.

1.6 TARGET COMPUTERS USED WITH THIS BOOK

To enhance the learning experiences with this book **three** different computer configurations are covered. (Details on their operation, purchase, and the like, are given in Appendix D.) The first configuration is done with software only, utilizing two programs created by Pseudo Corporation. The *PseudoSam Assembler Program* allows creation of programs that would be transferred to any 68000 for execution. The *PseudoMax 68K Simulator Program* will act as a 68000 system, eliminating the need to purchase hardware. If actual interfacing to external devices is needed, a working 68000 computer system would be

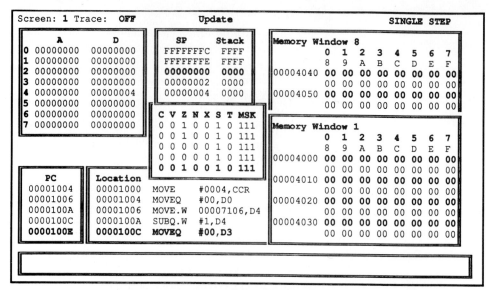

Figure 1-1   PseudoMax Simulator screen dump.

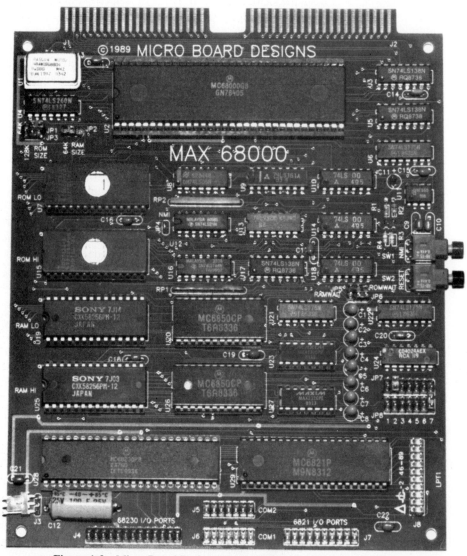

Figure 1-2   Micro Board Designs MAX 68000 microcomputer system.

needed, of course. A disk containing student versions of these programs is available. A typical screen dump from the simulator is shown in Figure 1-1. Its setup can be varied extensively.

Next, the *MAX 68000 Microcomputer System* by **Micro Board Designs** will be used. Programs will be created using the PseudoSam Assembler Program mentioned above, then transferred to the MAX 68000 for actual execution. A photograph of the economical and versatile MAX 68000 is shown in Figure 1-2.

Finally, the *Motorola MC68000 ECB* (**Educational Computer Board**) will be covered (see Figure 1-3). It was the first system available, offered at attractive prices by the Technical Training Department of Motorola, Inc.

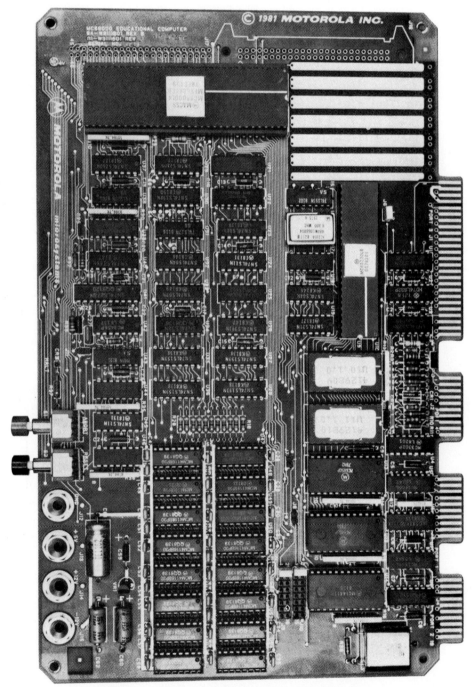

Figure 1-3 Motorola MC68000 ECB computer system.

1.7 OTHER MEMBERS OF THE 68XXX FAMILY

Since the introduction of the MC68000 in 1979, several other members of the family have been introduced.

1. 68008 (1982): an economical version of the 68000, having an 8-bit data bus. Internally identical to the 68000, it allows simple interfacing with 8-bit memory. It comes in a 48-pin package, limiting it to 20 address lines. This gives a memory addressing range between 00000_{16} and $FFFFF_{16}$, allowing a megabyte (1,048,576 locations) of memory. With minor hardware differences, the 68008 is completely software compatible with the 68000. It runs slower than the 68000 since all operations are done a byte at a time. Those interested in construction of a 68000 system from scratch should consider starting with the simpler 68008. A typical system is given in a later chapter.

2. 68010 (1982): an upgraded verison of the 68000 that supports a concept called **virtual memory,** which gives the programmer what appears to be unlimited memory access. It does this by swapping areas of memory to/from a hard disk whenever needed. Upwardly compatible, the 68010 contains additional registers to perform the virtual memory functions.

3. 68012 (1985): similar to the 68010, giving a gigabyte of addressing.

4. 68020 (1984): a true 32-bit μP, having a 32-bit data bus, 32-bit address bus, and a 32-bit arithmetic/logic unit (ALU). Its overall organization resembles the 68000 and it contains all of the 68000 instructions as a subset. It is *upwardly compatible,* meaning that any code written for a 68000 will run without modifications. It supports virtual memory and a very fast 256-byte internal memory system called **cache** that speeds up delivery of data or instructions to the μP. It has a different pin configuration than the 68000 and can address a memory range from 00000000_{16} to $FFFFFFFF_{16}$, a 4.3-gigabyte (4,294,967,295 locations) range. The 68020 runs all 68000 programs and has several new and enhanced instructions and addressing modes, the most obvious being improved multiply and divide instructions. The 32-bit multiply can provide 64-bit results, and 64-bit numbers can be used with the divide instruction. Until an official name is developed for a 64-bit number, we can call it a *double-longword* or a *quad-word,* and this data type can be handled by the 68020. It can easily be interfaced with the 68881 coprocessor for high-speed floating-point arithmetic.

5. 68030 (1987): upwardly compatible with its predecessors, the 68030 has two separate 256-byte caches, one for instructions and the other for data. Other sophisticated enhancements are incorporated, some of which allow multiple internal operations to be done simultaneously. The net effect of the enhancements is the higher rate at which instructions are executed.

6. 68HC000 (1986): a low-power-consumption version of the 68000, consuming only a fraction of the power of a 68000.

7. 68881 (1984): a specialized IC, called a *coprocessor,* that performs mathematical calculations of floating-point numbers. Used in conjunction with the 68020 or 68030 μP.

8. 68040 (1990): recently announced, again upwardly compatible with its predecessors, the 68040 is typically three times faster than the 68030.

chapter 2
a simplified look at the hardware
from a programmer's viewpoint

The hardware components described in this chapter are shown in Figure 2-1.

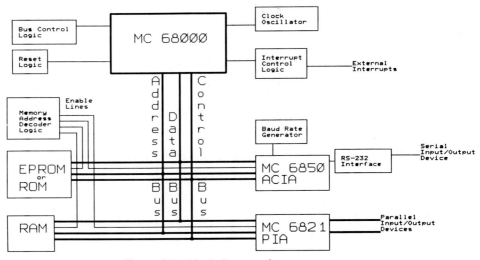

Figure 2-1   Block diagram of a computer system.

2.1 CENTRAL PROCESSING UNIT ARCHITECTURE

The Motorola 68000, introduced in 1979, was introduced when integrated circuit packaging was in transition. A 40-pin IC was considered big back then. The 68000 came out in a very unusual 64-pin design. Figure 2-2 shows the pins and their names, and Figure 2-3 is a photograph of the various packages of the 68000. A prefix of MC is often used when referring to the Motorola family of computer devices. Motorola also allows other companies to produce their devices, under their own name, and the MC prefix is usually dropped. We refer to them without the prefixes for simplicity.

One of the most innovative but no longer unique approaches to introducing the role of the microprocessor chip is to view it as a super-fast, super-efficient *postmaster.* We view the memory addresses as the mailing addresses, which are the mailboxes that we ''read from'' and ''write to.'' Each µP chip differs in its structure, called architecture, and each has a different set of instructions that it is capable of carrying out. Figure 2-4 is a

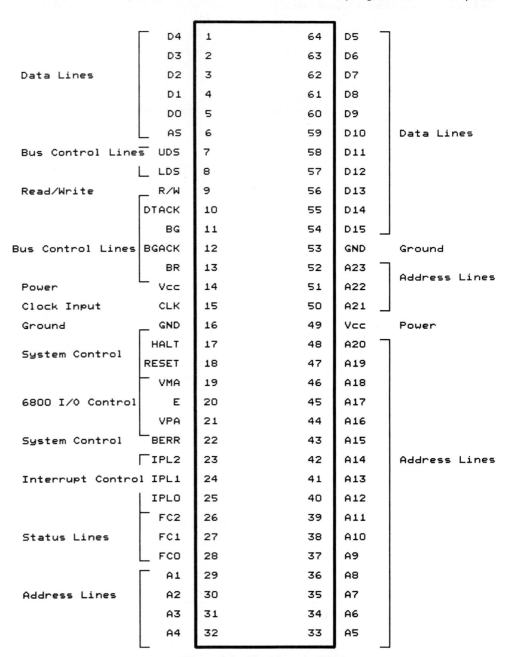

Figure 2-2   The Motorola 68000.

simplified sketch of what the 68000 *postmaster* has on its desk. Look at the registers, Program Counter, and Stack Pointers as individual *notepads,* handy for keeping track of numbers.

The ***Arithmetic/Logic Unit*** (**ALU**) is the *calculator*. With it, the 68000 can add, subtract, divide, and multiply. Square root, trigonometric functions, and other more complex functions are not available, and most 8-bit μPs do not even have the multiply and divide capability. The ALU can also do logical AND, OR, exclusive OR, and NOT operations. The ALU is used automatically by the postmaster; the programmer does not load or use it directly.

The ***Program Counter*** is used to keep track of the program execution **AND ALWAYS CONTAINS THE ADDRESS OF THE NEXT INSTRUCTION TO BE EXECUTED.**

Figure 2-3   Physical appearance of the 68000.

The **Data Registers** are temporary *scratchpads* used to hold data. Data can be stored in single 8-bit *(byte)*, 16-bit *(word)*, or 32-bit **(longword)** form. Operations can also be performed on individual bits.

The **Address Registers** are used to hold addresses used as **pointers** to the source or destination of data to be processed. Addresses are stored as either 16-bit words or 32-bit longwords.

The **Stack Pointers** point to regions of memory used to store temporary information.

The **Status Register** provides information about the results of instructions as they are executed. A zero result, for a mathematical operation, for instance, sets one of the bits in this register (to a "1"), which can be used to alter program execution.

Not shown in Figure 2-4 is the postmaster's "brain," which consists of several THOUSAND different instructions that it knows how to read and execute. This invisible brain is more correctly called an **instruction decoder.**

Each of these instructions performs the hardware tasks of reading from memory, moving data to registers or to the calculator, moving data to memory, and so on. The internal operations of the instructions or of the µP itself are not our main concern. We will discuss how many memory locations the instructions occupy and how long it takes them to be executed, two very important aspects to programming.

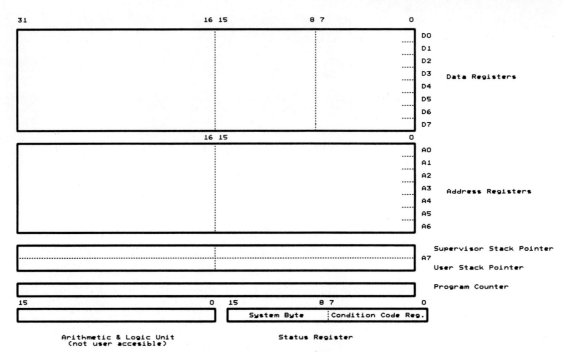

Figure 2-4   Programmer's model of the 68000.

In the memory *mailboxes* we will always find:

1. An **INSTRUCTION** telling the μP what to do next

or

2. DATA (numbers or letters) needed to accomplish a task

or

3. An **ADDRESS** telling the μP where to get some data

How does the μP know where to start in the task of executing instructions? On the μP chip itself is a *reset* pin. It is tied to circuitry that allows an orderly startup upon application of power (or upon pressing of a reset button).

When a reset occurs, the μP puts the 32-bit longword contents of memory addresses 00000000 through 00000003 into the Supervisor Stack Pointer (SSP), puts the 32-bit longword contents of memory addresses 00000004 through 00000007 into the Program Counter (PC), and begins program execution at the address stored in the PC. It then acts as a very efficient postmaster, going from one mailbox to the next reading instructions and performing the operations requested. Needless to say, these two addresses have to have been stored in the computer's memory BEFORE hitting the reset button. The normal approach is to store them in an EPROM or ROM IC. An important fact worth repeating: *The PROGRAM COUNTER ALWAYS CONTAINS THE ADDRESS OF THE NEXT INSTRUCTION TO BE EXECUTED.*

The μP then reads the contents of the memory location pointed to by the PC, expecting to see an INSTRUCTION. It reads the instruction, looks it up in the instruction decoder to see how many memory locations that particular instruction occupies, adds that value to the existing value in the PROGRAM COUNTER, stores the result in the PROGRAM COUNTER, and THEN executes the instruction just read.

After that instruction is executed, the process is repeated, with the reading of the PROGRAM COUNTER to see where to go next to get an instruction. Unless we use the category of instructions called *CONTROL AND BRANCHING,* the postmaster would continue through memory, executing one instruction after another. If memory chips existed for all addresses, it would eventually make it to address FFFFFF ($16,777,215_{10}$), then to 000000 and start all over. This cycle is not usually done, however.

Looking at the program execution steps in a little more detail we have the following sequence of events:

1. The memory address of the first instruction is loaded into the PC.
2. The μP places that memory address onto the address bus in order to get the instruction.
3. The memory circuitry decodes the address to find the specific address requested.
4. The memory sends the instruction back to the μP over the data bus.
5. The instruction is placed into the Instruction Register, an invisible (to the programmer) register in the μP.
6. The μP decodes the instruction, using knowledge internal to the μP, to determine the action to take.
7. The PC is incremented the proper number of bytes, depending on the size of the instruction.
8. The instruction action is taken, be it a move of data, arithmetic operation, and so on.
9. Resume at step 2. Some instructions may alter the PC contents, allowing operation of instructions not located next in memory. Steps 2 to 5 are considered the *fetch* portion and steps 6 to 8 are the *execute* portion.

2.2 MEMORY ORGANIZATION: THE MEMORY MAP

Rather than oversimplifying using 123 Oak Street, 4325 Elm Avenue, and so on, for our mailbox addresses, let's use **hexadecimal numbers** from 000000 to FFFFFF, a total of 16,777,216 (16 MB) addresses. If binary and hexadecimal numbers are totally new to you, refer briefly to Chapter 4.

Next, looking at the programming model, the registers you see are 32 bits wide, indicating that it can do 32-bit operations. However, due to pin limitations, the chip has only 16 *DATA lines*. These lines are used to write data to memory addresses and to read from memory. Also, note that there are only 23 *ADDRESS lines*. [Actually, there are 24, due to the way that even and odd addresses are selected, using the UDS/LDS (7,8) pins.] The address lines are used selectively to address specific memory addresses. This limits our addressing range from 000000 to FFFFFF, which can be represented by 24 binary bits.

The 68000 was initially dubbed a 16/32-bit computer. It was a 32-bit computer to its friends and a 16-bit computer to its foes. It does have instructions that tell it to fill up the 32-bit registers completely with data, and it can do 32-bit operations. It has to do two memory reads to perform this operation. The ''argument'' has since been resolved with the introduction of the 68020 32-bit μP, which has a full complement of 32 data lines and 32 address lines, capable of addressing from 00000000 to FFFFFFFF, for a total of 4,294,967,730 memory locations (4.3 gigabytes).

In comparison, the popular 8-bit Zilog Z-80 has 8 DATA lines, 16 ADDRESS lines, and 8/16-bit internal registers. It addresses memory from 0000 to FFFF, a total of 65,536 (decimal) memory locations. It had to do two memory reads when 16 bits of data were needed for its limited 16-bit operations. Memory locations for an 8-bit μP are visualized as 8 bits wide, each containing a binary number (address, or data, or instruction) between 00 and FF (Figure 2-5). Memory can be displayed either progressing downward or upward.

Visualization of memory in a 68000 system is complicated slightly by the fact that not only can single 8-bit bytes be handled, but 16-bit WORDS and 32-bit LONGWORDS can also be utilized. Since the 68000 has a restriction that all **instructions** START in an EVEN address, and a single instruction can span from two to ten 8-bit memory locations, it is easier to view the memory in a 16-bit format (Figure 2-6). Eight-bit **data** can be stored at either an odd or an even address.

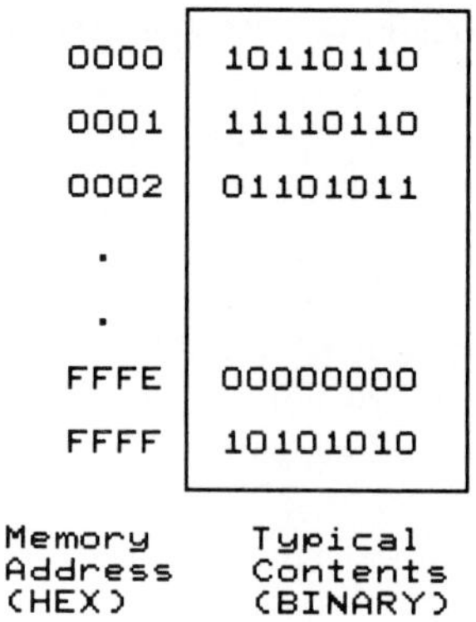

```
        0000 | 10110110
        0001 | 11110110
        0002 | 01101011
             |    .
             |    .
        FFFE | 00000000
        FFFF | 10101010
```

Memory   Typical
Address   Contents   **Figure 2-5** Eight-bit memory map.
(HEX)   (BINARY)

```
  000000 | 10110110 ┆ 10101010 | 000001
  000001 | 11110110 ┆ 00000000 | 000003
  000002 | 01101011 ┆ 00100101 | 000005
         |    .     ┆          |
         |    .     ┆          |
  FFFFFC | 00000000 ┆ 11110101 | FFFFFD
  FFFFFE | 10101010 ┆ 00011100 | FFFFFF
```

Even   Typical   Odd
Memory   Memory   Memory
Address   Contents   Address
(HEX)   (BINARY)   (HEX)

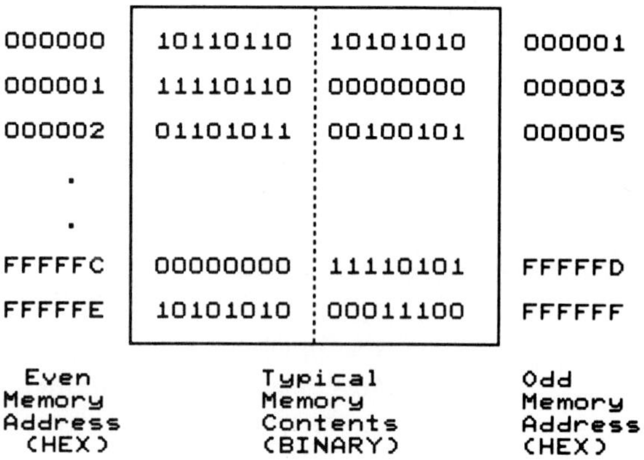

Figure 2-6 Sixteen-bit memory map.

Perhaps you could look at each group of four memory locations as a quad-plex apartment, with the first EVEN address being the **main** mailing address. If the postmaster wants to deliver an 8-bit (byte) "letter," it is deposited in location 000000_{16}, for example (Figure 2-7). That 8-bit letter, for example, can represent a single unsigned integer between 0_{10} and 255_{10} (00_{16} to FF_{16}). Similarly, bytes can be deposited in odd addresses, such as 000001_{16}.

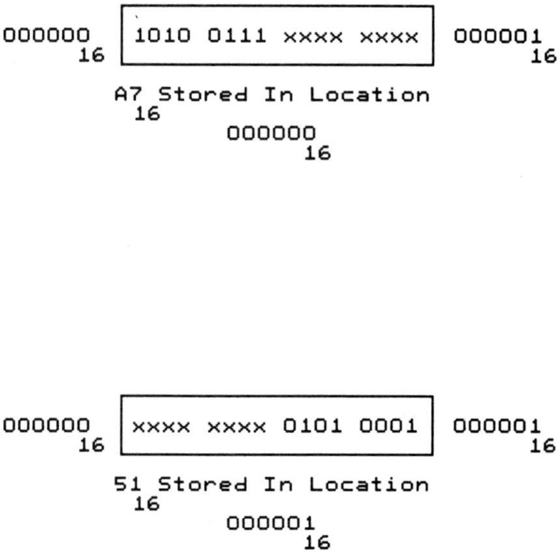

```
  000000 | 1010 0111 xxxx xxxx | 000001
      16 |                     |     16
```
           A7 Stored In Location
              16
                      000000
                          16

```
  000000 | xxxx xxxx 0101 0001 | 000001
      16 |                     |     16
```
           51 Stored In Location
              16
                      000001
                          16

Figure 2-7 Byte storage.

If a 16-bit (word) "letter" is to be deposited in location 000000_{16}, the postmaster uses locations 000000_{16} and 000001_{16} to store the data. **Words ARE NOT stored starting at an ODD address.** If the data were 1234_{16}, it would be stored as shown in Figure 2-8. Similarly, if the 32-bit longword "letter" 12345678_{16} is to be deposited in location 000000_{16}, it would be located as shown in Figure 2-9. This ordering sequence is not followed by all μPs, some of which store the bytes in reverse order (backward).

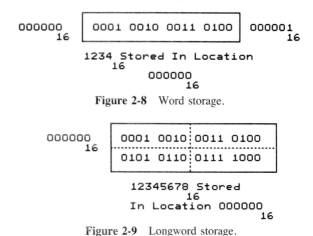

```
000000        0001 0010 0011 0100    000001
      16                                   16

         1234 Stored In Location
             16
                    000000
                          16
```

Figure 2-8   Word storage.

```
000000         0001 0010 : 0011 0100
      16       -----------:-----------
               0101 0110 : 0111 1000

           12345678 Stored
                   16
           In Location 000000
                             16
```

Figure 2-9   Longword storage.

Since the 68000 has different instructions for byte, word, and longword operations, writing a byte to a single location does not affect nearby locations. Writing a word affects only the two locations. Of course, reading from any location(s) does not affect the memory contents.

    TWO IMPORTANT CONSIDERATIONS (or quirks) WHEN WE ADDRESS MEMORY:

1. **All instructions must begin on an even address.**
2. **All WORD and LONGWORD address references must be to an even address. We can, however, refer to a byte located at an odd address.**

    With a little care, these restrictions will not surface often, and subsequent examples will explain how to avoid problems. Some assemblers take care of this potential problem for us.

2.3   INPUT AND OUTPUT

Considering next how we are able to get useful work out of a 68000 system, we need to connect input and output devices to the system. Since 16 MB of memory locations is at our disposal, circuitry is added to our system to allow some of the memory locations to act as input or output interfaces to the external world.

    Typical **inputs** to a computer system:

Keyboards

Switches

ON/OFF sensors

Serial or sequential digital data

Variable (analog) voltages

Variable (analog) pressures

Variable (analog) temperatures

Variable (analog) audio signals

Variable (analog) video images

    The possible types of inputs are limited only by the human imagination. Because of their speed and memory requirements, the last two, audio and video, are perhaps the driving force for the popularity and need for 32-bit computing systems. The first three types of inputs—keyboards, switches, and ON/OFF sensors—can be connected in a DIGITAL or ON/OFF manner to provide the necessary 0- or 5-volt (V) Transistor-Transistor-Logic (TTL)-level signals. Examples in later chapters show how to hook these devices to an existing computer system. Serial data, coming in from an external source a bit at a time over a single pair of wires, can easily be processed with a serial-to-parallel converter IC to provide an 8-bit digital simultaneous or parallel input to a memory location.

We will learn in later examples how to accommodate variable or analog inputs and convert them to digital values so that they can be processed directly by a digital computer.

Typical **outputs** for a computer system:

Video displays (analog or digital)

Lights (analog or digital)

Relays (digital)

Actuators (analog)

Motors (analog)

Valves (analog or digital)

Audio devices (analog or digital)

Servo positioning devices (analog)

Serial data output devices (digital)

The digital outputs are easily processed to provide a DC 0- or 5-V level signal which can be converted to the appropriate higher voltage or AC voltage. The analog signals are processed by a digital-to-analog (D/A) device. Serial data are output easily with a parallel-to-serial IC, and sent out a bit at a time over a pair of wires.

Interfacing a new or unique input or output device to an existing computer system that already contains the necessary memory address decoding circuitry and buffer (protection) ICs is well within the capability of the eager beginner. Several examples are shown in later chapters. Extreme care must be exercised when dealing with 110-V AC voltages, however.

The input or output work to be performed could be categorized according to their associated speed requirements. *Slow-speed I/O* typically provide (or receive) new data not more than once a second. In such cases, the program need only check/update their status every millisecond or so. *Medium-speed I/O* devices may process data at rates from 1 to 10,000 bits per second. In other words, they may output over 1000 bytes of data per second. *High-speed I/O* processes data at well over 10,000 bits per second. As we get into programming tasks, we will see how we have the computer *ready* to receive data that may come in at unpredictable times, how a *stream* of data can be output, and how we are able to maintain the rates needed. Realize that the actual operation to transfer a single byte of data takes just a few MICROSECONDS.

In summary, we have the ability to deposit INPUT information into memory locations, and the ability to receive OUTPUT information from memory locations, to accomplish an assigned task. We create a computer program, deposit it into memory, hit the reset button, and let the postmaster "run" or "execute" our program. In later chapters we will see how/why data and addresses are stored in the postmaster's "internal scratchpads," called Data and Address Registers.

chapter 3

programming a computer:
getting a taste of the software

Before a programming task can be undertaken, we of course must have knowledge of the tools at our disposal. Techniques for organizing and visualizing an assignment will be presented here before tackling the 68000 instruction set. Problem-solving skills need to be developed before attempting complicated problems. *Beginner's luck* and *common sense* are all it takes for the beginner to write some interesting introductory programs, however-er.

3.1 VISUALIZING AN ASSIGNED TASK FOR A COMPUTER SOLUTION

Once you are armed with the knowledge of computer programming fundamentals, you are ready to accept a work assignment that you could attempt to complete by utilization of a computer. The procedures you use to visualize or lay out the task before you are very similar to the techniques you would use if a computer was not available. The programming tools are new and the algorithms may be slightly different, but the basic process is the same.

One important consideration should be made before you start. A computer solution to an assignment is SUPPOSED to provide a time savings over a noncomputer solution. Is the time you are spending creating or modifying a computer program going to be *recouped* in the future with the repeated application of your program? I suppose that if you are simply a hired computer programmer, you would not worry about this. Some assignments can still be done a LOT faster manually, without anything more than a piece of paper, a pencil, and a $10 calculator! For example, let's prepare a computer program to help us complete our yearly income tax forms. I can assure you that there will be no time savings or *ease of completion* in this case! However, such programs are commercially available and may be used by companies that prepare a lot of income tax returns.

Let's take the following job assignment and see if it can be *computerized*. (It will be one of our programming examples in a later chapter.)

Assignment

Set up an organized, efficient method for computation (and display) of the following statistics for an array of data. An *array of data* is simply a collection of numbers, hopefully representing something useful. Find:

1. The average
2. The highest
3. The lowest
4. The number of data points obtained

3.2   GATHERING THE KNOWNS AND SPECIFIC REQUIREMENTS

The noncomputerized way of setting this up would be to gather (1) pencils, (2) a tablet with useful columnar divisions, (3) a calculator, (4) a typewriter, and (5) some typing paper. Next we would go back to gather some *details* on what was *known* information and to determine any specific requirements. Suppose that we were told the following:

The data represent temperature measurements.
The allowable range is between 0 and 255 degrees.
As many as 32,767 measurements may be made.
The statistics requested are to be displayed.

3.3   ORGANIZING THE INPUTS, OUTPUTS, AND MEMORY REQUIREMENTS

Setting up the initial layout is simple at first sight. Our report or display could look like:

Temperature Data Analysis Project

Highest temperature is _____ Average temperature is _____
Lowest temperature is _____ Based on _____ measurements

Once the input data format and desired data output are determined, some thought can be given to creation of the program. A few tips:

1. Divide the task into small, logically independent tasks.
2. Use graphical aids as much as possible.
3. Document your thoughts and actions as you create the program.

3.4   SKETCHING THE PLAN OF ATTACK

There are three popular ways to attack the assignment. Some are inclined to take nothing more than the information given above and plunge into the project, using the brute-force, trial-and-error method until done. Others at least scribble down a few notes and perhaps a sketch or two before diving in. Then there is a small group that has the foresight to spend a little more time in preparation by developing algorithms and a readable sketch called a **flowchart.**

3.5   DEVELOPING ALGORITHMS

Algorithm is a fancy buzzword often used by seasoned programmers. Recall, it is simply a systematic, efficient, generalized procedure that is developed for the solution of a problem. One more detail—an algorithm is supposed to have a FINITE number of steps. Beginners generally develop awkward but finite algorithms; seasoned programmers develop elegant, efficient ones. Experience is the key to improvement. By tackling increasingly difficult problems you begin to synthesize new ideas and improved algorithms. These creative endeavors are the keys to success as a skilled programmer.

3.6   CREATING FLOWCHARTS

A *flowchart* is nothing more than a *map* showing the flow of a program as it performs the various procedures of an algorithm. There are several approaches to the creation of a flowchart depicting a solution to an assignment. We can describe each step in a program fairly completely or describe it more generally in terms of minitasks, called subprograms or subroutines. We can create the flowchart utilizing words and commands specific to the particular programming language we are using, or it can be created in *plain English,* which allows use of the flowchart with almost any programming language.

Perhaps some of the reluctance in the utilization of flowcharts is due to the excessive amount of time it takes to create them when done manually. Templates, drawing supplies, and so on, were put to the task in times past. If a minor change in the flowchart were needed, it was "back to the drawing board." Most students put off the creation of flowcharts until AFTER a program had been debugged and running. However, today a variety of computer programs are available that do the hard work for us. Shown in Appendix G is an approach using a shareware program called FLOWCHARTER. Its use is as simple as creating a text file with a few specific commands, and the program does the flowchart creation for you. Figure 3-1 is a simple flowchart outlining preliminary steps to be done before we begin to write the program.

As we progress through learning the 68000 assembly language, we will learn the relatively simple chores of moving data, doing arithmetic, and the making of logical decisions. We must however, first cover the techniques of **branching,** called **loop structures** and **decision structures** among computer professionals. Basically, a computer starts running (or executing) a program from some location in memory and continues systematically down through its memory, doing each instruction in turn. This is exactly true if we are considering programming at the lower assembly level. It is also basically true as far as we are concerned while programming in a high-level language, such as BASIC, which executes each program line sequentially. See, for example, Figure 3-2, which describes a handy but not very powerful program that shows only the computer's speed and accuracy.

We can also put what are called **labels** on our program lines and instruct the computer to JUMP to that particularly labeled line and resume operation from there. This is called an **unconditional jump structure.** The example of Figure 3-2 can be modified as shown in Figure 3-3 to allow continuous operation.

This program is a little more useful than its predecessor, but we must have provisions for stopping eventually. A loop structure as above is well accepted and easy to follow. However, if we use the JUMP instruction to go to another portion of our program and then on to another, and so on, we soon create an unwieldy sort of program. Excessive use of this instruction is frowned upon and violates our goal of creating what could be classified as a "structured program." A program that jumps all around becomes impossible to understand even by its creator a few weeks after the ink has dried. A program that "flows" is much easier to follow, and hopefully, to modify or improve when needed. If it is large, it should be written as smaller *modules* each having a single point of entry and a single exit point. A *main* program then links the modules together. We will find that 68000 assembly language supports this concept very well.

When we add the ability to *exit* or terminate a continuous loop structure, we create what is called a **conditional loop structure** or, more precisely, a **decision structure.** Aside from its ability to do arithmetic and to accomplish tasks in short order, a computer's ability to make a **decision** (an accurate one at that) is perhaps its most valuable asset.

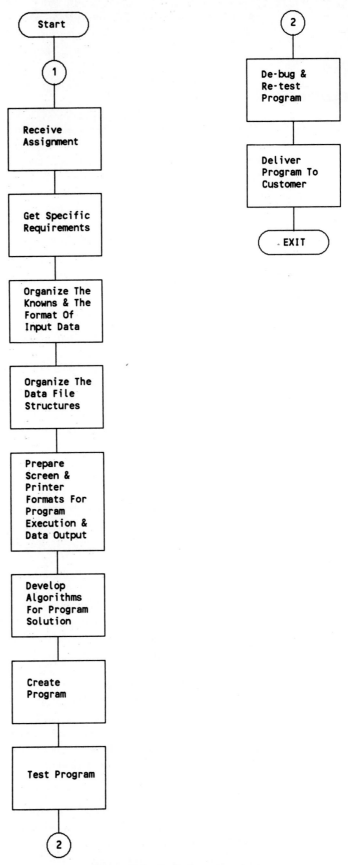

Figure 3-1   Preliminary flowchart.

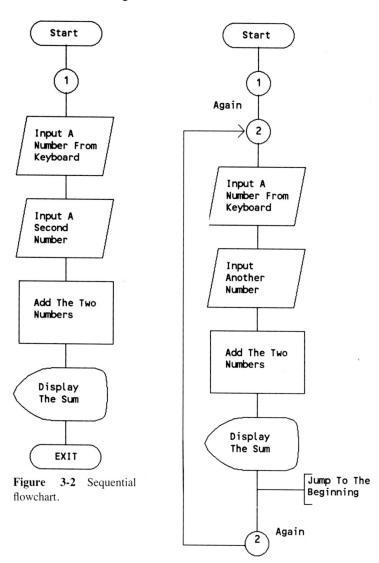

Figure 3-2 Sequential flowchart.

Figure 3-3 Unconditional jump structure flowchart.

The computer can compare the already known values for numbers *X* and *Y,* and if the result is TRUE (i.e., if *X IS* less than *Y*), it will perform an alternative instruction. If the result is FALSE, it will ignore the alternative instruction and continue on to the next instruction (see Figure 3-4).

Although the example shown in the figure is considered as a **single-alternative** decision structure, we could easily add an instruction into the NO branch and would then have a **double-alternative** decision structure. The YES branch is performed if the arithmetic comparison is TRUE and the NO branch if it is FALSE. A structure such as this is called **IF-THEN-ELSE** in the BASIC programming language.

Next, we can combine this decision-making ability with the looping ability and create what is called a **WHILE LOOP.** Combining the previous examples; we have the flowchart shown in Figure 3-5.

This example would continue to add and display numbers AS LONG AS we entered in a value for *X* that was less that the value entered for *Y*. If *X* is greater than *Y*, it would **exit** the loop and resume with the following instruction. In other words, we **DO** the loop **WHILE** *X* is LESS THAN *Y*.

Another variation is the **REPEAT UNTIL LOOP,** where the decision (or loop test) is at the bottom of the loop (Figure 3-6).

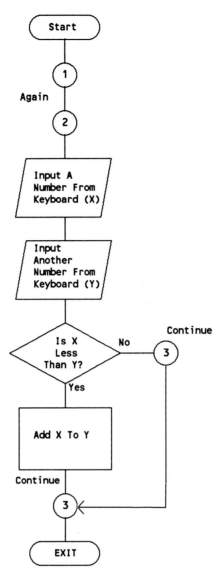

Figure 3-4 Conditional loop structure flow-chart.

This example would input, add, and display the numbers AS LONG AS *X* **IS NOT LESS THAN** *Y*. In other words, it will **REPEAT UNTIL** *X* WAS GREATER THAN *Y*.

When programs begin to get complicated, a more appropriate approach is to create **SUBPROGRAMS,** often called **SUBROUTINES** (see Figure 3-7). They serve two basic purposes. If we have a series of instructions that appear in one portion of the program and appear in similar or identical fashion elsewhere in the program, we put these instructions into a SINGLE subroutine. Our main program flows *down as normal* and then CALLS the subroutine. The subroutine's instructions are completed and then the program flow is resumed at the instruction immediately following the CALL in the main program. This approach saves typing and reduces the size of our program. Another reason to use this concept would be to break a large and perhaps hard-to-follow program down into a series of smaller subroutines, each performing a *stand-alone* function.

An added bonus of subroutines is that they can be individually tested and debugged. A ''find the square root'' subroutine, for example, would be provided with the location or value of the number whose square root is needed, and the subroutine will provide the result to a desired location (a memory location or Data Register). If properly written, its execution will not disturb the main program in any way.

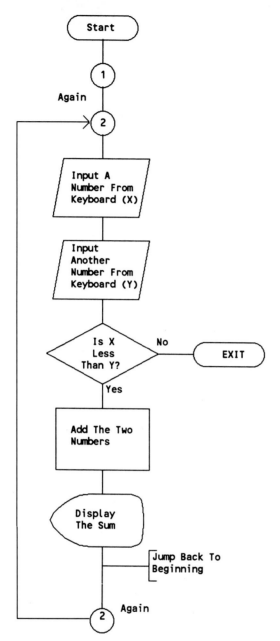

Figure 3-5 Double alternative, Do While loop flowchart.

An Important Note: Subroutines ALWAYS appear separate or at the end of a main program. In other words, the main program must have a STOP or some type of LOOPing instruction to prevent the program execution from flowing downward into the subroutine. If it did, when the RETURN instruction was encountered, the program would have no idea where to return to, since the subroutine had not been CALLed. Since the RETURN is also considered as a *dead-end,* we can put further subroutines after it as well.

Not shown in the looping and decision structures is a way in which we can make a **multiple-choice** decision. Programming at the assembly level, we have to realize that the computer is only a binary system capable of making *true–false* decisions. If you have programmed in a high-level language such as BASIC, you know instructions are available that allow the computer to do a specific task based on several or many choices. At the assembly level we will be limited to making the decisions one at a time (i.e., making one

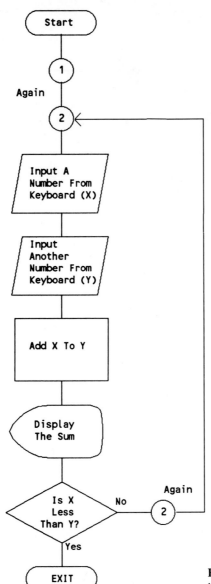

Figure 3-6 Double alternative, Repeat Until loop flowchart.

comparison between two values at a time). When multiple choices are to be made, we will use the basic double-alternative decision structure along with innovative algorithms.

The classic game of trying to guess a number chosen by someone shows this approach. If I am thinking of a number between 0 and 100 (say 67) and you are to guess it, and my only answers are "too high," "too low," or "correct," we have two (at least) approaches to guess the number. You could start at 0 and keep getting "too low" responses until you get the correct number at 67. Another way would be to guess 50 (halfway between 0 and 100), find that it was too low, guess 75 (halfway between 50 and 100), find that it was too high, guess 62 (or 63), find that it was too low, guess 66, find that it was too low, guess 71, find that it was too high, guess 68, find that it was too high, guess 67, and get a CORRECT answer.

To program this, giving a triple-alternative response (too high, too low, correct), we need to use two double-alternative decision structures. It would look as follows:

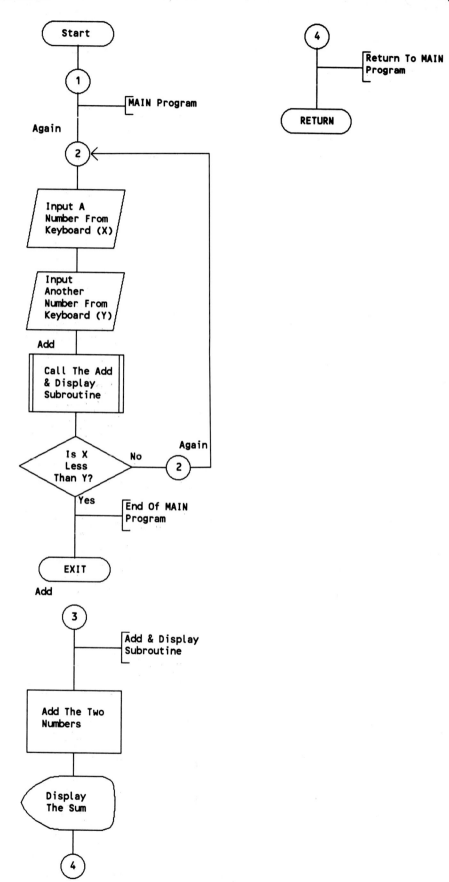

Figure 3-7 Flowchart utilizing a subroutine.

Compare the guess with the number

Is the number greater than the guess → yes → give *too high* report

   ↓ no

Is the number equal to the guess → yes → give *correct* report

   ↓ no

Give *too low* report

Another alternative to flowcharting a task is a technique called **pseudocode.** Pseudocode is simply a generalized statement of each program step, written line by line. Boxes can be drawn around the steps but are not required. Each line will specify a task to be done using decision structures or subroutines.

Flowcharts and pseudocode are much more effective if written INDEPENDENT of the specific computer language to be used. More important, the more time spent in creating an accurate flowchart, the less time that will be needed in the creation and debugging of the computer program.

Efficiency in programming is increased through the use of subroutines, or a modular approach, and through use of the loop structures described above. They enhance creation of what is called STRUCTURED PROGRAMS, those that have a single point of entry and a single exit point. Emphasis should be placed on creation of structured programs, and the result will be more systematic, easier to understand, and better organized programs.

In summary, to solve a computer problem, we:

1. Analyze the situation, gathering the requirements, constraints, etc.
2. Synthesize the algorithms needed for the solution.
3. Create, run, and evaluate the program.
4. Refine the program until it completely meets all of the requirements and perhaps add a few additional frills.

A good general approach to take in creation of the program is to:

1. Initialize parameters and conditions.
2. Enter a programming loop that does repeated operations.
3. While looping, check for error conditions or sign of program completion, taking appropriate action.
4. End the program.

Before we formally tackle the assignments, let's get familiar with the way the 68000, or any binary computer, processes its instructions and stores its data.

chapter 4

how a computer handles instructions, addresses, and data

4.1 NUMBER SYSTEMS

Of the two basic types of computers, the BINARY computer has far surpassed the ANALOG computer, with the latter being almost extinct. An *analog computer* uses variable voltage levels, resistors, and capacitors to perform its work. A *binary computer*, on the other hand, utilizes only two voltage levels and a maze of integrated circuits. An "ON" or binary "1" is a high level (5 volts DC on some computers), and an "OFF" or binary "0" is a low level (typically, almost zero volts DC on some computers). Before we can begin to give a computer input or expect output, we must know how it handles numbers in a binary fashion. Similarly, we utilize binary (or hexadecimal) representations for the computer instructions and for the memory addresses we read from and write to.

4.1.1 Binary and Hexadecimal Numbers

To see how a computer (binary, from now on) holds these binary levels, let's first review how we represent numbers in general.

The **decimal** number 1011 can be represented as

$$1 \text{ times } 1000 + 0 \text{ times } 100 + 1 \text{ times } 10 + 1 \text{ times } 1$$

or as

$$1 \times 10^3 + 0 \times 10^2 + 1 \times 10^1 + 1 \times 10^0$$

The *base* of this number system is **10.** The digits allowable are 0, 1, 2, 3, 4, 5, 6, 7, 8, 9. Numbers can be created using any base, however. We will limit discussion to **binary** (base **2**) and **hexadecimal** (base **16**), the two primary ones used in computer work.

Considering a **binary** number system, the allowable digits are 0 and 1. The **binary** number 1011 can be represented as

$$1 \text{ times } 8 + 0 \text{ times } 4 + 1 \text{ times } 2 + 1 \text{ times } 1$$

or as

$$1 \times 2^3 + 0 \times 2^2 + 1 \times 2^1 + 1 \times 2^0$$

The **hexadecimal** number system's digits consist of 0, 1, 2, 3, 4, 5, 6, 7, 8, 9, A, B, C, D, E, F. Note that we have to continue with letters for digits greater than 9. The **hexadecimal** number 1011 can be represented as

$$1 \text{ times } 4096 + 0 \text{ times } 256 + 1 \text{ times } 16 + 1 \text{ times } 1$$

or as

$$1 \times 16^3 + 0 \times 16^2 + 1 \times 16^1 + 1 \times 16^0$$

We have taken similar sets of digits and represented them in three different number systems. To avoid confusion, a subscript representing its base is often included. We would have 1011_{10}, 1011_2, and 1011_{16}. With a little arithmetic for the example above, we see that:

1011_{10} is equal to 1,011 in our decimal system.
1011_2 is equal to 11 in our decimal system.
1011_{16} is equal to 4,113 in our decimal system.

We next consider how to convert from one number system to another.

4.1.2   Converting from Decimal to Binary to Hexadecimal and Back

Several techniques are used to do these conversions, and fortunately, once the basic concepts are understood, we can usually rely on the computer to do this work for us. Converting 1011_{10} to a binary number, we create an aide:

Bit:	15	14	13	12	11	10	9	8	7	6	5	4	3	2	1	0
	32,768	16,384	8192	4096	2048	1024	512	256	128	64	32	16	8	4	2	1
Result:	—	—	—	—	—	—	—	—	—	—	—	—	—	—	—	—

We then take our decimal number, 1011, and subtract from it the LARGEST number possible from the sequence above. The number 512 looks like the proper choice to start with.

$$
\begin{array}{r}
1011 \\
-\ 512 \\
\hline
499 \\
-\ 256 \\
\hline
243 \\
-\ 128 \\
\hline
115 \\
-\ 64 \\
\hline
51 \\
-\ 32 \\
\hline
19 \\
-\ 16 \\
\hline
3 \\
-\ 2 \\
\hline
\end{array}
$$

We put a 1 below the 512 in the aide above

Continuing
We put a 1 below the 256

We put a 1 below the 128

We put a 1 below the 64

We put a 1 below the 32

We put a 1 below the 16

We put a 1 below the 2

$$\begin{array}{r} 1 \\ -\ 1 \\ \hline 0 \end{array} \qquad \text{We put a 1 below the 1}$$

We put zeros in all the other columns. Our result: $0000\ 0011\ 1111\ 0011_2$. The digits are grouped for ease of reading and for subsequent conversion to hexadecimal. To check our work (skipping the leading zeros):

$$1 \times 512 + 1 \times 256 + 1 \times 128 + 1 \times 64 + 1 \times 32 + 1 \times 16 + 1 \times 2 + 1 \times 1 = 1,011_{10}$$

Next, let's look at the following binary numbers and their corresponding decimal and hexadecimal equivalents:

Binary	Decimal	Hexadecimal	Binary	Decimal	Hexadecimal
0000	0	0	1000	8	8
0001	1	1	1001	9	9
0010	2	2	1010	10	A
0011	3	3	1011	11	B
0100	4	4	1100	12	C
0101	5	5	1101	13	D
0110	6	6	1110	14	E
0111	7	7	1111	15	F

With these equivalents it is now easy to convert a **binary** number to a **hexadecimal.** The $0000\ 0011\ 1111\ 0011_2$ becomes $03F3_{16}$.

So far we have converted a decimal number to binary and then to hexadecimal. Going the other way, the conversion from hexadecimal to binary is almost instantaneous. Looking at the conversion table above: $03F3_{16} = 0000\ 0011\ 1111\ 0011_2$. Converting this binary number to decimal, using the previous aide, we put the corresponding 0's and 1's in place:

32768	16384	8192	4096	2048	1024	512	256	128	64	32	16	8	4	2	1
(0	0	0	0)	(0	0	1	1)	(1	1	1	1)	(0	0	1	1)
		0				3				F				3	

Note how we group four binary bits for each hex character. Adding the numbers that have "1 multipliers," we get

$$512 + 256 + 128 + 64 + 32 + 16 + 2 + 1 = 1011_{10}$$

If you want to convert directly from **hexadecimal** to **decimal,** we use

$$MNOP_{16} = M \times 16^3 + N \times 16^2 + 0 \times 16^1 + P \times 16^0$$

which is equivalent to

$$M \times 4096 + N \times 256 + O \times 16 + P \times 1$$

where M, N, O, and P represent any hex digits. Our example becomes

$$03F3_{16} = \mathbf{0} \times 4096 + \mathbf{3} \times 256 + \mathbf{F} \times 16 + \mathbf{3} \times 1 = 1011_{10}$$

If you do not have a calculator capable of multiplying F times 16, simply multiply its equivalent (15) by 16.

4.1.3 Simple Arithmetic Operations with Binary and Hexadecimal Numbers

An understanding of binary and hexadecimal arithmetic techniques will help in the development of programming techniques. Again we rely on the computer to do most of the work. The techniques for binary and hexadecimal addition and subtraction are identical to decimal arithmetic.

Adding some binary numbers, we have to realize that $1 + 1 = 10_2$ (a zero with a carry). Adding hex numbers $F + 1$, we get 10_{16}. Adding decimal numbers $9 + 1 = 10_{10}$. Adding two numbers is fairly straightforward, but keeping up with the carries is awkward for larger series of numbers.

Example:

$$
\begin{array}{cccccc}
\text{Add (binary)} & 1011 = & \textbf{(decimal)} & 11 = & \textbf{(hex)} & B \\
& +\,0111 = & & +\ 7 = & & +\ 7 \\
\hline
& 1\,0010 = & & 18 = & & 12 \\
\end{array}
$$

Subtraction can be described similarly. To subtract two numbers, we take what is called the *TWO's complement* of the subtrahend and then add the two numbers. Taking the two's complement changes the sign of the number, and the process is likened to ordinary subtraction where the sign of the subtrahend is changed and the two numbers then added.

Example:

$$
\begin{array}{cccccc}
\text{Subtract (binary)} & 1011 = & \text{or decimal} & 11 = & \text{or (hex)} & B \\
& -0111 = & & -\ 7 = & & -\ 7 \\
\hline
& 0100 = & & 4 = & & 4 \\
\end{array}
$$

To get the two's complement of 0111, we first take the *ONE's complement* and then add one. The one's complement is obtained by complementing the 0's (changing them to 1's) and complementing the 1's (changing them to 0's). This gives $0111 >> 1000$. Adding 1 to get the two's complement, we have $1000 + 1 = 1001$. Now adding this to 1011, we get

$$
\begin{array}{r}
1011 \\
+\ 1001 \\
\hline
0100 \\
\end{array}
$$

which agrees with our decimal answer of 4.

Fortunately, we rarely get involved with manual binary division and multiplication, leaving those chores to the computer. Some interesting programming examples will be given later. Perhaps the most valuable operations done with a computer involve the **COMPARING** of two numbers using the LOGICAL operations AND and OR. We can also compare two numbers by subtracting them and looking for a zero result, a carry, or a borrow.

4.1.4 Logical AND, OR, Exclusive OR, and NOT Operations

The logical AND, OR, exclusive OR, and NOT operations are basic and simple to master. Operations are done on a **bit-by-bit** basis, with no carries being used.

And

A 1 **AND** a 1 gives a 1, indicating they **BOTH** are 1. A 1 **AND** a 0 gives a 0, indicating they both are NOT 1.

Or

When we **OR** a 1 and a 1 we get a 1, indicating that either the first OR the second bit is a 1. When we **OR** a 1 and a 0 we get a 1, indicating **AT LEAST ONE** of them was a 1. Of course, **OR**ing two 0's gives a 0, indicating that neither of them are 1.

EOR

The exclusive OR function (called XOR in some texts) gives a 1 when ONE but not BOTH of the bits are 1's. That is the only difference between the **EOR** and the **OR** functions.

Not

The NOT function simply ''inverts'' the bits of a single byte, word, or longword, performing a one's complement on them. A 0 becomes a 1, and a 1 becomes a 0.

Examples:

AND	OR	EOR	NOT
1011	1011	1011	1011
1001	1001	1001	
1001	1011	0010	0100

Handy tables for AND, OR, and EOR logic, showing all possible outputs:

AND			OR			EOR		
In	In	Out	In	In	Out	In	In	Out
0	0	0	0	0	0	0	0	0
0	1	0	0	1	1	0	1	1
1	0	0	1	0	1	1	0	1
1	1	1	1	1	1	1	1	0

The operations considered so far have all been positive integers, that is, numbers greater than or equal to zero. When we cover the actual arithmetic instructions of the 68000, we will also consider negative integers and real numbers (positive or negative), which can have decimal portions (i.e., 32.765).

4.1.5. Representing Ordinary Text in Binary Form

In addition to the manipulation of numbers (data), instructions, and addresses, we also need representation for alphabetical characters, which really are a form of data. The most common method is to represent them in what is called ASCII (American Standard Code for Information Interchange) format. A partial ASCII chart, including equivalent decimal and hexadecimal equivalents, is shown in Figure 4-1. Note that these representations do not take the full 8 bits, which is the normal size of a single memory location. A space 20_{16} is 100000_2, or 5 bits, and the highest character, an unprintable null $7F_{16}$, is 1111111_2 or 7 bits. In some situations, the values from 00_{16} to 19_{16} and those from 80_{16} to FF_{16} are used for special control and graphics symbols. Elaborate packing and conversion routines are sometimes used to conserve memory space.

BITS B4 B3 B2 B1 \ B7 B6 B5	0 0 0 CONTROL CHARACTER (HEX / DEC)	0 0 1 CONTROL CHARACTER (HEX / DEC)	0 1 0 NUMBERS & SYMBOLS (HEX / DEC)	0 1 1 NUMBERS & SYMBOLS (HEX / DEC)	1 0 0 UPPER CASE (HEX / DEC)	1 0 1 UPPER CASE (HEX / DEC)	1 1 0 LOWER CASE (HEX / DEC)	1 1 1 LOWER CASE (HEX / DEC)
0 0 0 0	0 NUL 0	10 DLE 16	20 SP 32	30 0 48	40 @ 64	50 P 80	60 ` 96	70 p 112
0 0 0 1	1 SOH 1	11 DC1 17	21 ! 33	31 1 49	41 A 65	51 Q 81	61 a 97	71 q 113
0 0 1 0	2 STX 2	12 DC2 18	22 " 34	32 2 50	42 B 66	52 R 82	62 b 98	72 r 114
0 0 1 1	3 ETX 3	13 DC3 19	23 # 35	33 3 51	43 C 67	53 S 83	63 c 99	73 s 115
0 1 0 0	4 EOT 4	14 DC4 20	24 $ 36	34 4 52	44 D 68	54 T 84	64 d 100	74 t 116
0 1 0 1	5 ENQ 5	15 NAK 21	25 % 37	35 5 53	45 E 69	55 U 85	65 e 101	75 u 117
0 1 1 0	6 ACQ 6	16 SYN 22	26 & 38	36 6 54	46 F 70	56 V 86	66 f 102	76 v 118
0 1 1 1	7 BEL 7	17 ETB 23	27 ' 39	37 7 55	47 G 71	57 W 87	67 g 103	77 w 119
1 0 0 0	8 BS 8	18 CAN 24	28 (40	38 8 56	48 H 72	58 X 88	68 h 104	78 x 120
1 0 0 1	9 HT 9	19 EM 25	29) 41	39 9 57	49 I 73	59 Y 89	69 i 105	79 y 121
1 0 1 0	A LF 10	1A SUB 26	2A * 42	3A : 58	4A J 74	5A Z 90	6A j 106	7A z 122
1 0 1 1	B VT 11	1B ESC 27	2B + 43	3B ; 59	4B K 75	5B [91	6B k 107	7B { 123
1 1 0 0	C FF 12	1C FS 28	2C , 44	3C < 60	4C L 76	5C \ 92	6C l 108	7C	124
1 1 0 1	D CR 13	1D GS 29	2D - 45	3D = 61	4D M 77	5D] 93	6D m 108	7D } 125
1 1 1 0	E SO 14	1E RS 30	2E . 46	3E > 62	4E N 78	5E ^ 94	6E n 109	7E ~ 126
1 1 1 1	F SI 15	1F US 31	2F / 47	3F ? 63	4F O 79	5F _ 95	6F o 110	7F DEL 127

Figure 4-1   ASCII chart.

34

4.2 PUTTING A COMPUTER PROGRAM INTO MEMORY

Before we can begin to understand how a computer runs (or *executes*) a program, we must consider how the program is created and how it is deposited into the computer's memory. Let's take a quick trip back in history to see how programs used to be deposited and run in a computer. The front panel of most personal computers, introduced in the late 1970s, were fully decorated with an array of light-emitting diodes (LEDs) and switches. A typical setup:

```
(Memory Addresses)                          (Memory Contents)
A15 - - - - - - - - - - - - - - - - - - - - A0     D7 - - - - - - - - D0
LEDs
 *  *  *  *  *  *  *  *  *  *  *  *  *  *  *  *      *  *  *  *  *  *  *  *
SWITCHES
 + + + + + + + + + + + + + + + +      + + + + + + + +

READ          GO
  +            +
WRITE        HALT
```

Programs were created *manually* using mnemonics, and prepared for the computer using a reference card giving the corresponding hexadecimal values. Taking a fictitious instruction that had a hex code of 3E, we next picked an address [i.e., 0000_{16} that we knew contained usable memory and started depositing the instruction into memory (manually)]. We depressed all 16 address switches to select an address of 0000_{16} or 0000000000000000_2, then set the data switches (up for a 1, down for a 0) for the $3E_{16}$ instruction (00111110_2). This was followed with the momentary pressing of the WRITE switch to deposit the instruction into memory. This process was continued, selecting the next address 0001_{16}, and so on, until the program was finally loaded into memory. We then selected the starting address of the program with the address switches, depressed the run switch, and then the µP executed our program.

Today the process is much improved. Rather than describing the process here, the student anxious to experiment with the floppy disk (available from your instructor) can jump to Example 6.1 and do a *hands-on* exercise now.

how the 68000 works
from a software viewpoint

5.1 A SOFTWARE MODEL OF THE 68000

Programming at the assembly language level requires an intimate knowledge of the internal operation of the specific computer being used. The software programming model (Figure 5-1) gives us an idea of what the postmaster has at its disposal during execution of the program. Look at the following as a series of *scratchpads* (called *registers*) on the postmaster's (mobile) desk. Remember that the postmaster will be on the road a lot, going through the computer's memory, reading one mailbox after another. Hopefully, this over-simplification was of some value.

The first register, perhaps the most important one, is the **Program Counter (PC).** It is 32 bits wide with the bits identified as 0 through 31 going from right to left. THE PROGRAM COUNTER ALWAYS CONTAINS THE MEMORY ADDRESS OF THE

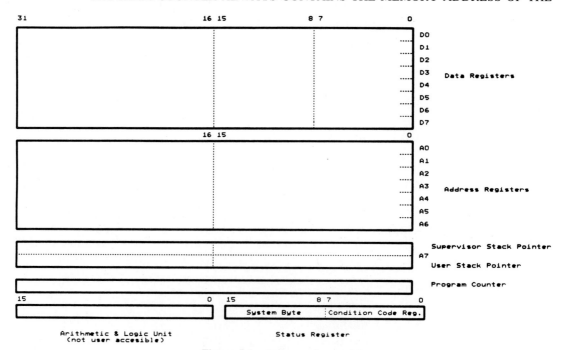

Figure 5-1 Programming model.

NEXT INSTRUCTION TO BE EXECUTED. The 68000's addressing capability is limited to 24 bits, so the top 8 bits are not used. The 68020 and later versions have full 32-bit capability.

The **Status Register (SR)** consists of two independent bytes, the **system** byte and the **user** *byte,* which is called the **Condition Code Register (CCR).** The system byte deals with advanced topics and is not addressed in this book. The **CCR** is another important register, consisting of individual bits that are set or cleared during program execution. The condition of these bits allows decision making and deviation from the normal execution sequence, which is to progress sequentially through memory. It will be covered in depth when programming examples are given.

The **Data Registers,** numbered **D0 to D7,** are identical in nature. In other words, none are reserved or required for specific operations. They can be loaded with a **byte** in the lower 8 bits, with a **word** in the lower 16 bits, or with a **longword** for all 32 bits. Loading a byte does not disturb the remaining 24 bits; loading a word does not disturb the high 16 bits. A **byte** can only be loaded into the **lower** 8 bits, and a **word** can only be loaded into the **lower** 16 bits. Various arithmetic, logical, shifting, and testing operations allow manipulation and testing of individual bits, bytes, and so on, and their operations are covered in the examples. The Data Registers will be referenced as **Dn** in general (*n* between 0 and 7). The Data Registers can be used to hold **addresses** as well, but adequate **Address Registers** are available and should be used for this purpose.

The **Address Registers,** numbered **A0 to A7,** are identical in nature except for A7. There are actually two A7 registers, with only one of them being available at a time. The Address Registers are used to hold **memory addresses,** which are used as **POINTERS.** A special notation using parentheses, such as (A1), is used to indicate the memory **contents pointed to** by Address Register A1. If A1 contained the address ABCD and memory location ABCD contained 12, we are referring to the contents, which is 12. A specific series of instructions are available for the use of the Address Registers as memory pointers. They can be used to hold data as well, but again, we have adequate Data Registers for this purpose. The high 8 bits of the Address Registers are not used on the 68000, as is the case of the Program Counter.

To discuss the operation of **A7,** the **SUPERVISOR MODE** and **USER MODE** must be explained. These comprise a *sophisticated* feature, not necessary for introductory coverage of the 68000. The 68000 was designed for potential use in multitasking or multiuser operations. We will be in *complete control* of the 68000 systems used here, so we will be running in the **supervisor mode.** If we crash the system, we have no one to blame but ourselves. Refer to advanced texts if you are developing programs for use on a system where you *are not in complete control.* The two A7 registers are sometimes referred to as A7 and A7' or as the **USER STACK POINTER (USP)** and the **SUPERVISOR STACK POINTER (SSP).** We will stick to the SSP, with a complete explanation of stack pointer operations coming in due time.

As mentioned earlier, there are other ''invisible'' registers, but they will not be our concern during programming. Next, let's consider the instruction set at our disposal.

5.2 INSTRUCTION FORMAT

Recall that a computer instruction is simply a series of binary values stored in memory and later *executed.* The 68000's instructions range from 2 to 10 bytes in length. Looking at a 2-byte instruction, we realize that with 2 bytes we could have hex numbers from 0000 to FFFF. We could have 65,536 different instructions. For example, the NOP, or No OPeration (''do nothing''), instruction is 4E71. Other instructions, such as ''MOVE the contents of one data register to another data register,'' can be represented by 2 bytes. ADD instructions that add the contents of registers or memory addresses being pointed to by

Address Registers can also be represented by 2 bytes. These 2-byte instructions completely describe the operation to be performed.

But if we want to move a NUMBER (i.e., 1234_{16}) to a register or to a memory location, our instruction must include not only the instruction but also the number; similarly, if a specific MEMORY ADDRESS is needed for an instruction. If we want to move the contents of one MEMORY location to another, our instruction includes a 2-byte instruction for the move operation, plus up to 4 bytes for the source of the number, and up to 4 bytes for its destination address, bringing the total length to 10 bytes. Figure 5-2 shows typical instructions and their hexadecimal equivalents for the examples above.

Typical Instructions	Hexadecimal Equivalent
NOP	4E71
MOVE.L D0,D1	2200
ADD.L D0,D1	D280
ADD.L D0,(A0)	D190
MOVE.W #$1234,D0	303C1234
MOVE.W #$1234,$7000	31FC12347000
MOVE.B $90000,$A00000	13F90090000000A00000

Figure 5-2   Instruction syntax.

Programming at the assembly level, the binary (or hex) instructions are represented by mnemonics. The format of an instruction becomes

OPERATION-CODE   OPERAND, OPERAND

All instructions contain an **operation code (opcode),** and depending on the instruction, contain **one** or **two operands.** Operands are either **memory addresses** or **data** to be used with the instruction.

Since we have a wide variety of ways to address memory, a general form for the instruction is written as

OPCODE   <Sea>, <Dea>

Where **<Sea>** represents the **Source Effective Address** and **<Dea>** represents the **Destination Effective Address.** When it is clear concerning the source and destination, **<ea>** or **Effective Address** is used. The term "effective address" is used since there are several places from which data can be obtained, along with quite a few ways of addressing (pointing to) these locations. When DATA follows the opcode, it is also considered as an *effective address* since it is located in the addresses immediately following the opcode.

Once an *assembler source program* is created, it is then processed by an assembler program. Each assembler has its own *rules and regulations* and special features. Three of these, introduced at this time, are:

   $ used to denote a **hexadecimal number.** Instead of $ABCD_{16}$, we write it as $ABCD for the assembler.

   #   used to denote **DATA,** or a number, as opposed to an ADDRESS. To move or add the hex number ABCD, we use #$ABCD.

   (An)   used to refer to the *MEMORY CONTENTS being **pointed to** by Address Register An*. If we want to refer to the actual contents of Address Register An, we use An; similarly for the Data Register Dn.

Attention must be given to spaces and commas used in the construction of the instructions. Details are given in the examples.

5.3 *THE INSTRUCTION SET*

The 68000 utilizes only 56 basic instructions, but with variations of some, the total exceeds 300. When the addressing modes of both the source effective address and the destination effective address are taken into account (see the 14 addressing modes below), the total possible different instructions reaches an astronomical number. A quick reference card containing the exact hexadecimal opcodes is impossible. Instead, the reference card contains information about what each instruction does, along with information on how to manually construct the instructions, which range from 2 to 10 bytes in length. For this reason alone, instruction entry with hex keypads on a 68000 system is rarely done, and almost all work even at the educational or beginner level is done at the assembly level. Assemblers are economical and interfacing with a personal computer is simple, so there is no reason to begin at the hex keypad level. An understanding of the 56 instructions together with some of the addressing modes is all that is required to begin programming.

Let's look at two approaches to learning the instructions and addressing modes of the 68000:

1. Discuss the syntax of each instruction, either in alphabetical order or by functional grouping, one by one. Provide examples having no practical application periodically for some of the instructions, sometimes not until all the instructions are covered. The references at the end of the book offer this approach.
2. Get a brief description of the functional grouping of the instructions and addressing modes, just to get a general understanding of what we are undertaking. Then tackle an example, picking the appropriate instructions and addressing modes needed to accomplish the task. Reasons for picking the particular instructions and addressing modes are given together with the discussion of programming techniques.

Comparing these two techniques: Suppose that you are given an extensive toolbox, loaded with many tools. Then you are given a complete description of what each tool does, "all at once," before any tasks are assigned. Then when the assignments arrive, we are expected to remember the function of all the tools.

The other approach would be to take the toolbox and be given a brief description of the types of things possible with the tools. Then a simple task is assigned, along with a description of the needed tools and how to use them. Then we tackle increasingly complex tasks, which require more and different tools along with more programming techniques.

Learning will be enhanced greatly utilizing the second approach. However, if you wish to progress through this book utilizing the first approach, the following brief discussion of the instructions and addressing modes include reference to applicable examples in subsequent chapters, and can be referenced when desired.

5.4 *OPERATIONS WITHIN THE CAPABILITY OF THE 68000*

In Table 5-1 we summarize the 56 basic types of instructions in alphabetical order.

TABLE 5-1

68000 Basic Instruction Set

Mnemonic	Description	Mnemonic	Description
ABCD	Add Decimal with Extend	MOVE	Move Source to Destination
ADD	Add Binary	MULS	Signed Multiply
AND	Logical AND	MULU	Unsigned Multiply
ASL	Arithmetic Shift Left	NBCD	Negate Decimal With Extend
ASR	Arithmetic Shift Right	NEG	Negate
Bcc	Branch Conditionally	NOP	No Operation
BCHG	Bit Test and Change	NOT	One's Complement
BCLR	Bit Test and Clear	OR	Logical OR
BRA	Branch Always	PEA	Push Effective Address
BSET	Bit Test and Set	RESET	Reset External Devices
BSR	Branch to Subroutine	ROL	Rotate Left without Extend
BTST	Bit Test	ROR	Rotate Right without Extend
CHK	Check Register with Bounds	ROXL	Rotate Left with Extend
CLR	Clear Operand	ROXR	Rotate Right with Extend
CMP	Compare	RTE	Return from Exception
DBcc	Decrement and Branch Conditionally	RTR	Return and Restore
DIVS	Signed Divide	RTS	Return from Subroutine
DIVU	Unsigned Divide	SBCD	Subtract Decimal with Extend
EOR	Exclusive OR	Scc	Set Conditionally
EXG	Exchange Registers	STOP	Stop
EXT	Sign Extend	SUB	Subtract Binary
ILLEGAL	Illegal Instruction	SWAP	Swap Data Register Halves
JMP	Jump to Effective Address	TAS	Test And Set Operand
JSR	Jump to SuBroutine	TRAP	Trap
LEA	Load Effective Address	TRAPV	Trap on Overflow
LINK	Link Stack	TST	Test
LSL	Logical Shift Left	UNLK	Unlink
LSR	Logical Shift Right		

The instructions can be grouped by the basic function they perform. Quick reference to this grouping is easier than referring to an alphabetical listing when a *tool* is needed to complete a function in a program. The instructions are outlined briefly in the following sections, with an index given to the location of an example describing each instruction.

Notes concerning notation in the tables that follow:

@−		Decrement
@+		Increment
(31:16)		Refers to the upper 16 bits of a register or location
(15:0)		Refers to the lower 26 bits of a register or location
→		Is moved to
(Source)		CONTENTS of the Source location
Data Register		CONTENTS of the Data Register
(Destination)		CONTENTS of the Destination Address
CCR		Condition Code Register (lower byte of Status Register)
PC	Program Counter	C　Carry bit of CCR
SP	Stack Pointer	Z　Carry bit of CCR
SR	Register	∧　Logical AND operation
USP	User Stack Pointer	∨　Logical OR operation
SSP	Supervisor Stack Pointer	%　Exclusive OR operation
X	Extend bit of CCR	~　One's complement

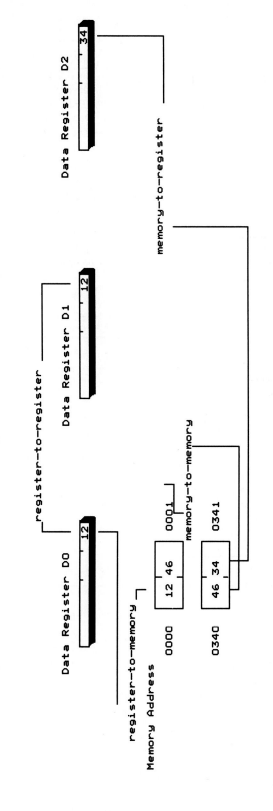

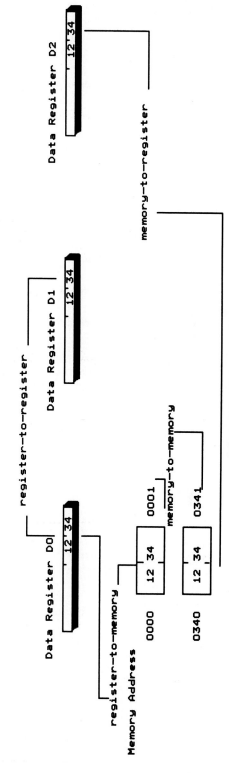

Figure 5-3 Movement of data.

5.4.1 Data Movement or Transfer Instructions

The **MOVE** instruction is the basic method of data acquisition. Data can be stored and loaded as part of our program, or obtained by using the MOVE instruction from a memory location (attached to an input device) during program execution. The MOVE instruction allows DATA movements of bytes, words, and longwords from **memory to memory, memory to** a **register, register to memory,** and **register to register** (Figure 5-3). Memory-to-memory transfers are not typically available on 8-bit μPs. MOVE instructions involving ADDRESSES, not the memory contents, allow movements of words, and longwords *only*. Several other movement or transfer instructions are available.

When data is moved into an ADDRESS REGISTER, data sizes are restricted to word and longword sizes. Byte and word sizes to a DATA REGISTER do NOT alter the HIGHER portions of the Data Register. However, word-sized moves to an ADDRESS REGISTER DO alter the HIGHER portion of the Address Register. The most significant bit of the word-sized data being moved into the lower word of the Address Register is COPIED into bits 16 to 31 of the Address Register. This is called SIGN EXTENSION. We will show later that the most significant bit of a SIGNED number is used to denote its sign; a 1 represents a negative number, a 0 represents a positive number.

In Table 5-2 we show the data MOVEment instructions in alphabetical order together with examples that show their use.

TABLE 5-2

Data Movement Instruction Example List

Mnemonic	Examples
EXG	
LEA	6.19a, 6.21, 6.22, 6.23, 6.25, 6.26, 6.27; 7.6, 7.7, 7.10, 7.11
LINK	
MOVE	6.1, 6.2, 6.3, 6.4, 6.6, 6.10, 6.11, 6.12, 6.14, 6.15, 6.16, 6.17, 6.18, 6.19, 6.20, 6.21, 6.22, 6.23, 6.24, 6.25, 6.26, 6.27; 7.1, 7.2, 7.3, 7.4, 7.5, 7.6, 7.7, 7.8, 7.9, 7.10, 7.11; 8.0, 8.1, 8.2, 8.3; 9.0
MOVEA	6.5, 6.15, 6.16, 6.17, 6.19, 6.20, 6.23
MOVEM	7.10; 8.2; 9.0
MOVEP	
MOVEQ	6.14, 6.15, 6.16, 6.17, 6.18, 6.19, 6.20, 6.21, 6.22, 6.24
PEA	
SWAP	
UNLK	

Table 5-3 provides the data MOVEment instructions together with a brief description of their operation and the allowable operand sizes.

TABLE 5-3

Data Movement Instructions (Parts 1, 2, and 3)

Mnemonic	Description	Operand Size (B, W, L)	Operation
EXG	Exchange Registers	L	Register X ↔ Register Y
LEA	Load Effective Address	L	Destination → Address Register
LINK	Link and Allocate	—	Address Register → Stack Pointer Stack Pointer → Address Register Stack Pointer + displacement → Stack Pointer
MOVE	Move Data from Source to Destination	B, W, L	(Source) → Destination

TABLE 5-3 (continued)

Data Movement Instructions (Parts 1, 2, and 3)

Mnemonic	Description	Operand Size (B, W, L)	Operation
MOVEA	Move contents of Source to Address Register	W, L	(Source) → Address Register
MOVEM	Move Multiple Registers	W, L	Registers → Destination or (Source) → Registers
MOVEP	Move Peripheral Data	W, L	(Source) → Destination
MOVEQ	Move Quick	L	Immediate Data → Destination
PEA	Push Effective Address	L	Destination → Stack Pointer @-
SWAP	Swap Register Halves	L	Register (31:16) ↔ Register (15:0)
UNLK	Unlink	—	Address Register → Stack Pointer Stack Pointer @+ → Address Register

5.4.2 Integer Arithmetic Instructions

The integer arithmetic instructions perform mathematical operations on byte, word, or longword data that represent signed or unsigned integers. A separate group are dedicated to arithmetic with binary-coded-decimal (BCD) numbers. Also in this group is one of the most important instructions, the **CMP (CoMPare)**, useful to compare two quantities. Program branching can be controlled by the result of the comparison.

In Table 5-4 are listed the integer arithmetic (and compare) instructions in alphabetical order.

TABLE 5-4

Integer Arithmetic and Compare Instruction Example List

Mnemonic	Examples
ADD	6.6, 6.7, 6.8, 6.9, 6.14, 6.15, 6.16, 6.17, 6.20, 6.25; 7.7, 7.11; 8.0
ADDA	6.15, 6.22; 7.7
ADDI	
ADDQ	7.4, 7.6, 7.7, 7.10
ADDX	
CLR	6.20, 6.21, 6.22, 6.24, 6.25, 6.26, 6.27; 7.4, 7.6, 7.7, 7.10, 7.11
CMP	6.21, 6.22, 6.27; 7.3, 7.5, 7.6, 7.7, 7.11; 8.2
CMPA	6.22, 6.27; 7.7
CMPI	6.22; 7.10; 8.0, 8.2
CMPM	6.23
DIVS	6.25; 8.0
DIVU	6.12; 7.7
EXT	6.25
MULS	
MULU	6.11; 8.0
NEG	
NEGX	
SUB	6.10, 6.20; 7.4
SUBA	
SUBI	8.0
SUBQ	6.21, 6.22, 6.23, 6.24, 6.26; 7.10, 7.11; 8.2
SUBX	
TAS	
TST	

In Table 5-5 we list the integer arithmetic (and compare) instructions together with a brief description of their operation and the allowable operand sizes.

TABLE 5-5

Integer Arithmetic and Compare Instructions

Mnemonic	Description	Operand Size (B, W, L)	Operation
ADD	Add <ea> to Data Register	B, W, L	(Source) + Data Register → Data Register
ADD	Add Data Register to <ea>	B, W, L	(Destination) + Data Register → Destination
ADDA	Add <ea> to Address Register	W, L	(Source) + (Address Register) → Address Register
ADDI	Add a constant to <ea>	B, W, L	Immediate Data + (Destination) → Destination
ADDQ	Add a constant (1 to 8) to <ea>	B, W, L	Immediate Data + (Destination) → Destination
ADDX	Add <Sea> + Extend Bit to <Dea>	B, W, L	(Source) + (Destination) + X → Destination
CLR	Clears <ea> to zero	B, W, L	0 → Destination
CMP	Compare <Sea> with Data Register	B, W, L	(Destination) − (Source) sets appropriate CCR bits
CMPA	Compare <Sea> with Address Register	W, L	(Address Register) − (Source) sets appropriate CCR bits
CMPI	Compare a constant with <Dea>	B, W, L	(Destination) − Immediate Data sets appropriate CCR bits
CMPM	Compare two memory contents	B, W, L	(Destination) − (Source) sets appropriate CCR bits
DIVS	Divide (signed) Data Register by <Sea>	L/W → L	(Destination)/(Source) → Destination
DIVU	Divide (unsigned) Data Register by <Sea>	L/W → L	(Destination)/(Source) → Destination
EXT	Extend the Sign bit	W, L	(Destination) sign extended → Destination
MULS	Multiply (signed) Data Register by <Sea>	W × W → L	(Source) × (Destination) → Destination
MULU	Multiply (unsigned) Data Register by <Sea>	W × W → L	(Source) × (Destination) → Destination
NEG	Two's complement of <ea>	B, W, L	0 − (Destination) → Destination
NEGX	Two's complement with Extend Bit of <ea>	B, W, L	0 − (Destination) − X → Destination
SUB	Subtract a Data Register from <ea>	B, W, L	(Destination) − Data Register → Destination
SUB	Subtract <ea> from a Data Register	B, W, L	Data Register − (Source) → Data Register
SUBA	Subtract <Sea> from Address Register	W, L	(Address Register) − (Source) → Address Register
SUBI	Subtract a constant from <ea>	B, W, L	(Destination) − Immediate Data → Destination
SUBQ	Subtract a constant (1 to 8) from <ea>	B, W, L	(Destination) − Immediate Data → Destination
SUBX	Subtract <Sea> + Extend Bit from <Dea>	B, W, L	(Destination) − (Source) − X → Destination
TAS	Test and Set Bits of <ea>	B	(Destination) tested, bits → CCR then 1 → bit 7 of Destination
TST	Test <ea> for Negative or Zero	B, W, L	(Destination) tested, bits → CCR

5.4.3 BCD Arithmetic Instructions

Binary-coded-decimal numbers are often used for information displays and for some mathematical operations. These numbers consist simply of the decimal digits 0 to 9. Since a computer does its work in a binary fashion, with a hexadecimal notation being used by us for simplicity, a special set of instructions is needed to allow us to add BCD numbers 6 + 4 to get 10 instead of the hex sum of A. We can also *pack* two BCD digits into a single byte.

The BCD arithmetic instructions are listed in alphabetical order in Table 5-6.

TABLE 5-6

BCD Arithmetic Instruction Example List

Mnemonic	Examples
ABCD	6.18, 6.19, 6.19a, 6.20, 6.21
SBCD	6.21
NBCD	

In Table 5-7 we list the BCD arithmetic instructions together with a brief description of their operation and the allowable operand sizes.

TABLE 5-7

BCD Arithmetic Instructions

Mnemonic	Description	Operand Size (B, W, L)	Operation
ABCD	Add BCD with Extend Data Registers	B	(Source) + (Destination) + X → Destination
ABCD	Add BCD with Extend Memory Locations	B	(Source) + (Destination) + X → Destination
SBCD	Subtract BCD with Extend Data Registers	B	(Destination) − (Source) − X → Destination
SBCD	Subtract BCD with Extend Memory Locations	B	(Destination) − (Source) − X → Destination
NBCD	Negate BCD (Ten's Complement) of <ea>	B	0 − (Destination) − X → Destination

5.4.4 Logical Instructions

The logical instructions allow the AND, OR, exclusive OR, and NOT logical functions to be performed down to the bit level, adding to the versatility of the 68000.

The logic instructions are given in alphabetical order in Table 5-8.

TABLE 5-8

Logical Operations Instruction Example List

Mnemonic	Examples
AND	7.3, 7.5
ANDI	
OR	9.0
ORI	
EOR	9.0
EORI	
NOT	

In Table 5-9 the logical instructions are given together with a brief description of their operation and the allowable operand sizes.

TABLE 5-9

Logical Operations Instructions

Mnemonic	Description	Operand Size (B, W, L)	Operation
AND	Bitwise AND <ea> and Data Register	B, W, L	Data Register $\wedge$ (Destination) $\rightarrow$ Destination
AND	Bitwise AND Data Register and <ea>	B, W, L	(Source) $\wedge$ Data Register $\rightarrow$ Data Register
ANDI	Bitwise AND a constant with <Dea>	B, W, L	Immediate Data $\wedge$ (Destination) $\rightarrow$ Destination
OR	Bitwise OR <ea> and Data Register	B, W, L	Data Register $\vee$ (Destination) $\rightarrow$ Destination
OR	Bitwise OR Data Register and <ea>	B, W, L	(Destination) $\vee$ Data Register $\rightarrow$ Data Register
ORI	Bitwise OR a constant with <Dea>	B, W, L	Immediate Data $\vee$ (Destination) $\rightarrow$ Destination
EOR	Exclusive OR a Data Register with <ea>	B, W, L	Data Register % (Destination) $\rightarrow$ Destination
EORI	Exclusive OR a constant with <Dea>	B, W, L	Immediate Data % (Destination) $\rightarrow$ Destination
NOT	One's Complement of <ea>	B, W, L	$\sim$ (Destination) $\rightarrow$ Destination

5.4.5 Shift and Rotate Instructions

The shift and rotate instructions give us a variety of ways to further manipulate numbers to perform programming tasks. They are best explained with examples. The shift and rotate instructions are listed in alphabetical order in Table 5-10.

TABLE 5-10

Shift and Rotate
Instructions Example List

Mnemonic	Example
ASL	7.2
ASR	7.4
LSL	6.24
LSR	
ROL	
ROR	
ROXL	6.24
ROXR	

In Table 5-11 we list the shift and rotate instructions together with a brief description of their operation and the allowable operand sizes.

TABLE 5-11

Shift and Rotate Instructions

Mnemonic	Description	Operand Size (B, W, L)	Operation	
ASL	Left shift Data Register by a constant	B, W, L	Operand	
ASL	Left shift Data Register by Data Register	B, W, L	Operand	

TABLE 5-11 (continued)

Shift and Rotate Instructions

Mnemonic	Description	Operand Size (B, W, L)	Operation	
ASL	Left shift <ea> by one bit	B, W, L	Operand	
ASR	Right shift Data Register by a constant	B, W, L	Operand	
ASR	Right shift Data Register by Data Register	B, W, L	Operand	ASR
ASR	Right shift <ea> by one bit	B, W, L	Operand	
LSL	Logic left shift Data Register by a constant	B, W, L	Operand	
LSL	Logic left shift Data Register by Data Register	B, W, L	Operand	LSL
LSL	Logic left shift <ea> by one bit	B, W, L	Operand	
LSR	Logic right shift Data Register by a constant	B, W, L	Operand	
LSR	Logic right shift Data Register by a Data Register	B, W, L	Operand	LSR
LSR	Logic right shift <ea> by one bit	B, W, L	Operand	
ROL	Rotate left Data Register by a constant	B, W, L	Operand	
ROL	Rotate left Data Register by a Data Register	B, W, L	Operand	ROL
ROL	Rotate left <ea> by one bit	B, W, L	Operand	
ROR	Rotate right Data Register by a constant	B, W, L	Operand	
ROR	Rotate right Data Register by Data Register	B, W, L	Operand	ROR
ROR	Rotate right <ea> by one bit	B, W, L	Operand	
ROXL	Rotate left with Extend Data Register by a constant	B, W, L	Operand	
ROXL	Rotate left with Extend Data Register by Data Register	B, W, L	Operand	ROXL
ROXL	Rotate left with Extend <ea> by one bit	B, W, L	Operand	
ROXR	Rotate right with Extend Data Register by a constant	B, W, L	Operand	
ROXR	Rotate right with Extend Data Register by Data Register	B, W, L	Operand	ROXR
ROXR	Rotate right with Extend <ea> by one bit	B, W, L	Operand	

5.4.6 Bit Manipulation Instructions

The bit manipulation instructions allow access to individual bits in data, where testing for a 0 or 1 can be done, where the bit can be cleared to 0, can be set to 1, or can be changed from its previous value to the opposite. They offer an alternative approach to bit testing from the shift and rotate instructions. In Table 5-12 we present the bit manipulation instructions in alphabetical order.

TABLE 5-12

Bit Manipulation Instruction Example List

Mnemonic	Examples
BTST	7.3; 8.2
BSET	7.3, 7.5
BCLR	7.3, 7.5
BCHG	7.2, 7.3, 7.5

The bit manipulation instructions together with a brief description of their operation and the allowable operand sizes are given in Table 5-13.

TABLE 5-13

Bit Manipulation Instructions

Mnemonic	Description	Operand Size (B, W, L)	Operation
BTST	Bit test Data Register with <ea>	B, L	~ (Bit of Data Register) → Z
BTST	Bit test a constant with <ea>	B, L	~ (Bit of Destination) → Z
BSET	Test and SET bit Data Register with <ea>	B, L	~ (Bit of Data Register) → Z 1 → Bit of Data Register
BSET	Test and SET bit a constant with <ea>	B, L	~ (Bit of Destination) → Z 1 → Bit of Destination
BCLR	Test and CLEAR bit Data Register with <ea>	B, L	~ (Bit of Data Register) → Z 0 → Bit of Data Register
BCLR	Test and CLEAR bit a constant with <ea>	B, L	~ (Bit of Destination) → Z 0 → Bit of Destination
BCHG	Test and invert bit Data Register with <ea>	B, L	~ (Bit of Data Register) → Z ~ (Bit of Data Register) → Bit of Data Register
BCHG	Test and invert bit a constant with <ea>	B, L	~ (Bit of Destination) → Z ~ (Bit of Destination) → Bit of Destination

5.4.7 Program Control Instructions

This group of instructions allow decision making based on a mathematical or comparison operation.

The **Bcc** branch instructions are executed if the bit conditions (contained in the Status Register) are satisfied, and are simply skipped if the conditions are not met. A byte or word displacement is added to the Program Counter to determine the address where execution is to resume.

The **cc** general notation is used to denote the last two letters of the various branch instructions. (There are BCC and DBCC instructions as well.)

The **DBcc** decrement and branch instructions are simply a "combination instruction" performing two operations.

The **Scc** set conditionally instructions allow setting of all of the bits in an effective address, based on the bit conditions in the Status Register. (The previous bit manipulation instructions set individual bits unconditionally.)

The **BSR** and **JSR** branch and jump to subroutine instructions allow temporary deviation from the normal main program execution to allow execution of individual or stand-alone modules, with return to the next main program instruction after completion of the subroutine.

The **RTR** and **RTS** return from subroutine instructions are put at the end of a subroutine to signify their end.

In Table 5-14 the program control instructions are listed in alphabetical order.

TABLE 5-14

Program Control Instruction Example List

Mnemonic	Examples
BCC	6.17, 6.19, 6.19a, 6.21, 6.27; 7.7
BCS	6.19, 6.24; 7.4, 7.7
BEQ	6.21, 6.22; 7.3, 7.4, 7.5, 7.10, 7.11; 8.0
BGE	
BGT	
BHI	6.22; 7.6
BLE	6.22
BLS	6.22
BLT	6.22
BMI	
BNE	6.23, 6.24, 6.27; 7.7, 7.10, 7.11; 8.2
BPL	
BVC	
BVS	
BRA	6.19a, 6.20, 6.21, 6.22, 6.23, 6.24, 6.25, 6.26, 6.27; 7.1, 7.2, 7.3, 7.4, 7.6, 7.7, 7.10, 7.11; 8.0, 8.2
DBCC	6.25
DBCS	
DBEQ	
DBGE	
DBGT	
DBHI	
DBLE	
DBLS	
DBLT	
DBMI	
DBNE	
DBPL	
DBRA (DBF)	
DBVC	
DBVS	
DBF	6.16, 6.17, 6.19, 6.19a, 6.20, 6.21, 6.22, 6.23, 6.25, 6.26; 7.1, 7.11
DBT	
SCC	
SCS	
SEQ	
SGE	
SGT	
SHI	
SLE	
SLS	
SLT	
SMI	
SNE	
SPL	
SVC	
SVS	
SF	
ST	
BSR	7.7, 7.10, 7.11; 8,0, 8.2
JSR	
JMP	6.13, 6.15, 6.16, 6.17, 6.18, 6.19
RTR	
RTS	7.7, 7.10, 7.11; 8.0, 8.2

The program control instructions together with a brief description of their operation and the allowable operand sizes are given in Table 5-15.

TABLE 5-15

Program Control Instructions

Mnemonic	Description	Displacement Size (B, W, L)	Operation
Bcc		B, W	If condition is true, then PC + displacement → PC; otherwise, execute next instruction
BCC	Branch if C bit is CLEAR		
BCS	Branch if C bit is SET		
BEQ	Branch if Z bit is SET		
BGE	Branch if N and V bits are either both SET or both CLEAR		
BGT	Branch if N and V bits are both SET and Z bit is CLEAR or if N, V, and Z bits are ALL CLEAR		
BHI	Branch if C and Z bits are both CLEAR		
BLE	Branch if the Z bit is SET or if the N bit is SET and the V bit is CLEAR or if the N bit is CLEAR and the V bit is SET		
BLS	Branch if either the C or Z bits are SET		
BLT	Branch if the N bit is SET and the V bit is CLEAR or if the N bit is CLEAR and the V bit is SET		
BMI	Branch if the N bit is SET		
BNE	Branch if the Z bit is CLEAR		
BPL	Branch if the N bit is CLEAR		
BVC	Branch if the V bit is CLEAR		
BVS	Branch if the V bit is SET		
BRA	Branch ALWAYS		
DBcc		W	
DBCC	Terminate if C bit is CLEAR		
DBCS	Terminate if C bit is SET		
DBEQ	Terminate if Z bit is SET		
DBGE	Terminate if N and V bits are either both SET or both CLEAR		
DBGT	Terminate if the N and V bits are both SET and the Z bit is clear or if the N, V, and Z bits are ALL CLEAR		
DBHI	Terminate if the C and Z bits are both CLEAR		
DBLE	Terminate if the Z bit is SET or if the N bit is SET and the V bit is CLEAR or if the N bit is CLEAR and the V bit is SET		
DBLS	Terminate if EITHER the C or Z bit is SET		
DBLT	Terminate if the N bit is SET and the V bit is CLEAR or if the N bit is CLEAR and the V bit is SET		
DBMI	Terminate if the N bit is SET		
DBNE	Terminate if the Z bit is CLEAR		
DBPL	Terminate if the N bit is CLEAR		
DBRA	Terminate by countdown of Dn only (Decrement and Branch)		
DBVC	Terminate if the V bit is CLEAR		
DBVS	Terminate if the V bit is SET		
DBF	Terminate by countdown of Dn only (equivalent to DBRA)		
DBT	Always terminate, no loop at all		
Scc		B	If condition is true, then 1 → Destination; otherwise, 0 → Destination
SCC	Set <ea> if the C bit is CLEAR		
SCS	Set <ea> if the C bit is SET		
SEQ	Set <ea> if the Z bit is SET		
SGE	Set <ea> if the N and V bits are either both SET or both CLEAR		
SGT	Set <ea> if the N and V bits are both SET and the Z bit is CLEAR or if the N, V, and Z bits are ALL CLEAR		
SHI	Set <ea> if the C and Z bits are both CLEAR		
SLE	Set <ea> if the Z bit is SET or if the N bit is SET and the V bit is CLEAR or if the N bit is CLEAR and the V bit is SET		
SLS	Set <ea> if either the C or Z bit is SET		

TABLE 5-15 (continued)

Program Control Instructions

Mnemonic	Description	Displacement Size (B, W, L)	Operation
SLT	Set <ea> if the N bit is SET and the V bit is CLEAR or if the N bit is CLEAR and the V bit is SET		
SMI	Set <ea> if the N bit is SET		
SNE	Set <ea> if the Z bit is CLEAR		
SPL	Set <ea> if the N bit is CLEAR		
SVC	Set <ea> if the V bit is CLEAR		
SVS	Set <ea> if the V bit is SET		
SF	Never set <ea>		
ST	Always set <ea>		
BSR	Branch to <ea> (SubRoutine)	B, W	PC → − (SP), then PC + displacement → PC
JSR	Jump to <ea> (SubRoutine)		PC → − (SP), then Destination → PC
JMP	Jump to <ea>		Destination → PC
RTR	Return and Restore		(SP) + → CCR, then (SP) + → PC
RTS	Return from Subroutine		(SP) + → PC

5.4.8 System Control Instructions

The system control instructions perform a variety of operations, with examples of some of them in later chapters. If we make a serious programming error, the TRAP will let us know. The Motorola ECB has one *good* TRAP instruction that will allow us to use some of the useful subroutines located in the TUTOR monitor in our programs. In Table 5-16 we list the system control or other instructions in alphabetical order.

TABLE 5-16

System Control Instructions Example List

Mnemonic	Examples
CHK	
ILLEGAL	
MOVE USP	
NOP	7.1, 7.7, 8.0
RESET	
RTE	9.0
STOP	
TRAP #0	
TRAP #1	
TRAP #2	
TRAP #3	
TRAP #4	
TRAP #5	
TRAP #6	
TRAP #7	
TRAP #8	
TRAP #9	
TRAP #10	
TRAP #11	
TRAP #12	
TRAP #13	
TRAP #14	7.10, 8.0
TRAP #15	
TRAPV	

The system control instructions together with a brief description of their operation and the allowable operand sizes are listed in Table 5-17.

TABLE 5-17

System Control Instructions

Mnemonic	Description	Displacement Size (B, W, L)	Operation
CHK	Check register against bounds	W	If Data Register <0 or Data Register> <ea>, then TRAP
ILLEGAL			
MOVE USP	Move user stack pointer to/from Address Register	L	If in supervisor state, the USP → Address Register and Address Register → USP; else TRAP
NOP	No operation		Does nothing, PC advanced by one word
RESET	Reset external devices		If in supervisor state, then assert RESET line; else TRAP
RTE	ReTurn from Exception		If in supervisor state, then (SP)+ → SR, (SP)+ → PC; else TRAP
STOP	Stop (enable and wait for interrupts)		If in supervisor state, then Immediate Date → SR, STOP; else TRAP
TRAP			PC → -(SSP), SR → -(SSP), (Vector) → PC
TRAP #0	Trap to address 00080		
TRAP #1	Trap to address 00084		
TRAP #2	Trap to address 00088		
TRAP #3	Trap to address 0008C		
TRAP #4	Trap to address 00090		
TRAP #5	Trap to address 00094		
TRAP #6	Trap to address 00098		
TRAP #7	Trap to address 0009C		
TRAP #8	Trap to address 000A0		
TRAP #9	Trap to address 000A4		
TRAP #10	Trap to address 000A8		
TRAP #11	Trap to address 000AC		
TRAP #12	Trap to address 000B0		
TRAP #13	Trap to address 000B4		
TRAP #14	Trap to address 000B8		
TRAP #15	Trap to address 000BC		
TRAPV	Trap to address 0001C if V bit is set		If V flag is a logic 1, then TRAP

5.5 HOW MEMORY IS ADDRESSED

Recalling that our computer program, together with any stored or computed data, are stored somewhere in the 68000's 16-megabyte (MB) memory map, we need some method of addressing each location. The idea of using 123 Oak St. and 456 Elm St. gets cumbersome in a hurry. The majority of the computation involves MOVEMENT of data from one location to another. A few arithmetic operations and decisions are made along the way as well. Temporary data are also stored in the 68000's internal Data and Address Registers. Our instruction set must have a way of *pointing to* or identifying both the SOURCE and DESTINATION of the data being transferred or operated on.

5.5.1 Overview of the Addressing Modes

The 68000 utilizes six basic types of addressing modes, with variations of some, bring the total to 14 different addressing modes.

The operands of each instruction may reside in computer memory or in the internal registers. By using the 14 different addressing modes, an EFFECTIVE ADDRESS <ea> is created that identifies the location of the operands. When the operand is a number or data immediately following the opcode, it will be designated by a # sign. It is not imperative to learn the names of the various addressing modes.

By practicing with the examples, ideas on how to point to and retrieve the <ea> will be given, and a *general* sense of the capability of the modes will develop. When creating a program, pick the mode you think will do best, then check the specific instruction specification page to see if it exists. If it does, you can get the proper syntax and use that mode. If that mode is not available for that specific instruction, another mode or another instruction must be chosen to perform the task at hand. It is also not imperative to learn ALL 14 modes. Examples in this book will use nine of them at most, leaving the others for the programming experts or for programming tasks beyond the complexity needed here.

In the descriptions below, remember that some instructions have both a SOURCE EFFECTIVE ADDRESS <Sea> and a DESTINATION EFFECTIVE ADDRESS <Dea>. The addressing mode used for <Sea> DOES NOT have to be the same as that chosen for <Dea>. In Table 5-18 we present the complete name for each addressing mode; the notation used; its short name, which is handy during the example discussions; and the example number in which it was first introduced.

TABLE 5-18

Addressing Modes and Example List

Complete name/short name	Notation	Introductory example
1. Direct Data Register Addressing **Data Direct**	**Dn**	6.2
Direct Address Register Addressing **Address Direct**	**An**	6.5
2. Direct Memory Addressing (Short Address) **Direct Short**	**$s**	6.1
Direct Memory Addressing (Long Address) **Direct Long**	**$l**	
3. Address Register Indirect Memory Addressing **Address Indirect**	**(An)**	6.5
Address Register Post-increment Indirect Memory Addressing **Address Post-increment**	**(An)+**	6.5
Address Register Pre-decrement Indirect Memory Addressing **Address Pre-decrement**	**−(An)**	6.10
Address Register Indirect Memory Addressing with Displacement **Address Indirect w/Disp**	**$w(An)**	
Address Register Indirect Memory Addressing with Index and Displacement **Address Indirect w/Indx+Disp**	**$b(An,Xn)**	7.6
4. Implied Register Addressing—Status Register **Implied-SR**		
Implied Register Addressing—Condition Code Register **Implied-CCR**		

TABLE 5-18 (continued)

Addressing Modes and Example List

Complete name/short name	Notation	Introductory example
5. Program Counter Relative Addressing with Displacement **PC Relative w/Disp**	**$w(PC)**	
Program Counter Relative Addressing with Index and Displacement **PC Relative w/Indx+Disp**	**$b(PC,Xn)**	
6. Immediate Data Addressing **Immediate Data**	**#$q**	**6.4**
Immediate Data Addressing—Quick **Quick Immediate**	**#p**	**6.4**

Notes:

Dn	D0 to D7
An	A0 to A7
Xn	Address (An) or Data Register (Dn) used as an Index Register
b	8 bits (byte)
w	16 bits (word)
l	32 bits (longword) (upper byte not used on 68000)
q	8, 16, or 32 bits (b, w, or l)
p	A number between 1 to 8 ($0 = 8, 1 = 1, 2 = 2, \ldots, 7 = 7$)
s	A short address in the range $0000 to $7FFF
$	Denotes a hexadecimal number
#	Denotes data (immediately following the opcode)
<Sea>	Source Effective Address
<Dea>	Destination Effective Address
<ea>	Effective Address

5.5.2 Immediate Data Addressing Mode

The immediate data addressing mode is definitely the most important one since it allows us to *load* the data and address registers with data or addresses, and to *deposit* numbers (or data) into memory locations. There are two variations of this mode:

1. **Immediate:** denoted by **#$q**. The <ea> is the number (or data) immediately following the instruction opcode. The **#** denotes this mode, and **$** is used to denote a hexadecimal number. If the $ is omitted, most assemblers will assume the value to be a decimal number. Some assemblers allow use of binary numbers as well. The q represents a number between $00000000 and $FFFFFFFF.

2. **Quick Immediate:** denoted by **#p**. Identical to the Immediate mode except that the value must be between 1 and 8 (a 0 is used to represent 8 in this case). This allows the value to be embedded in the instruction, greatly reducing the size of the instruction. $ is not needed since decimal numbers between 0 and 7 are the same as their hex equivalents.

5.5.3 Direct Register Addressing Mode

There are two variations of the direct register addressing mode:

1. **Data Register Direct:** denoted as **Dn**. This mode uses one of the Data Registers (D0 to D7) for the <ea>. Taking the MOVE instruction, for instance, we can MOVE into (or from) a Data Register: (a) the contents of another Data Register, (b) the contents of one of the Address Registers, (c) a number (or data), or (d) the contents of a memory location being pointed to by an Address Register.

2. **Address Register Direct:** denoted as **An**. This mode uses one of the Address Registers (A0 to A7) for the <ea>. The other <ea> can be as described for the Data Register Direct mode.

5.5.4   Direct Memory Addressing Mode

There are two variations of the direct memory addressing mode:

1. *Absolute Short:* denoted as **$s**. The <ea> for this mode is a memory location between $00000000 and $00007FFF. The address is short and can be defined with one word (with the four leading zeros omitted), conserving memory space for the instruction.
2. *Absolute Long:* denoted as **$l**. The <ea> for this mode is any memory address between $00000000 and $FFFFFFFF. It takes two words to define <ea> in this case.

5.5.5   Indirect Memory Addressing Mode

There are five variations of the indirect memory addressing mode, the first three being the most commonly used.

1. *Register Indirect:* denoted as **(An)**. The <ea> for this mode is the memory address being indirectly pointed to by the Address Register An. For example, if A1 contains the address $00001000, an instruction using (A1) would load or use the contents of MEMORY LOCATION $00001000. The contents of A1 is not affected.
2. *Post-increment Register Indirect:* denoted as **(An)+**. This mode is identical to the preceding one except that the contents of the Address Register is incremented (increased by one) AFTER the instruction is executed. It is simply a shortcut, doing two instructions at once.
3. *Pre-decrement Register Indirect:* denoted as **-(An)**. This mode is identical to the Register Indirect mode except that the contents of the Address Register is decremented (decreased by one) BEFORE execution of the instruction. Another shortcut instruction.
4. *Register Indirect with Displacement:* denoted as **$w(An)**. This mode operates similar to Register Indirect except that the hex number ($w) is added to the Address Register value to determine the <ea>. The Address Register itself is not altered.
5. *Register Indirect with Index and Displacement:* denoted as **$b(An,Xn)**. Similar to the preceding mode except that the hex number ($b) is 8 bits in length, plus the contents of another Address Register or Data Register (denoted as Xn) is added to the Address Register (An) to determine the <ea>.

5.5.6   Implied Register Addressing Mode

The <ea> for the implied register addressing mode is implied by its name, which can be the Status Register **(SR)**, the User Stack Pointer **(USP)**, the Supervisor Stack Pointer **(SSP)**, or the Program Counter **(PC)**.

5.5.7   Program Counter Relative Addressing Mode

There are two variations of the program relative addressing mode:

1. *Program Counter Relative with Displacement:* denoted as **$w(PC)**. Similar to the Address Register Indirect with Displacement, except that the Program Counter (PC) is used.
2. *Program Counter Relative with Displacement:* denoted as **$b(Pc,Xn)**. Similar to the Address Register Indirect with Index and Displacement, with the PC being used.

chapter 6

getting started with some simple programs

Armed with a general idea of how our 68000 system works, and a general understanding of the types of instructions available for our work, let's begin with some examples. They will increase in complexity, and some will have more practical application than others. Most will be appropriate for entry into the Motorola Educational Computer Board, the Pseudo-Sam and PseudoMax cross-assembler and simulator, or the Micro Board Designs MAX 68000 computer board.

To get to a point where meaningful examples can be presented, we cover three basic functions:

1. *Moving* data from one location to another
2. *Adding* two numbers
3. Program *jumping,* which diverts program execution to a new location

From there we can leapfrog into more useful examples as we cover new instructions. As the examples become more meaningful, the programs increase in complexity and must be thought out before creation. We will develop algorithms, or a plan of attack, sometimes described best by a graphical flowchart, BEFORE we attempt to write the computer program itself. The skill level of a programmer is more accurately determined by his or her ability to develop and use sound algorithms to accomplish the task at hand than by the person's knowledge of the computer's instruction set.

Important Note: From now on all addresses and data will be assumed to be in hexadecimal unless otherwise noted. They will be prefixed by $ when being used with the assembler, and occasionally where needed for clarity.

The creation of a program involves several steps, enough in fact to warrant a flowchart for the process (see Figure 6-1). Although the process is the same for development on almost any computer, in this book we address the specifics for use with a MS-DOS-compatible personal computer. Speaking generally for now, the steps identified by the flowchart in Figure 6-1 are as follows:

1. Turn on the computer, insert the proper disk, and "boot up" the computer.
2. Load the appropriate editor program that will create an ASCII file.
3. Identify the source program that you plan to create. (A .ASM filename extension is typically used to identify the source program.)
4. Create the program (USING YOUR PREVIOUSLY CREATED FLOWCHART), save it to the disk, and return to the computer operating system.

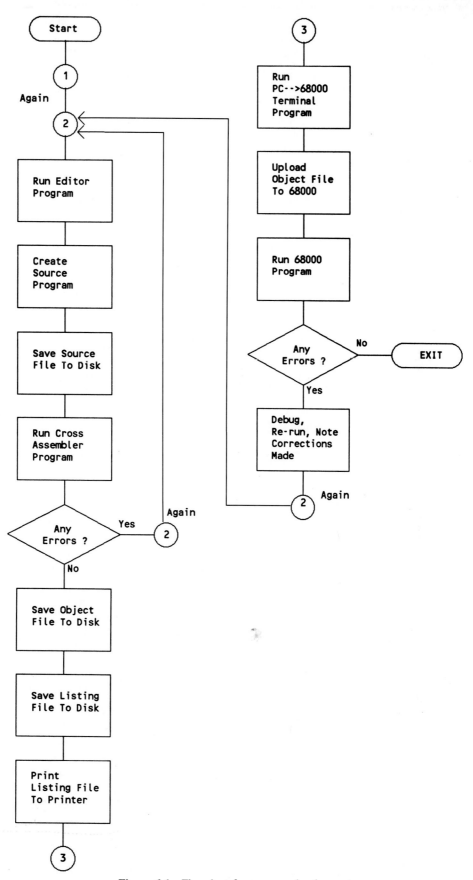

Figure 6-1 Flowchart for program development.

5. Load and run the cross-assembler program, identifying your source program, along with desired options. You will be returned to the operating system upon completion of the assembly.

6. Inspect any error messages provided by the assembler. Return to step 2 if necessary.

7. Upon elimination of errors, check your disk for the appropriate OBJECT code file (it has an .OBJ filename extension). This file contains the actual hex codes that will be loaded into the 68000 computer for execution. The file created is usually in readable ASCII form as a long series of hexadecimal numbers.

8. Check your disk for an OUTPUT file (it has an .LST extension) that contains both your original source code and the hex codes created by the cross-assembler. Copy it to your printer; it will be needed for debugging and execution of the program on the 68000.

9. Load and run the appropriate *terminal* program that allows communication between the personal computer and the 68000 system. After the 68000 system is operable, use the terminal program to UPLOAD your OBJECT code to the 68000.

10. Initialize the 68000 and enter the appropriate instruction to begin execution of the program.

11. Using the available debugging capabilities of the 68000 system, eliminate any programming errors, making note of the changes. Return to step 2 to modify your original source code, to provide a finished product.

When using the PseudoMax simulator in step 9, the personal computer acts as the 68000 computer system. The actual steps for performing steps 9 to 11 are given in Chapter 10.

6.1 MOVING DATA FROM ONE LOCATION TO ANOTHER

Now our first example.

EXAMPLE 6.1

If address 7000 represents an 8-bit INPUT device and address 7010 an 8-bit OUTPUT device, create a program that MOVES a byte from the input to the output.

> **The formats for the MOVE instruction:**
> MOVE.B   <Sea>,<Dea>
> MOVE.W   <Sea>,<Dea>
> MOVE.L   <Sea>,<Dea>

The .B, .W, and .L denote the size of the data to be moved (byte, word, or longword).

The purpose of the MOVE instruction, of course, is to move data from a SOURCE EFFECTIVE ADDRESS, denoted as <Sea>, to a DESTINATION EFFECTIVE ADDRESS, denoted as <Dea>. The general term "effective address" is used since there are several places from which the data can be obtained, along with quite a few ways of addressing (pointing to) these locations.

In addition to the <ea> format to denote an effective address, Motorola 68000 assemblers use the $ sign to denote **hexadecimal** numbers. The **#** sign is used to denote **data** when it is part of the instruction. If the **$ is omitted,** the number will be interpreted as a **DECIMAL** number.

Not all addressing modes are available with this specific MOVE instruction, and Table 6-1 shows the allowable addressing modes and the notation used for <Sea> and

<Dea> for the MOVE instruction. The specific ones available are marked with asterisks (*) or with numbers representing the first example in which they were introduced.

Each example contains the complete instruction description for the newly introduced instructions. Appendix B contains them all, in alphabetical order.

TABLE 6-1

MOVE Instruction

MOVE   **Move Source to Destination**

Assembler Syntax: MOVE.B <Sea>,<Dea>
                      .W
                      .L

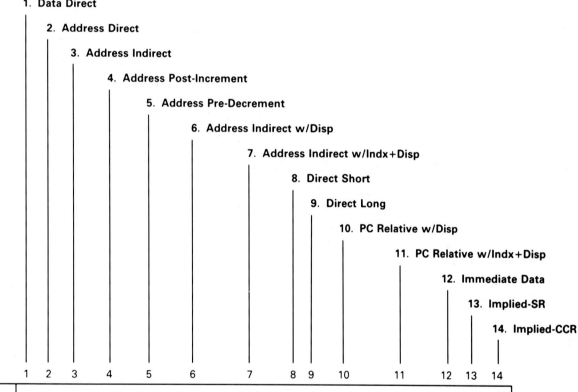

1. Data Direct
2. Address Direct
3. Address Indirect
4. Address Post-Increment
5. Address Pre-Decrement
6. Address Indirect w/Disp
7. Address Indirect w/Indx+Disp
8. Direct Short
9. Direct Long
10. PC Relative w/Disp
11. PC Relative w/Indx+Disp
12. Immediate Data
13. Implied-SR
14. Implied-CCR

<Sea> Source Effective Address	<Dea> Destination Effective Address													
	Dn	An	(An)	(An)+	-(An)	$w (An)	$b (An, Xn)	$s	$l	$w (PC)	$b (PC, Xn)	#$q	SR	CCR
Dn	*		*	*	*	*	*	*	4					
An	*		*	*	*	*	*	*	*					
(An)	*		*	*	*	*	*	*	*					
(An) +	*		*	*	*	*	*	*	*					
- (An)	*		*	*	*	*	*	*	*					
$w (An)	*		*	*	*	*	*	*	*					
$b (An, Xn)	*		*	*	*	*	*	*	*					
$s	*		*	*	*	*	*	1	*					
$l	*		*	*	*	*	*	*	*					
$w (PC)	*		*	*	*	*	*	*	*					
$b (PC, Xn)	*		*	*	*	*	*	*	*					
#$q	3		*	*	*	*	*	2	*					
SR														
CCR														

TABLE 6-1 (continued)

MOVE Instruction

1, 2, 3, and 4 denote modes used for Examples 6.1, 6.2, 6.3, and 6.4, respectively.

Action: <Sea> → <Dea>

Status Register Condition Codes: X   N   Z   V   C
                                      -   *   *   0   0

X is not affected.
N is set if the result is negative. Cleared otherwise.
Z is set if the result is zero. Cleared otherwise.
V always cleared.
C always cleared.

Notes:

1. MOVEA is used when the destination is an Address Register.
2. MOVEM allows moving of a group of registers to/from memory.
3. MOVEP is used for moving data to 8-bit peripheral devices.
4. MOVEQ is used for moving a BYTE, which gets sign-extended to 32 bits.
5. The CCR bits ARE affected by this instruction, unlike some 8-bit microprocessors.
6. $s denotes a short (hex) address in the range $0000 to $7FFF.
7. $l denotes a long (hex) address in the range $8000 to $FFFFFF.
8. $b denotes an 8-bit (hex) number.
9. $w denotes a 16-bit (hex) number.
10. $q denotes an 8-, 16-, or 32-bit (hex) number.

NOTE: **Use of the Status Register Condition Codes is introduced in Example 6.17.**

Addressing Modes Used

Referring to Table 6-1, we can use the **Direct Short** mode, denoted by $s, since our source and destination addresses are below $8000. The instruction is "directly" or "absolutely" defined by the source and destination addresses, which are "short."

To MOVE our **byte** of data from 7000 to 7010, we would use:

```
; Example 6.1
; Move a byte from one memory location to another (7000 to 7010)
;
MOVE.B   $7000,$7010
```

Note that we have added comments to our instructions. They provide clarity to our program, making the program easier to follow the next day. Most assemblers assume them to begin with the ; character.

Referring to Figure 6-2, we see that the contents of only the 8-bit locations at 7010 is affected. Remember that we wanted to move the contents of location 7000 to location 7010. Our program itself (the MOVE instruction) would be located at some other memory area, and location $7000 is somehow previously "loaded" with some meaningful data. Later examples will show this procedure.

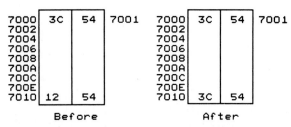

Figure 6-2   Execution of Example 6.1.

NOTE: This example shows a powerful or unique feature of the 68000. The contents of a memory location can be moved from one location to another, WITHOUT processing it through a Data Register. Typical microprocessors require the contents to be loaded into a register and then moved to the new location.

Hands-On Execution of This Example

To develop a good feeling toward assembly language programming, you should consider actually executing this (and subsequent) example(s). If you have purchased the **Motorola Educational Board (ECB),** refer to Section 10.1 at this time for general operation of this useful development tool. Then refer to Section 10.1.2 for details on the execution of Example 6.1.

If you have obtained the floppy disk containing the **PseudoSam and PseudoMax assembler and simulator,** refer to Section 10.2 at this time for general operation of these programs. Then refer to Section 10.2.2 for details on execution of this example.

If you are using the **Micro Board Designs MAX 68000 Board,** refer to Section 10.3 for general operation of this useful development tool. Then refer to Section 10.3.2 for details on execution of this example.

Bit-by-Bit Construction of an Instruction

Although not important when programming at the assembly level, the bit-by-bit construction of an instruction is informative. A foldout card with the hex codes for each of the Motorola's instructions is not practical. If you have entered and executed the example above, perhaps you saw the following bytes in memory for the instruction: 11F8 7000 7010. The last two words, of course, represent the source address and destination address. The MOVE instruction itself is arranged as follows in a bit-by-bit fashion:

$$0\ 0\ a\ a\ b\ b\ b\ c\ c\ c\ d\ d\ e\ e\ e$$

0 0	The identifying most significant bits of a MOVE instruction
a a	Represents the operation size (01 = byte, 11 = word, 10 = longword)
b b b	Represents the register effective address specification for the destination
c c c	Represents the mode used for the destination
d d d	Represents the mode used for the source
e e e	Represents the register effective address specification for the source

Our example has 00 01 000 111 111 000 (equivalent to 11F8), indicating a byte operation using absolute short addressing for both the source and destination.

EXAMPLE 6.2

Put the hex number A6 into memory location $7010.
To MOVE the **byte** of data into location $7010, we would use

```
; Example 6.2

; Put a byte of data into memory location 7010

;

MOVE.B   #$A6, $7010
```

The <Sea> is "immediate data," data that immediately follow the operation code for the MOVE instruction. It must be preceded by a # sign. The <Dea> is a short memory location (one that is less than $8000). The addressing modes used are denoted with a 2 in Table 6-1. The .B extension is used with the opcode to denote movement of a byte. Execution of this example is shown in Figure 6-3.

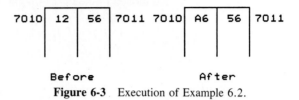

Figure 6-3 Execution of Example 6.2.

EXAMPLE 6.3

Put the hex number A6 into a Data Register.
　　　To MOVE the **byte** of data into Data Register D0, we would use

> ; Example 6.3
>
> ; Put a byte of data into Data Register D0
>
> ;
>
> **MOVE.B   #$A6, D0**

Again, the <Sea> is *immediate data,* data that immediately follow the operation code for the MOVE instruction. The <Dea> is the Data Register D0. The addressing modes used are denoted with a 3 on Table 6-1. The .B extension is used with the opcode to denote movement of a byte. Note that a $ is NOT used in front of D0 for the <Dea>. D0 represents Data Register D0, while $D0 represents a hex number D0. Errors with the #'s and $'s will cause a little discomfort for the beginning programmer. See Figure 6-4 for execution of this example.

Figure 6-4 Execution of Example 6.3.

EXAMPLE 6.4

Move the least significant byte in Data Register D0 to memory location $9000.
　　　To MOVE our **byte** of data from Data Register D0 to $9000, we would use

> ; Example 6.4
>
> ; Put the byte contents of a Data Register into memory location 9000
>
> ;
>
> **MOVE.B   D0, $9000**

The <Sea> is a data register. The <Dea> is a long memory address. The addressing modes used are denoted with a 4 on Table 6-1. The .B extension is used with the opcode to denote movement of a byte. Again, note that a $ is NOT used in front of D0 for <Sea>. (See Figure 6-5.)

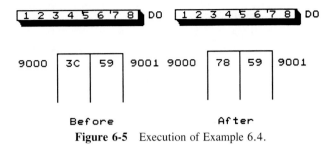

Figure 6-5 Execution of Example 6.4.

EXAMPLE 6.5

(a) Load an Address Register with the **word**-sized contents of a **short** memory location.

(b) Load an Address Register with the **longword**-sized contents of a **short** memory location.

(c) Load an Address Register with the **word**-sized contents of a **long** memory location.

(d) Load an Address Register with the **longword**-sized contents of a **long** memory address.

(e) Load an Address Register with a value representing a **short** address.

(f) Load an Address Register with a value representing a **long** address.

Referring to Table 6-1, we see that the <Dea> cannot be an Address Register for the MOVE instruction. A variation of this instruction, the MOVEA (Table 6-2), is used when an Address Register is the <Dea>. Most assemblers will automatically change the MOVE instructions involving Address Registers to MOVEA.

TABLE 6-2

MOVEA Instruction

MOVEA   (To address register)

Assembler syntax: MOVEA.W <Sea>,An
                              .L

<Sea> Source Effective Address	<Dea> Destination Effective Address An	
Dn		
An		
(An)	*	
(An) +	*	
− (An)	*	
$w (An)	*	
$b (An, Xn)	*	
$s	5a and 5b	
$l	5c and 5d	
$w (PC)	*	
$b (PC, Xn)	*	
#$q	5e and 5f	
SR		
CCR		

TABLE 6-2 (continued)

MOVEA Instruction

5a, 5b, 5c, 5d, 5e, and 5f denote the addressing modes used in Examples 6.5a, 6.5b, 6.5c, 6.5d, 6.5e, and 6.5f, respectively.

Action: $<$Sea$>$ $\rightarrow$ An

Status Register Condition Codes: X   N   Z   V   C

                                                      -    -   -   -   -

No bits are affected.

Notes:
1. The $<$Sea$>$ is stored in An.
2. Only word and longword data sizes are allowed.
3. All 32 bits of the An are affected.
4. Word-sized data is sign-extended to 32 bits.
5. The CCR bits ARE NOT affected, unlike the MOVE instruction.

(a) ; **Example 6.5**
      ; **6.5a Load an Address Register with the word-sized CONTENTS of a short memory**
      ; **address**
      ;
      MOVEA.W $7000, A0

The contents of locations 7000 AND 7001 would be loaded into the lower two bytes of A0, and the two upper bytes of A0 would be *sign-extended*. This means that they would be cleared to 0000 if the word contained in $7000 was less than $7FFF; otherwise, the upper bytes would contain FFFF. In other words, the most significant BIT of the word located in location 7000 is copied into the upper 16 bits of the Address Register. .W is used to denote that a word-sized memory contents is to be used. The addressing modes used are denoted with a 5a in Table 6-2. Sign extension is discussed later. (See Figure 6-6.)

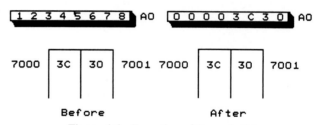

Figure 6-6   Execution of Example 6.5.

(b) ; **6.5b Load an Address Register with the longword-sized ; CONTENTS of a short**
      memory address
      ;
      MOVEA.L $7000, A0

The contents of locations 7000 to 7003 are copied into A0. No sign extension is done since the full 32 bits were provided from the memory locations by using the .L extension. (See Figure 6-7.)

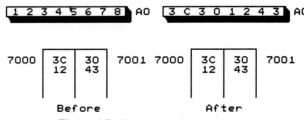

Figure 6-7   Execution of Example 6.5b.

(c) ; **6.5c Load an Address Register with the word-sized CONTENTS of a long memory**
     ; **address**
     ;

     MOVEA.W $8000, A0

This instruction acts just like Example 6.5a. It is the CONTENTS of the memory location that determines whether or not sign extension will be done when a word-sized <Sea> is used. (See Figure 6-8.)

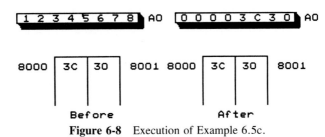

Figure 6-8   Execution of Example 6.5c.

(d) ; **6.5d Load an Address Register with the longword-sized CONTENTS of a long memory**
     ; **address**
     ;

     MOVEA.L $8000, A0

This instruction acts just like Example 6.5b.

(e) ; **6.5e Load an Address Register with a number representing a short memory address**
     ;

     MOVEA.L   #$7000, A0

The number 00007000 would be loaded into A0. If you accidentally use the .W mode, you will find the correct 00007000 to be loaded as well. (See Figure 6-9). **The .L mode should be used even though we are loading a short memory address.** The next example shows why.

Figure 6-9   Execution of Example 6.5e.

(f) ; **6.5f Load an Address Register with a number representing a long memory address**
     ;

     MOVEA.L   #$8000, A0

The number 00008000 would be loaded into A0. (See Figure 6-10.) If we used **MOVEA.W   #$8000,A0** instead, the $8000 will be sign-extended to become FFFF8000—NOT what we intended. *Use .L to load an Address Register with a memory address.*

Figure 6-10   Execution of Example 6.5f.

6.2 *DOING INTEGER ARITHMETIC ON DATA*

Next we add to our tools by doing some arithmetic on the data. You will recall from Chapter 5 that the binary data stored in memory can represent, among other types, signed integers and unsigned integers. In these simple examples we discuss unsigned integers. An

8-bit number can represent decimal values from 0 to 255, or 0 to FF hex; a 16-bit number represents decimal values from 0 to 65535, or 0 to FFFF hex; and a 32-bit number represents decimal values from 0 to 4,294,967,294, or 0 to FFFFFFFF hex. Refer to Section 5.4.2 for the available **ADD** instructions.

EXAMPLE 6.6

If addresses 7000 and 7002 represent 8-bit input devices, create a program that ADDS the bytes contained in these two locations and leaves the sum in Data Register D3.

The formats for the ADD instruction:

```
ADD.B   Dn, <Dea>
ADD.W   Dn, <Dea>
ADD.L   Dn, <Dea>
```

```
ADD.B   <Sea>, Dn
ADD.W   <Sea>, Dn
ADD.L   <Sea>, Dn
```

Recall, again, that .B, .W, and .L denote the size of the data to be moved (byte, word, or longword). We will be working with 8-bit integer data, representing decimal quantities from 0 to 255, so we use the .B extension.

Note: **We cannot directly ADD the contents of two memory locations.** (There is not an ADD <Sea>, <Dea> instruction.) We have to MOVE the contents of one of the addresses into a Data Register and then ADD it to the other location's contents. This approach is common to most or all other microprocessors, by the way.

The purpose of the ADD instruction, of course, is to add the data contained in the SOURCE EFFECTIVE ADDRESS, denoted as <Sea>, to a Data Register (D0 to D7), denoted as Dn, leaving the result in the Data Register. The contents of the <Sea> is not altered. It can also be used to add the data contained in a Data Register (Dn) to a DESTINATION EFFECTIVE ADDRESS, denoted as <Dea>, leaving the result in the <Dea>. The Data Register's contents are not altered.

Not all addressing modes are available with the ADD instructions and Tables 6-3 and 6-4 show the allowable addressing modes for the <Sea> and <Dea> for the ADD and ADDQ instructions.

TABLE 6-3

ADD Instructions

ADD   Add Binary

Assembler Syntax: **ADD.B   Dn,<Dea>**
                              .W
                              .L

<Sea> Source Effective Address	Dn	An	(An)	(An) +	– (An)	$w (An)	$b (An, Xn)	$s	$l	$w (PC)	$b (PC, Xn)	#$q	SR	CCR
Dn			*	*	*	*	*	9	*					

Note: Dn is added to the <Dea> and the result is stored in the <Dea>.

9 denotes the mode used for Example 6.9.

TABLE 6-3 (continued)

ADD Instructions

Assembler Syntax: ADD.B   \<Sea\>,Dn
                         .W
                         .L

\<Sea\> Source Effective Address	*Dn*	*\<Dea\>* Destination Effective Address
Dn	8	(source is word and longword only)
An	*	
(An)	*	
(An) +	*	
− (An)	*	
$w (An)	*	
$b (An, Xn)	*	
$s	6	
$l	*	
$w(PC)	*	
$b (PC, Xn)	*	
#$q	7	
SR		
CCR		

Note: The \<Sea\> is added to Dn and the result is stored in Dn.

6, 7, and 8 denote modes used for Examples 6.6, 6.7, and 6.8, respectively.

Action: \<Sea\> + \<Dea\> → \<Dea\>

Status Register Condition Codes: X N Z V C
                                                  * * * * *

X is set the same as the carry bit.
N is set if the result is negative. Cleared otherwise.
Z is set if the result is zero. Cleared otherwise.
V is set if an overflow is generated. Cleared otherwise.
C is set if a carry is generated. Cleared otherwise.

Notes:
1. ADDA is used when the destination is an Address Register.
2. ADDI and ADDQ are used when the source is immediate data, and are selected automatically by some assemblers.

Addressing Modes Used

Referring to Table 6-3, we can use the **Direct Short** mode, denoted by $s, since our source ADDRESS is below $8000. The **Data Direct** mode is used for the \<Dea\> destination. Since we want our result to be stored in Data Register D3, we can move one of the numbers to be added into that register. We then add the contents of the second address to the contents of D3, with the result being stored in D3.

To ADD the **bytes** of data in 7000 and 7002, leaving the result in D3:

```
; Example 6.6
; Add two bytes of memory (locations 7000 and 7002),
; leaving the sum in a Data Register
;
MOVE.B  $7000, D3
ADD.B   $7002, D3
```

TABLE 6-4

ADDQ Instructions

ADDQ   Add Binary (Quick {1–8} Data)

Assembler Syntax: ADDQ.B   <Sea>,<Dea>
                     .W
                     .L

<Sea> Source Effective Address	<Dea> Destination Effective Address														
	Dn	An	(An)	(An) +	− (An)	$w (An)	$b (An, Xn)	$s	$l	$w (PC)	$b (PC, Xn)	#$q	SR	CCR
#p	*	*	*	*	*	*	*	*	*					

**

Action: <Sea> + <Dea> → <Dea>

Status Register Condition Codes: X   N   Z   V   C
                                    *   *   *   *   *

X is set the same as the carry bit.
N is set if the result is negative. Cleared otherwise.
Z is set if the result is zero. Cleared otherwise.
V is set if an overflow is generated. Cleared otherwise.
C is set if a carry is generated. Cleared otherwise.

Notes:
1. The immediate data must be between 1 and 8 ($ not required); (0 = 8, 1 = 1, 2 = 2, . . . , 7 = 7).
**2. Only word and longword sizes are used when an Address Register is the destination. The source data is sign-extended to longword, and the condition codes are NOT affected for this addressing mode.
3. The 32-bit <Dea> is used regardless of the operation size.

Referring to the Figure 6-11, we see that the contents of only the lowest 8 bits of D3 is affected. Remember that we wanted to add the *contents* of location 7002 to the *contents* of D3. If we had prefixed the $7002 with #$7002, it would of course indicate an error, since this denotes a two-byte hex *number* 7002, not an address. A two-byte, or word-sized, number will not fit in a location designated byte size with the .B suffix.

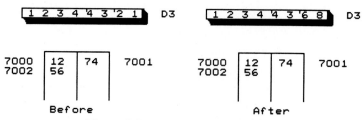

Figure 6-11   Execution of Example 6.6.

Since we specified a byte (.B) size for our operation, our sum will also be limited to byte size. If the numbers to be added are unsigned integers between 00 and FF, the maximum sum we can obtain is FF (255 decimal). We will learn in subsequent examples how to accommodate a possible sum exceeding byte size and how to add word- or longword-sized numbers.

Hands-On Execution of This Example

Refer to Section 10.1.3 for execution of this example for the **ECB,** to Section 10.2.3 for execution on the **PseudoSam/PseudoMax assembler/simulator,** or to Section 10.3.3 for the **MAX 68000.**

Next, here are a few short examples showing the other arithmetic functions: subtraction, multiplication, and division. These operations will, of course, be used in interesting application examples later.

EXAMPLE 6.7

Add the hex number C3 to a Data Register's contents.

To add the immediate data $C3 to D0's contents, we would use:

```
; Example 6.7
; Add a byte of data to the contents of a data register
;
ADD.B   #$C3, D0
```

Use of .B, #, and $ should be obvious by now. The addressing modes used are denoted with a 7 in Table 6-3. Figure 6-12 shows execution of this example.

Before   After

Figure 6-12   Execution of Example 6.7.

EXAMPLE 6.8

Add the least significant byte in Data Register D0 to the contents of D1.

To ADD the least significant byte of D0 to D1, we would use:

```
; Example 6.8
; Add the lower byte of one Data Register to another data register
;
ADD.B   D0, D1
```

The contents of D0 is unchanged. Since a .B extension was used, the addition will be a byte-sized addition. In other words, only the least significant byte of D1 will be affected. If the sum exceeds FF, an overflow would occur, setting the carry bit in the CCR, but the three upper bytes of D1 will not be affected. The addressing modes used are denoted with a 8 in Table 6-3. See Figure 6-13 for execution of this example.

Before   After

Figure 6-13   Execution of Example 6.8.

EXAMPLE 6.9

Add the least significant byte in Data Register D0 to the contents of memory location $7010.

To ADD the least significant byte of D0 to $7010, we would use

```
; Example 6.9
; Add the lower byte of a Data Register to the contents
; of memory location 7010
;
ADD.B   D0, $7010
```

The contents of D0 is unchanged. Since a .B extension was used, the addition will be a byte-sized addition to the previous contents of location 7010. Only location 7010 will be affected. The addressing modes used are denoted with a 9 in Table 6-3. (See Figure 6-14).

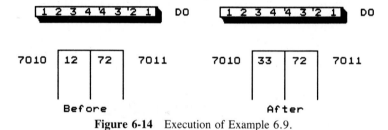

Figure 6-14   Execution of Example 6.9.

EXAMPLE 6.10

Suppose that memory location $7000 contains your salary, which can be represented by a 16-bit number, and location $7002 contains the amount of tax you owe. Write a program that stores, in locations starting at $7004, how much you will have remaining after taxes.

Referring to Table 5-5 to find the subtraction instructions, we find several that may be of value here (Table 6-5).

TABLE 6-5

SUB Instruction

SUB   Subtract Binary

Assembler Syntax: SUB.B   Dn,<Dea>
                            .W
                            .L

<Sea> Source Effective Address	Dn	An	(An)	(An)+	− (An)	$w (An)	$b (An, Xn)	$s	$l	$w (PC)	$b (PC, Xn)	#$q	SR	CCR
Dn			*	*	*	*	*	10	*					

TABLE 6-5 (continued)

SUB Instruction

___

Assembler Syntax: **SUB.B   <Sea>,Dn**
                            .W
                            .L

<Sea> Source Effective Address	*Dn*	*<Dea>* Destination Effective Address	
Dn	*		
An	*	(word and longword only)	
(An)	*		
(An) +	*		
– (An)	*		
$w (An)	*		
$b (An, Xn)	*		
$s	10		
$l	*		
$w (PC)	*		
$b (PC, Xn)	*		
#$q	*		
SR			
CCR			

10 denotes the addressing mode used in Example 6.10.

Action: <Dea> − <Sea> → <Dea>

Status Register Condition Codes: X   N   Z   V   C
                                          *   *   *   *

X is set the same as the carry bit.
N is set if the result is negative. Cleared otherwise.
Z is set if the result is zero. Cleared otherwise.
V is set if an overflow is generated. Cleared otherwise.
C is set if a borrow is generated. Cleared otherwise.

___

Notes:
1. SUBA is used when the destination is an Address Register.
2. SUBI and SUBQ are used when the source is immediate data.

___

As with addition, to use the **SUB** instruction, ONE of our <ea> effective addresses has to be one of the Data Registers. Note that for both forms of the SUB instruction, the operation performed is <Dea> − <Sea> → <Dea>. In other words, the data in the Source Effective Address is subtracted from the Destination Effective Address and the result is left in the Destination Effective Address. There are two ways to tackle this problem.

To SUBTRACT the word-sized data in 7002 from that contained in 7000, leaving the result in 7004:

```
; Example 6.10
; Subtract the contents of memory location 7002 from
; another (7000), leave result in another location (7004)
;
MOVE.W   $7000, D0      ; get the salary data
SUB.W    $7002, D0      ; subtract the tax from it
MOVE.W   D0, $7004      ; store the result
```

The addressing mode used is denoted by 10 in the upper part of Table 6-5. Figure 6-15 shows the operation.

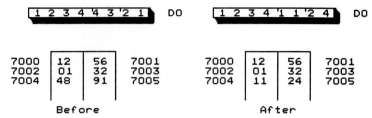

Figure 6-15   Execution of Example 6.10.

Another way:

```
; Example 6.10a
; Subtract the contents of memory location 7002 from
; another (7000), leave result in another location (7004)
;
MOVE.W   $7002, D0      ; get the taxes to be subtracted
SUB.W    D0, $7000      ; subtract them from the salary
MOVE.W   $7000, $7004   ; store the result
```

The addressing mode used is denoted by 10 in the lower part of Table 6-5. Figure 6-16 shows the operation.

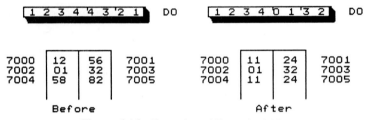

Figure 6-16   Execution of Example 6.10a.

Hands-On Execution of This Example

Refer to Section 10.1.4 for execution of this example for the **ECB,** to Section 10.2.4 for execution on the **PseudoSam/PseudoMax assembler/simulator,** or to Section 10.3.4 for the **MAX 68000.**

Addition and subtractions can be performed on 8-, 16-, and 32-bit data, with the result being the same size as the numbers being added or subtracted. If we were to add two large 8-bit numbers whose sum was more than 8 bits long, it would be **up to us** to take that into consideration in our program; similarly for 16- and 32-bit data. The 68000 handles multiplication and division slightly different. First a multiplication example.

EXAMPLE 6.11

Suppose that memory location $7000 contains a byte-sized number and you need to find its SQUARE. Let's store it in location $7004.

Referring to Table 5-5 to find the multiplication instructions, we find MULU (Table 6-6).

TABLE 6-6

MULU Instruction

MULU   Unsigned multiply

Assembler Syntax: **MULU.W <Sea>,Dn**

<Sea> Source Effective Address	<Dea> Destination Effective Address Dn
Dn	11
An	
(An)	*
(An) +	*
− (An)	*
$w (An)	*
$b (An, Xn)	*
$s	*
$l	*
$w (PC)	*
$b (PC, Xn)	*
#$q	*
SR	
CCR	

11 denotes the addressing mode used in Example 6.11.

Action: <Sea>*Dn → Dn

Status Register Condition Codes: X   N   Z   V   C
                                      -   *   *   0   0

Z is not affected.
N is set if the most significant bit of the result is negative. Cleared otherwise.
Z is set if result is zero. Cleared otherwise.
V is cleared.
C is cleared.

Note: The two 16-bit unsigned operands are multiplied using unsigned arithmetic, and 32 bits of the product are saved in Dn (no sign extension is done).

Our only choice for a <Dea> (Destination Effective Address) is a Data Register, and only 16-bit multiplication is available. This works fine for 8-bit numbers as well if we first clear out the upper bytes for the numbers to be multiplied; 32-bit multiplications will require a little extra programming effort.

Our program:

```
; Example 6.11
; Square a byte-sized number stored in memory location 7000,
; store the sum in another memory location (7004)
;
MOVE.L   #0,D0          ; clear out Data Register
MOVE.B   $7000,D0       ; get the number to be squared
MULU.W   D0,D0          ; multiply it by itself,
                        ; leave 32-bit result in D0
MOVE.L   D0,$7004       ; store the 32-bit result starting at $7004
```

Note: For our BYTE multiplication, only two locations are needed for the product, but we can allocate a longword for the product if we wish. The longword location starting at

7004 will have the proper value for an unsigned integer multiplication and will not be sign-extended.

The addressing mode used is denoted by 11 in Table 6-6. Figure 6-17 shows the operation. Note that $07_{16} \times 07_{16} = 31_{16} = 49_{10}$.

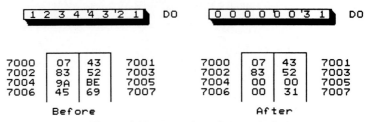

7000	07	43	7001	7000	07	43	7001
7002	83	52	7003	7002	83	52	7003
7004	9A	BE	7005	7004	00	00	7005
7006	45	69	7007	7006	00	31	7007

Before   After

Figure 6-17   Execution of Example 6.11.

Division works similarly, expecting a big number to start with and a small number after the division, the <Sea> is our 32-bit numerator, and the <Dea> is the 16-bit denominator. The result is 16 bits long, and the remainder is also preserved.

EXAMPLE 6.12

Write a program that divides the 16-bit number located in memory location $7000 by a 16-bit number located in $7002. Store the result in location $7004. Referring to Table 5-5 to find the division instructions we find DIVU (Table 6-7).

Our only choice for a <Dea> (Destination Effective Address) is a Data Register. The numerator can be 8, 16, or 32 bits long if the upper bytes are first cleared out. The denominator can be 8 bits if the upper byte is first cleared out.

Our program:

```
; Example 6.12
; Divide a 16-bit number stored in memory location 7000
; by another (in 7002),
; store the result in another location (7004)
;
MOVE.L  #0,D0          ; clear out the data register
MOVE.W  $7000,D0       ; get the numerator
DIVU.W  $7002,D0       ; divide it
MOVE.W  D0,$7004       ; store the 16-bit result,
                       ; ignoring the remainder
```

The addressing mode used is denoted by 12 in the table above. Figure 6-18 shows the operation. Note that the division instruction also stores a possible remainder in the upper word of <Dea>. Its use depends on the application, and it is up to the programmer to use it if necessary. To see how the remainder is obtained, note that $5678_{16} = 22136_{10}$ and $1326_{16} = 4902_{10}$. Dividing the hex numbers 5678/1326 gives 0004 with a remainder of C972. $C972_{16} = 51570_{10}$. Dividing the decimal equivalents 22136/4902 gives 4.515707874. Note that our fractional part cannot exceed $FFFF_{16}$ or 65535_{10}—hence the truncation of the least significant digits.

Now that we can do a little arithmetic on data, and move the data between registers, memory, and so on, let's learn a "looping" instruction so that we can come up with some less boring examples.

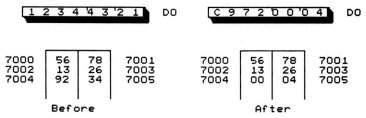

7000	56	78	7001
7002	13	26	7003
7004	92	34	7005

Before

7000	56	78	7001
7002	13	26	7003
7004	00	04	7005

After

Figure 6-18 Execution of Example 6.12.

TABLE 6-7

DIVU Instruction

DIVU   Unsigned Divide

Assembler Syntax: **DIVU.W <Sea>,Dn**

<Sea> Source Effective Address	<Dea> Destination Effective Address Dn	
Dn	*	
An		
(An)	*	
(An) +	*	
− (An)	*	
$w (An)	*	
$b (An, Xn)	*	
$s	12	
$l	*	
$w (PC)	*	
$b (PC, Xn)	*	
#$q	*	
SR		
CCR		

12 denotes the addressing mode used in Example 6.12.

Action: <Dea>/<Sea> → <Dea>

Status Register Condition Codes: X   N   Z   V   C
                                       -   *   *   *   0

X is not affected.
N is set if the quotient is negative. Undefined if overflow. Cleared otherwise.
Z is set if the quotient is zero. Undefined if overflow. Cleared otherwise.
V is set if division overflow occurs. Cleared otherwise.
C is cleared.

Notes:
1. The longword-sized <Dea> is divided by a word-sized <Sea>, and the result is stored in the <Dea>.
2. The operation is performed using unsigned arithmetic.
3. The quotient is stored in the lower word of the <Dea>.
4. The remainder is stored in the upper word of the <Dea>.
5. The sign of the remainder is the same as the dividend unless it is equal to zero.
6. Division by zero causes a TRAP.
7. Overflow may occur before completion of the instruction. If this occurs, the condition is flagged and the operands are unaffected.
8. Overflow occurs if the quotient is larger than a 16-bit unsigned integer.

6.3 *DIVERTING NORMAL PROGRAM EXECUTION WITH A JUMP (OR BRANCH) INSTRUCTION*

If you will recall, the *postmaster* starts execution of the program at its beginning and proceeds in a *straight line* down the program unless told to do otherwise. For our next example we consider what is called a JUMP (JMP) instruction. In other words, the Program Counter, which always contains the address of the NEXT instruction to be executed, is loaded with the *operand* (or <ea>) of the JMP instruction. Recall, the operands of an instruction follow the opcode, and in this case represents a memory address. Program execution resumes at that new location, not at the location following the JMP instruction. The JMP allows movement to ANY location within the 68000's addressing range. Another related instruction is the BRANCH (BRA) instruction. When the destination address is *nearby* in relation to the current Program Counter, the BRA (branch always) can be used. It is in the Bcc (branch conditionally) group and is covered in subsequent examples.

EXAMPLE 6.13

Write a program that causes the microprocessor to JUMP to location $4000. Refer to Section 5.4.7 for the **Program Control** instructions. One other JUMP instruction is available, the Jump to Subroutine, which is covered later.

The format for the JUMP instruction:

JMP <ea>

Not all addressing modes are available with the JMP instruction, and Table 6-8 shows the allowable addressing modes for <ea>.

TABLE 6-8

JMP Instruction

JMP   **Jump to effective address**

Assembler Syntax: **JMP <label> or JMP <ea>**

							<ea> Effective address							
Dn	An	(An)	(An) +	- (An)	$w (An)	$b (An, Xn)	$s	$l	$w (PC)	$b (PC, Xn)	#$q	SR	CCR
		*			*	*	13	*	*	*			

13 denotes the addressing mode used in Example 6.13.

Action: <ea> → Program Counter

Status Register Condition Codes: X N Z V C
                                 - - - - -

No bits are affected.

Note: Program execution resumes at the address specified by <ea>.

Addressing Modes Used

Referring to Table 6-8, we can use the **Direct Short** mode, denoted by $s, since our destination addresses is below $8000. To JUMP to 4000:

```
; Example 6.13
; Divert program execution to another address
;
JMP $4000
```

See execution of this example in Figure 6-19. Note that we do not, and should not, prefix the address $4000 with the # sign, which denotes a NUMBER instead of an ADDRESS. Since we are not dealing with data, the .B, .W, and .L notation is not used. Note also that when working with an assembler or cross-assembler, we can denote our <ea> with a **LABEL** instead of the actual hex address. We put the corresponding label at the beginning of the instruction we wish to jump to, and the assembler computes or finds the proper hex address for us. Later examples using the assembler show this technique.

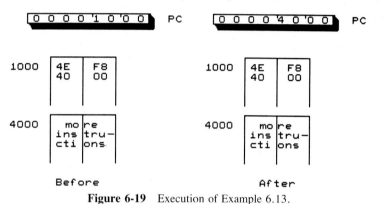

Figure 6-19   Execution of Example 6.13.

Now let's expand slightly on Example 6.13.

EXAMPLE 6.14

Suppose that addresses 7000, 7001, 7002, 7003, and 7004 represent five INPUT ports or addresses containing data (in byte format), and address 7010 is an 8-bit OUTPUT port or address. The input data could be obtained from a wide variety of devices, and the output address could be a visible digital display unit. Write a program that causes the microprocessor to add or total the contents of the input locations and put the sum in the output location.

Since this task is a bit more complicated, let's discuss our plan of attack. We can MOVE data from just about anywhere to anywhere. We can only ADD the contents of a Data Register to an effective address, or the contents of an effective address to a Data Register. The latter seems appropriate for this example. We can use a Data Register to *accumulate* the sum and then move the result to the output destination address. Let's use Data Register D0, and remembering that the ADD instruction *adds to the current contents* of D0, we will initialize it to zero. Refer to the MOVE table (Table 6-1) and the ADD instruction table (Table 6-3) to select appropriate instructions and addressing modes.

To put the sum of the contents of addresses 7000, 7001, 7002, 7003, and 7004 into location 7010:

```
; Example 6.14
; Sum the byte contents of five memory locations (starting at 7000), store sum in another
; memory address (7010)
```

```
;
MOVEQ.L #0,D0          ; clear out the data register
ADD.B $7000,D0         ; add the first number to D0
ADD.B $7001,D0         ; add the next number
ADD.B $7002,D0         ; add the next number
ADD.B $7003,D          ; add the next number
ADD.B $7004,D0         ; add the last number
MOVE.B D0,$7010        ; store the sum
```

A new instruction was introduced here. The **MOVEQ** is a shorter (2 bytes) and faster instruction for moving a small number. A little caution is needed when using it, however. Table 6-9 shows its characteristics.

TABLE 6-9

MOVEQ Instruction

MOVEQ   Move (Quick Data)

Assembler Syntax: MOVEQ.L #$q,Dn

<Sea> Source Effective Address	*<Dea>* Destination Effective Address	
	Dn	
#$q	14	

14 denotes the addressing mode used in Example 6.14.

Action: #$q → Dn

Status Register Condition Codes: X   N   Z   V   C
                                     -   -   -   -   -

No bits are affected.

Notes:
1. The immediate data is limited to byte size, allowing constants in the range −128 to +127 decimal.
2. The data is sign-extended to a longword and all 32 bits stored in the Dn.
3. Dn will contain a 32-bit representation of the signed number between −128 and +127.

MOVEQ is different from MOVE in that the entire 32 bits of the Data Register are affected. This will usually pose no unexpected problem. **MOVEQ #$7F,D0** would put 7F in the lower byte and would clear out the upper 24 bits. **MOVE.B #$7F,D0** would put 7F in the lower byte of D0 and leave the remaining 24 bits unaffected. If the number is greater than 7F, MOVE still does not affect the upper 24 bits, while MOVEQ would set them all to 1's. In other words, the MOVEQ sign extends the 8-bit number, ASSUMED to be a signed number, to a 32-bit signed number. This result could be used to advantage, or ignored if not needed. Another unusual characteristic of MOVEQ is that no bits are affected in the Condition Code Register.

In our program we clear out D0 initially, before accumulating our sum. The individual bytes from our input locations were added to the contents of D0 in turn, and finally the 8-bit sum was moved to the output address. This example is similar to a previous one, with two differences: The *odds* of our sum exceeding FF is larger, and the number of locations we wish to add is larger, requiring a longer program to be created. (See Figure 6-20.)

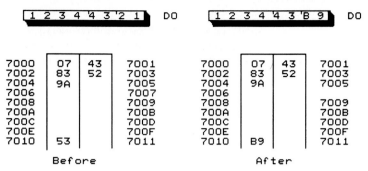

Figure 6-20   Execution of Example 6.14.

Addressing Modes Used

The new addressing mode used is denoted by 14 in Table 6-9. The first MOVE uses **Quick Immediate** addressing for the <Sea> and **Data Direct** for the <Dea>. The ADDs use **Direct Short** for the <Sea> and **Data Direct** for the <Dea>. The last MOVE uses **Data Direct** for the <Sea> and **Direct Short** for the <Dea>.

Hands-On Execution of This Example

Refer to Section 10.1.5 for execution of this example for the **ECB,** to Section 10.2.5 for execution on the **PseudoSam/PseudoMax assembler/simulator,** or to Section 10.3.5 for execution on the **MAX 68000.**

Now let's combine these three basic operations—MOVE, ADD, JMP—into an example.

EXAMPLE 6.15

Now, suppose that addresses 7000 to 70FF represent 256 INPUT ports or addresses containing data (in byte format), and address 7100 is an 8-bit OUTPUT port or address. Write a program that causes the microprocessor to add or total the contents of the input locations and put the sum in the output location.

We will attempt to write a program using the instructions covered so far, then seeing the flaws and limitations, cover some more instructions so that we can more efficiently satisfy the task.

Addressing Modes Used

Using techniques and instructions covered so far, we see a lot of typing ahead in the construction of this program. The program would be over 256 lines long using the techniques covered so far. Realizing that the Address Registers can be used to hold (or point to) memory addresses, let's use a new addressing mode to get the data we need to add. Referring to Section 5.5.1, we see the **Address Register Indirect Memory Addressing** mode, denoted by (An). The parentheses denote **THE ADDRESS POINTED TO,** hence the concept of indirectly pointing to an effective address.

Plan of attack:
1. Point to the beginning address of the input data.
2. Clear out Data Register D0.
3. Add the contents of the address being pointed to, to the contents of D0.
4. Increase the address pointer by one.
5. Repeat the process until done.

We use Address Register A0 as the Address Register pointer.

It sounds like we will have to resort to a graphical display of our plan of attack, more accurately called a flowchart, if the examples get any more sophisticated. Refer to previous tables to select the proper instructions and addressing modes, remembering to pick **Address Register Indirect** denoted by (An) for the <Sea> instruction used to point to the memory address of the start of the data.

To put the sum of the contents of addresses 7000 to 70FF into location 7000:

```
; Example 6.15 (first attempt)
; Sum the byte contents of 256 memory locations (starting at 7000),
; store result in another memory location (7100)
;
        MOVEQ.L #$0,D0          ; clear out the accumulator
        MOVEA.L #$7000,A0       ; point to the source of data
LOOP:   ADD.B (A0),D0           ; add to D0 the data pointed to
        ADDA.L #$1,A0           ; bump the address pointer up one
        MOVE.B D0,$7100         ; move the sum to the output address
        JMP LOOP               ; get the next data to add
```

Note that we have added what is called a LABEL called LOOP to our program. This identifies to the assembler a memory location (containing an instruction) that the program should jump to and resume program execution. Most assemblers require LABELS to start in the FIRST column, be 8 or fewer characters long, and end with a : (colon). If a source code line does not have a label, the OPCODE begins in any column AFTER column 1.

Even before you try to execute this program, you probably see its flaws. The continuous JUMP will work fine for the first 256 loops, BUT it will continue indefinitely. Although an UNCONDITIONAL JUMP appears useful in some applications, what we need now is a CONDITIONAL JUMP that knows when to quit.

We needed a new instruction, **ADDA,** in order to add one to our Address Register being used as a pointer (see Table 6-10). About the only thing different about this instruction is that no bits are affected in the Condition Code Register.

TABLE 6-10

ADDA Instruction

ADDA   Add Binary (To Address Register)

Assembler Syntax: ADDA.W <Sea>,An
                         .L

<Sea> Source Effective Address	<Dea> Destination Effective Address	
	An	
Dn	*	
An	*	
(An)	*	
(An) +	*	
− (An)	*	
$w (An)	*	
$b (An, Xn)	*	
$s	*	
$l	*	
$w (PC)	*	
$b (PC, Xn)	*	
#$q	15	
SR		
CCR		

TABLE 6-10 (continued)

ADDA Instruction

---

15 denotes the mode used in Example 6.15.

Action: <Sea>+<Dea> → <Dea>

Status Register Condition Codes: X   N   Z   V   C

                                              -   -   -   -   -

No bits are affected.

Notes:
1. The <Sea> is added to the <Dea>, and the result is stored in the <Dea>.
2. Word data is sign-extended to 32 bits.
3. The 32-bit <Dea> is used regardless of the operation size.
4. NO Condition Code Register bits are affected.

Addressing Modes Used

The second instruction uses the **Immediate Data** mode for the <Sea> and the **Address Direct** mode for the <Dea>. In other words, we "immediately" moved an address "directly" into an Address Register.

Before we tackle the conditional jump improvement, which will involve several new instructions, let's consider an improvement to Example 6.14 using a fancier addressing mode. Referring to Section 5.5.1, directly below the Register Indirect mode, we see **"Post-Increment Register Indirect,"** denoted by (An)+. Simply stated, this addressing mode is used to increment the contents of the ADDRESS REGISTER (not the memory location) *automatically* AFTER the instruction is executed. In effect, it eliminates the need for the **ADDA.L #$1,A0** instruction. Our new program would look as follows:

```
; Example 6.15 (second attempt)
; Sum the byte contents of 256 memory locations (starting at 7000),
; store the result in another memory location (7100)
;
          MOVEQ.L #$0,D0        ; clear out the accumulator
          MOVEA.L #$7000,A0     ; point to the source of data
LOOP:     ADD.B (A0)+,D0        ; add to D0 the data pointed to,
                                ; bump pointer
          MOVE.B D0,$7100       ; move the sum to the output address
          JMP LOOP             ; get the next data to add
```

If you choose to try this example, you will see that it performs the same function but occupies fewer memory locations.

Note also that A0 is incremented one memory location at a time since we are doing a BYTE operation. An Address Register can point to an ODD operation for byte operations, but can only point to EVEN addresses for WORD operations.

6.4 DIVERTING NORMAL PROGRAM EXECUTION WITH A CONDITIONAL BRANCH INSTRUCTION

The next level of sophistication is to learn how the *postmaster* can make comparisons that result in deviation of normal program execution. Instead of progressing normally through a program, addition of decision-making capability allows a lot more flexibility. Being a binary computer, only a *two-way* decision can be made. Programming techniques are used when a multiple-choice decision or selection is needed.

In the next example we use an instruction in the DBcc (conditional decrement and branch) group. Instructions in this group provide two functions with a single instruction, that of countdown or decrementing and conditional branching.

EXAMPLE 6.16

Perform Example 6.15 using new instructions to make it completely and correctly functional.

It looks like we need a CONDITIONAL jump (called a *BRANCH*) based on the results of a *countdown* from 256 to 0 (for the 256 input addresses), or perhaps on a comparison of the contents of an increasing Address Register with the value 70FF, which is the last input address to be used. We also need to look into how to accommodate a sum that exceeds FF.

Using Data Register D1 as the countdown register, we can load it with a countdown number (called a MOVE for the 68000, of course), perform an ADD instruction to add 2 bytes of input data, decrement the contents of D1 by one (subtract one from it), check it for a ZERO condition, and then branch back up to our *loop* instruction if the result is NOT ZERO. Careful consideration needs to be given to this to ensure that we get the correct number of loops to gather only the 256 locations for the input data.

The 68000 has a combination instruction that does a lot of the decrementing AND branching for us. Referring to the **Program Control** instructions in Section 5.4.7, we see the **DBF** (decrement and branch if false) instruction. It has two names, **DBRA** (decrement and branch) also being used by some assemblers. This instruction is generally defined as

DBcc Dn,<label>

The **cc** is replaced by various letters for several test conditions that are available. We want to use the **DBF** (false) instruction. This instruction checks the condition of the countdown register (Dn). If it is false (not equal to zero), the contents of Data Register Dn is decremented by one, and program execution is resumed at the address equivalent to the label.

If the check is true (equal to zero), the Data Register is not decremented, and program execution resumes at the next instruction following the DBcc instruction. One limitation is that the data register countdown is limited to a 16-bit value. Refer to Table 6-11 and previous tables to select the proper instructions and addressing modes for this example.

TABLE 6-11

DBcc Instruction

DBcc   **Decrement and Branch Conditionally**

Assembler Syntax: **DBcc.W Dn,<label>**   or   **DBcc.W Dn,displacement**

cc is replaced by one of the conditions below:

Action: If (condition is false), then $Dn - 1 \rightarrow Dn$;
               if Dn does not equal -1, then $PC+displacement \rightarrow PC$,
               else $PC+2 \rightarrow PC$ (resumes at next instruction)

DBCC	Terminate if C bit is CLEAR
DBCS	Terminate if C bit is SET
DBEQ	Terminate if Z bit is SET
DBGE	Terminate if N and V bits are either both SET or both CLEAR
DBGT	Terminate if the N and V bits are both SET and the Z bit is clear or if the N, V, and Z are ALL CLEAR
DBHI	Terminate if the C and Z bits are both CLEAR
DBLE	Terminate if the Z bit is SET or if the N bit is SET and the V bit is CLEAR or if the N bit is CLEAR and the V bit is SET

TABLE 6-11 (continued)

DBcc Instruction

DBLS	Terminate if EITHER the C or Z bit is SET
DBLT	Terminate if the N bit is SET and the V bit is CLEAR or
	if the N bit is CLEAR and the V bit is SET
DBMI	Terminate if the N bit is SET
DBNE	Terminate if the Z bit is CLEAR
DBPL	Terminate if the N bit is CLEAR
DBRA	Terminate on countdown of Dn only (Decrement and Branch)
DBVC	Terminate if the V bit is CLEAR
DBVS	Terminate if the V bit is SET
DBF	Terminate on countdown of Dn only; equivalent to DBRA
DBT	Always terminate; no loop at all

Status Register Condition Codes: X   N   Z   V   C

                                      -   -   -   -   -

No bits are affected.

Note: If the specified condition is not met, the **lower 16 bits** of Dn are decremented by one, and if it is not equal to -1, execution continues at <label>. Otherwise, execution resumes at the instruction following the branch instruction.

Our program looks like this:

```
; Example 6.16 (final version of Example 6.15)
; Sum the byte contents of 256 memory locations (starting at 7000),
; store the result in another memory location (7010)
; Exit (or loop) correctly when done
;
          MOVEQ.L #$0,D0        ; clear out the accumulator
          MOVE.W #$FF,D1        ; initialize the countdown register
          MOVEA.L #$7000,A0     ; point to the source of data
LOOP:     ADD.B (A0)+,D0        ; add to D0 the data pointed to,
                                ; bump the pointer
          DBF D1,LOOP          ; get the next data to add
          MOVE.B D0,$7100      ; move the sum to the output address
HERE:     JMP HERE             ; a way of stopping the program
```

Note that we move a WORD #$FF into D1, the countdown counter. It loads D1 as XXXX00FF. (The XXXX's indicate the previous contents of the upper word in D1, which is not altered.) We cannot use the MOVEQ instruction since the FF value exceeds $+127$ decimal. The DBF instruction decrements the word-sized (lower word) D1 register. **The FF gives us 256 times through the loop, not 255,** since the DBF does not fail until D1 is equal to -1. We will refer to this as the "DBF quirk," and hopefully, take it into account each time we use it.

The last instruction is but one way to *halt* a program. The program will continually loop on the same instruction, and the only way to regain control of the program would be to hit an abort button on the computer (red button on ECB). More elegant ways of regaining control are given in the individual explanations of the ECB, simulator, and MAX 68000.

Addressing Modes Used

No new addressing modes were covered.

Hands-On Execution of This Example

Refer to Section 10.1.6 for execution of this example for the **ECB,** to Section 10.2.6 for execution on the **PseudoSam/PseudoMax assembler/simulator,** or to Section 10.3.6 for the **Max 68000.**

Recapping, we can now gather data from up to FFFF + 1 (65536) input locations, add them, and output the 8-bit sum to an output memory location. Realizing now that our sum may indeed exceed FF, let's polish our program to allow for a much larger sum.

EXAMPLE 6.17

Now, suppose that addresses 7000 to 70FF represent 256 INPUT ports or addresses containing data (in byte format), and addresses 7100 and 7101 are considered together to provide a word-sized OUTPUT port. Write a program that causes the microprocessor to add or total the contents of the input locations and put the sum, which can be up to 16 bits long, into the output locations.

Before we start, a complete discussion of the **Status Register** (SR; Figure 6-21) and **Condition Code Register** (CCR; Figure 6-22) is in order. The 16-bit Status Register is used by the postmaster to record the status of various operations. The upper byte is used by the SYSTEM, in the supervisor mode for advanced operations, and the lower byte is the one we will use most. This USER byte is called the **Condition Code Register.**

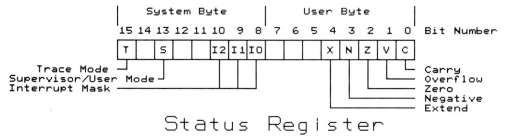

Figure 6-21   Status Register.

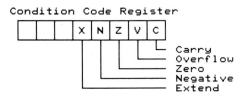

Figure 6-22   Condition Code Register.

A few brief words about the SYSTEM byte. The T bit, called the *trace bit,* is used for debugging a program. The S bit is called the *supervisor bit.* When it is set (1), the 68000 is operating in the supervisor mode; when clear (0), operation is in the user mode. These features are meaningful in larger systems; but for the two development boards used in this book, we can be both the user and supervisor, usually operating in the supervisor mode. When the system crashes, it is OUR fault. Bits 8 to 10, called the *interrupt mask,* are used to establish interrupt priorities.

For the present, let's consider only the lower 5 bits of this register. Denoted on an individual bit-by-bit basis as X, N, Z, V, and C, where X is the extend bit, N is the negative bit, Z is the zero bit, V is the overflow bit, and C is the carry bit.

The most often used bit is the Z bit. How it is set and reset is very basic. If we subtract two numbers and the result is **zero,** the **Z** bit is **set** to a **1** or **true** condition. Otherwise, the result is **not zero** and it is **cleared** to a **0** or **false** condition. Certain instructions activate or toggle these bits.

The next most popular bit is the C bit. Look at it as the ninth bit for a byte arithmetic operation or as the seventeenth bit for a word operation, or as the thirty-third bit for a longword operation. It is set when there is a carry (or borrow) during an arithmetic operation. It also receives bits during some shift and rotate instructions.

As for unsigned integers, we can use the Z bit to check for two equal numbers (or a zero result for arithmetic operations) or the C bit to find out which of two numbers is greater (or smaller). The V and N bits are used for operations with SIGNED numbers. The N bit is set if the most significant bit of a result is a 1, indicating a NEGATIVE signed number. It is cleared otherwise, indicating a POSITIVE number. The V bit is used to indicate that an operation corrupted the most significant bit of a result, which would in turn indicate the improper sign of the result. The X bit is used for multiprecision operations. Figure 6-23 shows how the bits of the Condition Code Register are affected by the various instructions.

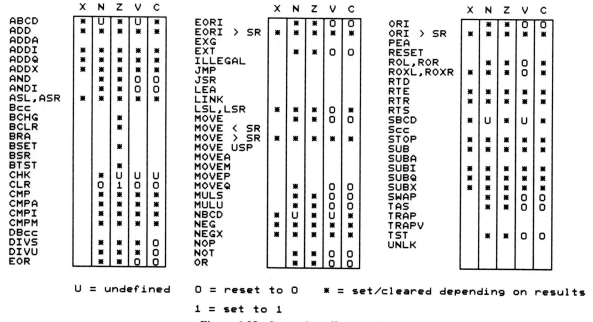

U = undefined   0 = reset to 0   * = set/cleared depending on results

1 = set to 1

Figure 6-23 Instruction effects on Condition Code Register bits.

Before proceeding with our arithmetic problem, let's consider the following binary addition examples that show the operation of the Condition Code Register bits.

		BINARY	UNSIGNED INTEGER (HEX)	SIGNED INTEGER (HEX)
(1)	Add	00110101	35	+35
	to	00100111	27	+27
		01011100	5C	+5C

If these numbers all represent unsigned integers, we are concerned only with the C and Z bits. Since we had no carry, C = 0. The result is not equal to zero, so Z = 0. Next, consider the numbers as signed integers, where the most significant (leftmost) bit is used to represent the sign of the number. A 0 is used for a positive number, a 1 for a negative. Our 7-bit signed numbers above represent a positive 35 hex and a positive 27 hex, whose sum is a positive 5C hex. Since the sum is a positive number, the N bit is not set to a 1 (N = 0). The overflow (V) bit is covered in Example 6-18.

		BINARY	UNSIGNED INTEGER (HEX)	SIGNED INTEGER (HEX)
(2)	Add	10000000	80	−00
	to	10000000	80	−00
		00000000	00 with a carry	+00

For unsigned integers, the carry makes C = 1; the zero result makes Z = 1. For signed numbers, we have added a negative zero to a negative zero, obtaining a positive zero for the sum. The N bit is cleared (N = 0). Adding these two negative numbers gave a POSITIVE sum. To indicate this error, the V bit is set (V = 1).

		BINARY	UNSIGNED INTEGER (HEX)	SIGNED INTEGER (HEX)
(3)	Add	01101011	6B	+6B
	to	01111100	7C	+7C
		11100111	E7	−67

For unsigned integers, C = 0 and Z = 0. For signed numbers, we have added a positive 6B hex to a positive 7C hex and obtained a negative 67 hex. Remember that the most significant bit is *stripped off* to be used for the sign. This leaves 110 0111, or 67 hex. The N bit is set (N = 1) to denote our negative sum, and V = 1 to indicate that our addition of two positive numbers gave a negative sum. (The bit adjacent to the most significant bit has *overflowed* into the most significant.)

Back to our original arithmetic problem! After an instruction is executed and the appropriate Status Register bits are set or cleared, it is up to the program to check and use the desired bits. In other words, the program execution continues normally UNLESS we USE the Bcc (BRANCH CONDITIONALLY) or DBcc (DECREMENT and BRANCH CONDITIONALLY) instructions.

For our example we will need to check the C or carry bit. As the BYTE size contents of D0 increases during the summing process, D0's contents go from 00000000 to 000000FF, then to 00000000, with the C bit being set to a 1, indicating that we had a carry condition. If our program had been set up for WORD-sized arithmetic, the sum stored in D0 would have been 00000100.

To perform our BYTE-sized arithmetic problem, we then use one of the BRANCH CONDITIONALLY instructions to *artificially* increment the second least significant nibble by one for each time we have a carry. Refer to Table 6-12 for the correct instruction.

TABLE 6-12

Bcc Instruction

Bcc   Branch Conditionally

Assembler Syntax: **Bcc.B <label>   or   Bcc.B displacement**
.W   .W

cc is replaced by one of the conditions below.

Action: If (condition is true), then PC+displacement → PC

BCC   Branch if C bit is CLEAR
BCS   Branch if C bit is SET
BEQ   Branch if Z bit is SET
BGE   Branch if N and V bits are either both SET or both CLEAR
BGT   Branch if N and V bits are both SET AND Z bit is CLEAR
           or
           if N, V, and Z bits are ALL CLEAR
BHI   Branch if C and Z bits are both CLEAR
BLE   Branch if the Z bit is SET or
           if the N bit is SET and the V bit is CLEAR or
           if the N bit is CLEAR and the V bit is SET
BLS   Branch if either the C or Z bit is SET
BLT   Branch if the N bit is SET and the V bit is CLEAR or
           if the N bit is CLEAR and the V bit is SET
BMI   Branch if the N bit is SET
BNE   Branch if the Z bit is CLEAR

TABLE 6-12 (continued)

Bcc Instruction

BPL	Branch if the N bit is CLEAR
BVC	Branch if the V bit is CLEAR
BVS	Branch if the V bit is SET

Status Register Condition Codes: X   N   Z   V   C

                                  -    -   -   -   -

No bits are affected.

Notes:
1. If the specified condition is met, program execution continues at the location equal to the sum of the Program Counter's contents plus the displacement. Otherwise, execution resumes at the instruction following the branch instruction.
2. The displacement, which is sign-extended to 32 bits, allows branching −128 to +127 bytes away for a byte displacement or −32,768 to +32,766 bytes away for a word displacement.
3. The displacement must always be an even number.
4. The label corresponds to an instruction, all of which start on an even address boundary.

We have selected the BCC instruction, but could have equally well selected the BCS (BRANCH if CARRY is SET) instruction by modifying our program slightly. Refer to Figure 3-4 for the Conditional Loop Structure. The decision point is entered from the top and if the decision is TRUE (YES), the *true code* is executed and then the loop is exited. IF the decision is FALSE (NO), the loop is exited.

For our example an instruction is executed at the entry that will set or clear the C (carry) bit in the Condition Code Register. It is the ADD.B (A0)+,D0 instruction that will provide a carry if the byte sum in D0 exceeds FF. The decision instruction (BCC NOPE) tests the C bit, and if it is CLEAR (TRUE condition), we exit the loop. IF the C bit is SET (FALSE condition), the *carry code* is executed (the ADD.W #$100,D0 instruction). The loop is then exited.

Our program would look as follows:

```
; Example 6.17
; Sum the byte contents of 256 memory locations (starting at 7000),
; store a word-sized result in another memory location (7100)
; Exit (or loop) correctly when done
;
            MOVEQ.L #$0,D0         ; clear out the accumulator
            MOVE.W #$FF,D1         ; initialize the countdown
                                  ; register to 256
            MOVE.W #$0,$7100      ; clean out the output display location
            MOVEA.L #$7000,A0     ; point to the source of data
LOOP:   ADD.B (A0)+,D0           ; add to D0 the data pointed to
                                  ; bump pointer
            BCC NOPE              ; skip over the carry operation
                                  ; if no carry
            ADD.W #$100,D0        ; we had a carry, increment the
                                  ; word nibble
NOPE:   DBF D1,LOOP              ; get the next data to add
            MOVE.W D0, $7100      ; move the word-sized sum
                                  ; to the output adr
HERE:   JMP HERE                 ; loop when done
```

Addressing Modes Used

No new addressing modes were introduced.

Two exercises for the student: (1) modify the program utilizing the BCS instruction; and (2) assuming that 256 word-sized numbers are to be added and that the sum can be up to 64 bits long, write the program.

Looking again at all of the BRANCHING instructions, it is often confusing for the beginner to choose the proper one for an operation. Table 6-13 should help, and examples are presented to show each.

TABLE 6-13

Choosing BRANCHING Instructions

Compare Condition	Unsigned Number Operation	Signed Number Operation
Greater than	BCC	BGE
Greater than or equal	BHI	BGT
Equal to	BEQ	BEQ
Not equal to	BNE	BNE
Less than or equal	BLS	BLS
Less than	BCS	BLT

6.5 PROCESSING BCD NUMBERS

A brief comparison of how the decimal digits from 0 to 9 can be represented is provided in Table 6-14.

TABLE 6-14

Decimal	Binary	Hexadecimal	Binary-Coded Decimal	ASCII (hex)
0	0000	0	0000	30
1	0001	1	0001	31
2	0010	2	0010	32
3	0011	3	0011	33
4	0100	4	0100	34
5	0101	5	0101	35
6	0110	6	0110	36
7	0111	7	0111	37
8	1000	8	1000	38
9	1001	9	1001	39

Binary-Coded-Decimal numbers have several applications. Most decimal displays, such as those used on clocks and calculators, require BCD inputs. Some input devices, such as keyboards, produce BCD outputs. (ASCII outputs are more common for keyboards.)

BCD numbers have several commercial applications. To represent a decimal number such as 1023.4573 accurately, BCD representation is often the best solution. Accounting applications, where both the input and output are in BCD format, can best be added, subtracted, and so on, by leaving the numbers in BCD format.

As you may have noticed, a BCD digit requires only 4 binary bits. Looking at a number such as 252 decimal:

DECIMAL	BINARY	HEXADECIMAL	BINARY-CODED DECIMAL
252	1111 1100	FC	0010 0101 0010

The spacing in the binary and BCD representations are for clarity only. Note how the number could be held in ONE 8-bit memory location in BINARY (or hex) format but will require 12 bits to be kept as a BCD number. Rather than storage of a single BCD digit (4 bits or a nibble) per 8-bit memory location, BCD digits are stored **two nibbles to a byte.** This is about the ONLY "humor" left to us by our programmer ancestors. The number 252 therefore takes 1½ bytes to store. Although not the most efficient form of storage, time is often saved in calculations by utilizing the BCD format.

Refer to Section 5.4.3 for the **BCD Arithmetic Instructions** at our disposal. Only addition and subtraction are available.

EXAMPLE 6.18

Suppose that addresses 7000, 7001, 7002, 7003, and 7004 represent five INPUT ports or addresses, each containing two BCD digits, and that address 7100 is an 8-bit OUTPUT port or address being used to store two BCD digits. The input data could be obtained from a wide variety of devices, and the output address could be a two-digit visible display unit. Write a program that causes the microprocessor to add or total the contents of the input locations and put the sum in the output location.

To keep this first BCD example *simple,* let's assume that the sum of the five input numbers will not exceed 99 decimal, which is the highest BCD number we can store in a single byte. Referring to Table 6-15, we see that only two modes are available. Since we cannot add the contents of two memory locations, we move one of the numbers into a Data Register and will use it as our <Sea> AND as <Dea>.

TABLE 6-15

ABCD Instruction

ABCD   Add Binary-Coded Decimal with Extend

Assembler Syntax;   ABCD.B <Sea>,<Dea>

<Sea> Source Effective Address	<Dea> Destination Effective Address												
	Dn	An	(An)	(An) +	− (An)	$w (An)	$b (An, Xn)	$s	$l	$w (PC)	$b (PC, Xn)	#$q	SR	CCR
Dn	18													
− (An)					*									

18 denotes the mode used in Example 6.18.

Action: <Sea>+<Dea>+extend bit → <Dea>

Status Register Condition Codes: X   N   Z   V   C
                                      *   U   *   U   *

X is set the same as the carry bit.
N is undefined.
Z is cleared if the result is nonzero. Unchanged otherwise.
V is undefined.
C is set if a BCD carry is generated from the most significant BCD nibble. Cleared otherwise.

Notes:
1. The LOW-order bytes of <Sea>, <Dea>, and the extend bit are added, and the sum is stored in the <Dea>.
2. The Pre-decrement addressing mode is used for adding multiple bytes in memory. Start at the highest address, which contains the least significant BCD nibble.
3. The Z bit is normally set with an instruction prior to execution of this instruction, allowing proper tests for a zero condition upon completion of multiple-precision operations involving long numbers.
4. The addition is done using binary-coded-decimal arithmetic.
5. Bits 8 to 31 of the <Dea> are unaffected.

Use of this instruction requires consideration of another bit in the Condition Code Register. The X or extend bit is used to indicate a carry (a sum greater than 99) during a BCD add. We will need to clear it (set it to 0) before doing our add process. It is also recommended to set the Z bit to allow for zero testing upon completion of multiple-precision operations.

Our crude but effective *first-cut* at this program would look like this:

```
; Example 6.18 (first attempt)
; Sum the BCD contents (two digits/byte) of five memory locations
; (starting at 7000),
; store result in memory location 7100
;
        MOVE.W #4,CCR        ; clear the X bit, set the Z bit
        MOVEQ.L #0,D0        ; clear out the sum storage
        MOVE.B $7000,D1      ; get the first number to be added
        ABCD.B D1,D0         ; add to the sum
        MOVE.B $7001,D1      ; get the second number
        ABCD.B D1,D0         ; add to the sum
        MOVE.B $7002,D1      ; get the next number
        ABCD.B D1,D0         ; add to the sum
        MOVE.B $7003,D1      ; get the next number
        ABCD.B D1,D0         ; add to the sum
        MOVE.B $7004,D1      ; get the next number
        ABCD.B D1,D0         ; add to the sum
        MOVE.B D0,$7100      ; store the sum
HERE:   JMP HERE             ; loop when done
```

Recall that the Condition Code Register bits are arranged as

$$— \quad — \quad — \quad X \quad N \quad Z \quad V \quad C$$

and a "4" will look like 0 0 0 0 0 1 0 0 to clear the X bit and set the Z bit.

Let's put a little polish on this example before executing it on a computer or simulator since it will give an incorrect answer if the sum exceeds 99.

EXAMPLE 6.19

Modify Example 6.18 to allow for a sum between 0 and 9999. The input data remain as before, 8 bits (two BCD digits) per (input) address. Output locations 7100 and 7101 will be used to provide a four-BCD-digit display.

To simplify the program greatly, we will *upgrade* it using techniques shown in Example 6.15. We also need to perform a CONDITIONAL branch to take care of a potential carry when the sum exceeds 99 decimal. If the sum, for example, is 1234 decimal, it will be stored in memory (hex) as

	12	34	
7100			7101
7102			7103

At first glance it looks like we need to keep track of our lower two BCD digits, whose sum could exceed 99, and store them in location 7101. Every time we have a carry, we need to add 1 to location 7100. We must use BCD arithmetic for this add as well; otherwise, we will accumulate a binary or hex number for the upper two BCD digits. The program would look like this:

```
; Example 6.19
; Sum the BCD contents (two digits/byte) of five memory locations
; (starting at 7000),
; store a result that can be between 0 and 9999 10 in two
; memory locations (7100 and 7101)
; in BCD format
;
        MOVE.W #4,CCR        ; clear the X bit, set the Z bit
        MOVEQ.L #0,D0        ; clear out the sum storage
        MOVEQ.L #4,D4        ; initialize the countdown
                            ; counter to one less
        MOVEQ.L #0,D3        ; clean out the carry sum
        MOVEQ.L #0,D2        ; to be used as the simple
                            ; carry incrementer
        MOVEA.L #$7000,A0    ; point to the beginning of the input data
LOOP:   MOVE.B (A0)+,D1      ; get the first number to be added,
                            ; bump pointer
        ABCD.B D1,D0         ; add to the sum
        BCC NOPE             ; take care of a sum > 99
                            ; otherwise proceed
        ABCD.B D2,D3         ; add one to the carry sum
NOPE:   DBF D4,LOOP          ; keep looping until done
        MOVE.B D3,$7100      ; store the upper two BCD digits
        MOVE.B D0,$7101      ; store the lower two BCD digits
HERE:   JMP HERE             ; loop when done
```

Most of the lines are self-explanatory. Remember in the third line to set the count to ONE LESS than the actual number of times we want to loop. D3 is used to accumulate our carries for each instance when the sum exceeds 99. Since we did not have an instruction to add one (BCD) *immediately* to D3, we used a combination of the X bit and the zero in register D2 to get the one needed for the BCD carry.

When the sum exceeds 99, both the X and C bits of the Condition Code Register are set. The BCC is chosen since it will branch when the C bit is **CLEAR.** There is not an instruction to test the X bit, but it is copied into the C bit, and tests for the C bit are done. The remaining instructions have been used in previous examples.

In summary, the program will add together five two-digit BCD numbers (between 0 and 99), and store the sum as four BCD digits (in two memory locations), allowing a sum between 0 and 9999.

While still in the mood for polishing up the program, let's introduce two new instructions that are very simple to use. They are the **BRA** (Table 6-16) and **LEA** (Table 6-17) instructions.

TABLE 6-16

BRA Instruction

BRA   Branch Always

Assembler Syntax: BRA.B <label>   or   BRA.B displacement
                          .W   .W

Action: The displacement is added to the contents of the Program Counter and program execution resumes from that location

TABLE 6-16 (continued)

BRA Instruction

Status Register Condition Codes: X   N   Z   V   C
                                        -   -   -   -   -

No bits are affected.

Notes:
1. The displacement is a two's-complement integer denoting the relative distance in bytes between the BEGINNING of the instruction following the BRA instruction and the next instruction to be executed.
2. Byte displacement give a branch range of −128 to +126 bytes away from the BRA instruction.
3. Word displacements give a branch range of −32,768 to +32,766 bytes away from the BRA instruction.
4. The displacement must always be EVEN, since instructions must begin on an even address.
5. Some assemblers use .S notation to denote .B (byte) displacements. Some do not require the .B or .W displacement size, calculating the proper size.

      The branch instructions (Bcc, BRA, and BSR) are unique in that they can take an ".S" size code. This suffix directs the assembler to assemble these as short branch instructions (i.e., one-word instructions with a range to −128 to +127 bytes). If the ".S" size code is used and the destination is actually outside this range, the assembler will print an error message. If the ".L" size code is used, the assembler will use a long branch, which is a two-word instruction with a range of −32,768 to +32,767 bytes. If neither size code is specified, the assembler will use a short branch if possible.

TABLE 6-17

LEA Instruction

LEA   **Load Effective Address**

Assembler Syntax: **LEA.L <Sea>,An**

<Sea> Source Effective Address	<Dea> Destination Effective Address	
	An	
Dn		
An		
(An)	*	
(An) +		
− (An)		
$w (An)	*	
$b (An, Xn)	*	
$s	19	
$l	*	
$w (PC)	*	
$b (PC, Xn)	*	
#$q		
SR		
CCR		

19 denotes the mode used in Example 6.19a.

Action: <Sea> → An

Status Register Condition Codes: X   N   Z   V   C
                                        -   -   -   -   -

No bits are affected.

Notes:
1. The <Sea> is loaded into An. All 32 bits of An are affected.
2. This instruction is used to write position-independent programs.
3. A constant can be added to An without altering the CCR flags using $w (An) <Sea>.

We replace **MOVE.W #$7000,A0** with **LEA $7000,A0** and **JMP HERE** with **BRA HERE. Note that we omit # since we are referring to an address and not to data.** You may want to try execution of the program before and after the changes to see their effect. LEA loads an effective address into an Address Register. The number of bytes used is the same as in MOVE. The main difference is that no bits of the Condition Code Register are affected, which will be of no consequence to us as beginners. The BRA instruction allows us to *branch* to another location instead of *jumping*. We can jump with equal ease to any of the 68000's memory addresses. BRA is limited to locations located within $-32,768$ to $+32,767$ locations from the present location. BRA requires 2 bytes, compared to JMP, which could use as many as 6 bytes. Most important, the use of BRA allows creation of **position-independent** programs, programs that can be loaded into and executed from any memory location. This is because the <Dea> is an OFFSET instead of an ABSOLUTE ADDRESS. Using a label with the assembler instead of an address with these instructions does not alter this.

Addressing Modes Used

No new addressing modes were introduced.

The enhanced program would look like this:

```
; Example 6.19a
; Sum the BCD contents (two digits/byte) of five memory locations
; (starting at 7000),
; store a result that can be between 0 and 9999 ₁₀ in two
; memory locations (7100 and 7101)
; in BCD format
;
              MOVE.W #4,CCR      ; clear the X bit, set the Z bit
              MOVEQ.L #0,D0       ; clear out the sum storage
              MOVEQ.L #4,D4       ; initialize the countdown counter
              MOVEQ.L #0,D3       ; clean out the carry sum
              MOVEQ.L #0,D2       ; to be used as the simple
                                 ; carry incrementer
              LEA $7000,A0        ; point to the beginning of the input data
LOOP:   MOVE.B (A0)+,D1     ; get the first number to be added,
                                 ; bump pointer
              ABCD.B D1,D0        ; add to the sum
              BCC NOPE            ; take care of a sum > 99
                                 ; otherwise proceed
              ABCD.B D2,D3        ; add one to the carry sum
NOPE:   DBF D4,LOOP         ; keep looping until done
              MOVE.B D3,$7100     ; store the upper two BCD digits
              MOVE.B D0,$7101     ; store the lower two BCD digits
HERE:   BRA HERE            ; loop here when done
```

Hands-On Execution of This Example

Refer to Section 10.1.7, 10.2.7, or 10.3.7 for execution of this example on the applicable systems.

Seeing the limitation of our example, let's modify it to accommodate a larger number of numbers and to accommodate six-digit BCD numbers (0 to 999,999 decimal).

EXAMPLE 6.20

Now suppose that addresses 7000 to 700F represent four longword-sized INPUT ports or addresses, each containing six-digit BCD data, and that addresses 7F00, 7F01, 7F02, and 7F03 are considered together to provide a longword-sized OUTPUT port for eight BCD digits. This would allow display of numbers from 0 to 99,999,999 decimal. The upper byte of each input number is not used. For simplicity only four BCD numbers are being added, but more could easily be accommodated. A sketch of the arrangement of our data is in order:

7000	00	12	7001	first BCD number 123456
7002	34	56	7003	
7004	00	99	7005	second BCD number 999999
7006	99	99	7007	
7008	00	11	7009	third BCD number 111111
700A	11	11	700B	
700C	00	22	700D	fourth BCD number 222222
700E	22	22	700F	

7F00	01	34	7F01	BCD sum 01,456,788
7F02	56	77	7F03	

7F04	00	00	7F01	BCD NUMBER OF NUM-BERS
7F06	00	04	7F03	

Addressing Modes Used

Refer to Examples 6.18 and 6.19 for hints on how to get started. To write this program effectively, we need to consider one of the *fancy* addressing modes of the 68000. We will use the **Address Register Pre-Decrement Indirect Memory Addressing** Mode for both <Sea> and <Dea>.

Referring to Table 6.15 and choosing a couple of address registers, the instruction we have is:

ABCD.B −(A1),−(A2)

Note in the sketch above that the *higher* BCD digits are stored in the LOWER memory locations. Rather than create a flowchart, let's look at the addition process *in slow motion:*

  1. We clear out the X bit since it is used for all BCD adds. It is the carry when the addition exceeds 99.
  2. Since we have the pre-decrement addressing mode instruction, we point to one location PAST the last BCD number for a starting point.

3. We use one Address Register as a pointer for the numbers being added, and one for the memory locations used for the sum.

4. The pointers are decremented BEFORE the operation and we then add the two numbers being pointed to, along with the X bit (in case there was a carry).

5. We repeat step 4 three more times, working toward the highest BCD digit pair. The carry condition is taken care of automatically with the combination of the addressing mode chosen and the BCD addition instruction.

6. So far we have added one of the six-digit BCD numbers to the previous sum (in location 7F00). We repeat the process four times (the number of numbers we have to add) to get the total sum. Our program looks like this:

```
; Example 6.20
; Add some six-digit BCD data stored in memory
; (starting at 7000), store sum in other
; memory locations (7F00 to 7F03) as an eight-digit BCD number
;
; $7000-7003 contains the first six-digit BCD number
; $7004-7007              second
; etc
; $7F00-7F03 contains the possible eight-digit sum
; $7F04-7F07 contains the NUMBER of scores to be added
;
.ORG $7000
START:    .RS 256              ; reserve space for the numbers
          .ORG $7F00
SUM:      .RS 4                ; reserve space for the SUM
NUM:      .DL 0004             ; the number of numbers is stored here
;
          .ORG $1000           ; the beginning of the program
;
          CLR.L D7             ; clear out the number counter
          MOVE.L NUM,D7        ; get the number of numbers to add
          MULU.W #4,D7         ; multiply by 4 since each
                               ; number is 4 bytes long
          ADD.L #START,D7      ; D7 now points to the address
                               ; PAST the last number
          MOVEA.L D7,A0        ; put the pointer in an Address Register
          MOVE.W #4,CCR        ; clear the X bit, set the Z bit,
                               ; for BCD arithmetic
          LEA.L SUM+4,A1       ; point to one past the sum
                               ; storage location
          CLR.L SUM            ; clear out the longword sum
                               ; storage location
          MOVE.L NUM,D2        ; initialize the countdown counter
          SUB.L #1,D2          ; correct for the DBF quirk, so
                               ; countdown will be correct
LOOP:     ABCD.B -(A0),-(A1)   ; decrement pointers, add the
                               ; least significant digits
          ABCD.B -(A0),-(A1)   ; repeat to add the middle two digits
          ABCD.B -(A0),-(A1)   ; repeat to add the most significant digits
          ABCD.B -(A0),-(A1)   ; all done, take care of potential carry
          LEA.L SUM+4,A1       ; reset the sum pointer
          DBF D2,LOOP          ; keep looping until all numbers are added
DONE:     BRA DONE
          .END
```

If you have used the assembler/simulator for any of the prior examples, you are familiar with the .ORG $1000 and .END lines. Called *pseudo-ops*, they are NOT 68000 instructions but are used by the assembler to perform its functions. This example introduces a few more of the pseudo-ops, which will make program creation a lot simpler. Since our numbers were (for this example) to be stored in locations starting at 7000, we *originated* at 7000 and then entered START: RS 256. This RESERVES 256 locations and gives the first location the label START. The memory did not actually need to be reserved in this case, but it is good programming practice. The label, however, makes programming smoother. If, for some reason, the numbers had to be stored elsewhere in memory, only the .ORG $7000 line would need to be changed since we are using the label START throughout the program. Similarly, we set up a longword location for the SUM. The next line, NUM: .DL 0004, is used to store the NUMBER of NUMBERS to be added. If the number of numbers to be added is changed, we simply change the 0004 located on that line.

Note that we introduced a new instruction: **CLR** (Table 6-18). The CLR is another way to MOVE a zero into a register or memory location.

TABLE 6-18

CLR Instruction

CLR   **Clear Operand**

Assembler Syntax: **CLR.B <Dea>**
                            .W
                            .L

											<Dea> Destination Effective Address									
	Dn	An	(An)	(An)+	–(An)	$w (An)	$b (An, Xn)	$s	$l	$w (PC)	$b (PC, Xn)	#$q	SR	CCR
	20		*	*	*	*	*	*	*					

20 denotes the mode used in Example 6.20.

Action: 0 → <Dea>

Status Register Condition Codes: X N Z V C
                                               - 0 1 0 0

X is not affected.
N is cleared.
Z is set.
V is cleared.

Note: A memory destination is read before it is written to.

The steps in the program are rather straightforward. The first five are but one way to show how to set the A0 pointer to the end of the array, automatically, calculated on the number of numbers. There are other ways of doing this with the assembler, by the way. Note that only the SUM location pointer (A1) is reset prior to each addition of a new number. The A0 pointer automatically follows down in the memory for each pair of BCD digits and for each complete number. Note how the labels NUM and START are used. To move the CONTENTS of memory location (0004 in this example) called NUM into D7, we use MOVE.L NUM,D7. To add the VALUE (the memory address) of the label START to the contents of a Data Register, we use ADD.L #START,D7. The # indicates a value, or *immediate data*. The CONTENTS of START would be the first BCD number to be added. See Figure 6-24 for execution of this program.

Hands-On Execution of This Example

Refer to Section 10.1.8, 10.2.8, or 10.3.8 for execution of this example on the applicable systems.

```
7000   00 │ 01   7001      7000   00 │ 01   7001
7002   23 │ 45   7003      7002   23 │ 45   7003
7004   00 │ 44   7005      7004   00 │ 44   7005
       55 │ 66                     55 │ 66

7F00   xx │ xx   7F01      7F00   00 │ 55   7F01
7F02   xx │ xx   7F03      7F02   79 │ 79   7F03
7F04   00 │ 00   7F05      7F04   00 │ 11   7F05
7F06   00 │ 02   7F07      7F06   00 │ 02   7F07
```

           Before   After

Figure 6-24   Execution of Example 6.20.

Let's try one more BCD example, one that would be helpful in calculating your course averages.

EXAMPLE 6.21

Find the AVERAGE for an array of numbers stored in BCD format. Originally, this example was to find the average of the rather large six-digit BCD numbers that were totaled in Example 6.20. Since we do not have an AVERAGE instruction in our toolbox, another innovative algorithm must be developed to accomplish this task. It was decided to take smaller two-digit numbers to show this, leaving the task of averaging six-digit BCD data to the student.

```
        ; Example 6.21
        ;
        ; Find the average value of an array of BCD data
        ;
        ; 7000 contains the first two-digit BCD number
        ; 7001 contains the second
        ; etc.
        ; 7100–7101 contains the SUM
        ; 7102–7105 contains the AVERAGE
        ; 7106–7107 contains the number of numbers (BCD format)
        ;         to be averaged
        ;
                .ORG $7000
START:  .RS 256              ; reserve space for the numbers
                .ORG $7100
SUM:    .RS 2                ; reserve space for the SUM (0000 to 9999 BCD)
AVER:   .RS 4                ; reserve space for the AVERAGE
NUM:    .DW 05               ; the number of numbers (BCD format)
                             ; is stored here
        ;
        ; First, the SUM of the array is determined (Example 6.19a used)
                .ORG $1000
        ; Example 6.19a
        ; Sum the BCD contents (two digits/byte) of five
        ;   memory locations (starting at 7000),
        ;   store a result that can be between 0 and 9999 in
        ;   two memory locations (7100 and 7101) in BCD format
        ;
                MOVE.W #4,CCR        ; clear the X bit, set the Z bit
                MOVEQ.L #0,D0        ; clear out the sum accumulator
                MOVE.W NUM,D4        ; initialize the countdown counter
                SUBQ #1,D4           ; take care of the DBF quirk
                MOVEQ.L #0,D3        ; clean out the carry part of the sum
```

```
              MOVEQ.L #0,D2        ; to be used as the simple
                                   ; carry incrementer
              LEA START,A0         ; point to the beginning of the input data
LOOP:         MOVE.B (A0)+,D1      ; get the first number to be added
                                   ; bump pointer
              ABCD.B D1,D0         ; add to the sum accumulator
              BCC NOPE             ; take care of a sum > 99
                                   ; otherwise proceed
              ABCD.B D2,D3         ; add one to the carry sum
NOPE:         DBF D4,LOOP          ; keep looping until done
              MOVE.B D3,SUM        ; store the upper two BCD digits
              MOVE.B D0,SUM+1      ; store the lower two BCD digits
              ;
              ; Find the AVERAGE
              CLR.L AVER           ; will contain the number of TIMES
                                   ; we subtract
                                   ; the number of numbers from the sum
              CLR.L D2             ; used with the X bit to take care
                                   ; of a carry
              CLR.L D0             ; used to total up the average
              MOVEQ #1,D3          ; used for our addition
              MOVE.W NUM,D1        ; get the BCD NUM to subtract
              MOVE.B SUM,D5        ; get the upper digits of SUM
              MOVE.B SUM+1,D6      ; get the lower
LOOP2:        SBCD D1,D6           ; subtract lower digits of NUM from
                                   ; SUM of lower digits
              SBCD D2,D5           ; subtract a possible borrow
                                   ; from upper digits
              CMP #$FF,D5          ; check to see if done
              BEQ EXIT             ; exit when done
              ABCD D3,D0           ; add one to the average (in D0 for now)
              BRA LOOP2            ; keep looping
                                   ; (BEQ EXIT will provide exit)
EXIT:         MOVE.L D0,AVER       ; store the average
DONE:         BRA DONE             ; loop forever
              .END
```

Techniques shown in Examples 6.19a and 6.20 were used in this example.

The algorithm used to compute an average is fairly simple to follow. Normally, you would take the sum of the numbers and divide it by the number of numbers. Since we do not have a BCD divide instruction in our toolbox, we must resort to another technique. We will SUBTRACT the number of numbers from the sum until we get zero and keep track of the number of times we subtract. This will be the average of the numbers (with the fractional part dropped).

Since we are doing a BCD subtraction, the NUMBER of NUMBERS must be stored in BCD format. If it had been obtained *automatically* by a device providing ordinary counts in binary (hex), it would have to be converted to BCD format first. If your program counted up the numbers as they were input into the memory locations, the count would also be a binary number and would need conversion to BCD. A binary-to-BCD conversion routine could easily be developed.

To make the countdown, or subtraction, go smoothly, two new instructions had to be introduced. The **SBCD** (Table 6-19) is used similar to the use of ABCD in Example 6.19a. The SBCD D2,D5 subtracts a zero (in D2) and the possible X bit from the contents of D5, which contains the upper BCD digit. However, when the upper BCD digit is zero and the X bit, which represents a borrow from the lower BCD digits, is subtracted, a result of FF is obtained for the upper BCD digit. No flags are set to indicate this undesired condition, which is really when we wish to exit from the repeated subtractions.

To check for this condition, we need to make a COMPARISON between D5 and the hex number FF. The **CMP** instruction is used, with its discussion delayed until the next section. In a nutshell, it sets the Z bit when D5 is equal to FF, and the BEQ instruction is executed, taking the program to the EXIT.

TABLE 6-19

SBCD Instruction

SBCD   Subtract Decimal with Extend

Assembler Syntax: **SBCD.B <Sea>,Dn**

<Sea> Source Effective Address	<Dea> Destination Effective Address													
	Dn	An	(An)	(An) +	– (An)	$w (An)	$b (An, Xn)	$s	$l	$w (PC)	$b (PC, Xn)	#$q	SR	CCR
Dn	21													
– (An)					*									

21 denotes the mode used in Example 6.21.

Action: Dn-<Sea>-extend bit → Dn

Status Register Condition Codes: X   N   Z   V   C
                                      *   U   *   U   *

X is set the same as the carry bit.
N is undefined.
Z is cleared if the result is nonzero. Unchanged otherwise.
V is undefined.
C is set if a BCD borrow is generated from the most
  significant BCD digit. Cleared otherwise.

Notes:
1. The LOW-ORDER bytes of the Data Registers are subtracted, and the result is stored in the destination Data Register.
2. The Pre-decrement addressing mode is used for subtracting multiple bytes in memory. Start at the highest address, which contains the most significant BCD digit.
3. The Z bit is normally set with an instruction prior to execution of this instruction, allowing proper tests for a zero condition upon completion of multiple-precision operations.
4. The subtraction is done using binary-coded-decimal arithmetic.

A variation of the **SUB** instruction, **SUBQ** (Table 6-20), was introduced. We can use it when the number to be subtracted is between **1 and 8.** Since we never have the need to subtract 0, SUBQ #0 subtracts 8. The other values, 1 to 7, are as expected. We can omit the $ used to denote hexadecimal.

Another new instruction, **BEQ**, is one of the Bcc series, and it branches if the Z bit in the CCR is equal to 1 (SET), indicating a zero result. Refer to Table 6-12 for information about the BEQ instruction.

TABLE 6-20

SUBQ Instruction

SUBQ   Subtract Binary (Quick {1-8} Data)

Assembler Syntax: SUBQ.B #$q,<Dea>
                         .W
                         .L

<Sea> Source Effective Address	Dn	An	(An)	(An) +	– (An)	$w (An)	$b (An, Xn)	$s	$l	$w (PC)	$b (PC, Xn)	#$q	SR	CCR
#$q	21	*	*	*	*	*	*	*	*					

21 denotes the mode used in Example 6.21.

Action: <Dea>-<Sea> → <Dea>

Status Register Condition Codes: X   N   Z   V   C
                                          *   *   *   *   *

X is set the same as the carry bit.
N is set if the result is negative. Cleared otherwise.
Z is set if the result is zero. Cleared otherwise.
V is set if an overflow is generated. Cleared otherwise.
C is set if a borrow is generated. Cleared otherwise.

Notes:
1. The immediate data must be between 1 and 8 ($ not required) (0 = 8, 1 = 1, 2 = 2, . . ., 7 = 7).
2. Only word and longword sizes are used for the Address Register Direct (An) mode; the source data is sign-extended to a longword, and the condition codes are not affected in this addressing mode.
3. The 32-bit destination Address Register is used regardless of the operation size.

6.6   *LEARNING HOW TO COMPARE AND TEST*

The Compare and Test instructions add to the power and versatility of the 68000 by allowing branching decisions based on comparisons and tests that **do not alter** either the Source Effective Address (<Sea>) or Destination Effective Address (<Dea>).

Since no computer text is complete without a classic bubble sort example, and since this is an excellent use of the compare instructions, it will be used to show how an array, or collection, of data can be sorted in memory. The bubble sort **algorithm** is not one a beginning programmer would conceive, but it is relatively easy to follow. A beginner would perhaps attempt some type of *insertion sort,* going about it in the natural way of simply taking a number and inserting it between the number above and below it. The bubble sort works differently.

To sort the numbers in ascending order, a comparison between the last two numbers of the array is done. If the last number is smaller than the one preceding it, the two numbers are exchanged. The smaller number is then checked with the one preceding it, and swapped if necessary. The smaller number *bubbles* its way to the top, moving one location at a time. Once the beginning of the array is reached, a check is done to see if a swap has occurred. If so, the pointers are reset at the bottom and the comparisons are made again, this time bubbling up the next smallest number. Eventually, a pass will be made through the array in which no swap takes place, and this will signify the end of the sorting process.

This is a very inefficient sorting algorithm since the numbers move only a single location at a time. Other sorting routines move the numbers over greater distances, completing the sort much faster. A flowchart of the process is shown in Figure 6-25. The following comments refer to the flowchart comment lines in the figure. We use Address Registers to point to memory locations, and Data Registers to contain data when needed.

1. Point to the first number with A0. The A0 pointer remains fixed.
2. Clear an indicator called a *swap flag,* which is simply Data Register D7. It is used to tell us when two numbers are exchanged during the sorting.
3. Point to the last number with A2. The A2 pointer begins at the bottom and moves up to the top.
4. We get the first (= bottom number initially) number for our comparison.
5. The variable pointer is decremented (moved toward the top) and is checked to see if it has reached the top. If so, check the swap flag.
6. The swap flag is not to the top of the array.
7. Get the next number for comparison. The two numbers are compared. If they are in order, get another number to compare.
8. The numbers need to be swapped to put the smaller number on top of the second number.
9. The swap flag is set; then get another number to compare.
10. If the swap flag is set, go to the beginning, clear the flag, and check the numbers again. If the flag is not set, exit the program; the numbers are sorted.

Before considering the program needed, let's discuss the compare instructions available in our toolbox. Refer to Table 5-5, where the compare instructions are included with the **Integer Arithmetic Instructions.** There are four COMPARE instructions:

CMP   Compare <Sea> with Data Register (Table 6-21)
CMPA   Compare <Sea> with Address Register (Table 6-22)
CMPI   Compare a constant with <Dea> (Table 6-23)
CMPM   Compare two memory contents (Table 6-24)

We can use the CMPI instruction to compare a specific *number* (immediate data) with the contents of a Data Register (the one used as our swap flag). We also want to compare the contents of two memory locations in step 8 of the flowchart, and to compare two address registers in step 6. At first glance, the CMPM instruction looks like a choice, but note in Table 6-24 that it is available in only one addressing mode—one that autoincrements our address registers in a way not useful for our problem. The alternative is to use the CMP instruction, used by first moving one of the numbers to be compared into a Data Register. We will use the CMPA for the Address Register comparisons.

In all cases the compare instructions *compare* two values and do not change them. The appropriate CCR bits are set or cleared. For example, if two numbers are equal to each other, the Z bit is set and a BEQ instruction could be used in this case. If the second number (the <Dea>) is LARGER than the first (the <Sea>), the C bit would be cleared (indicating NO CARRY), and a BCC instruction could be used.

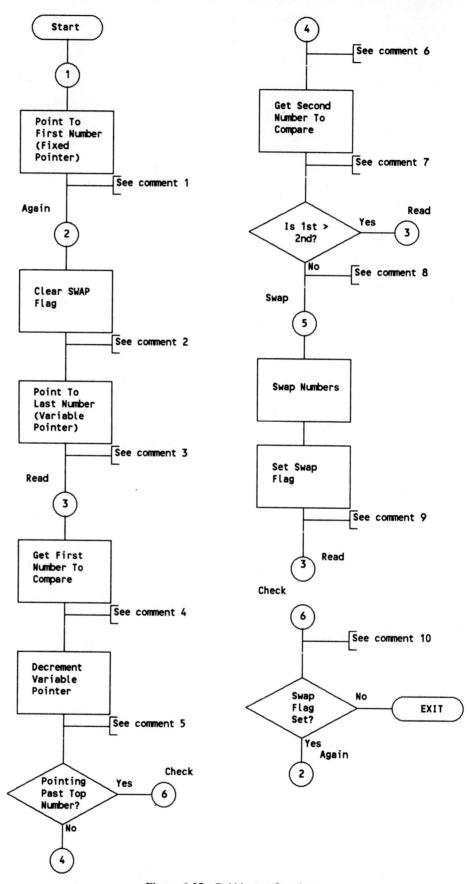

Figure 6-25   Bubble sort flowchart.

TABLE 6-21

CMP Instructions

CMP   Compare

Assembler Syntax: **CMP.B <Sea>, Dn**
                              .W
                              .L

<Sea> Source Effective Address	Dn	<Dea> *Destination Effective Address*
Dn	22	(word and longword only)
An	*	
(An)	*	
(An) +	*	
− (An)	*	
$w (An)	*	
$b (An, Xn)	*	
$s	*	
$l	*	
$w(PC)	*	
$b (PC, Xn)	*	
#$q	*	
SR		
CCR		

22 denotes the mode used in Example 6.22.

Action: The <Sea> is subtracted from the <Dea>, and the condition codes are changed accordingly

Status Register Condition Codes: X   N   Z   V   C
                                     -   *   *   *   *

X is not affected.
N is set if the result is negative. Cleared otherwise.
Z is set if the result is zero. Cleared otherwise.
V is set if overflow occurs. Cleared otherwise.
C is set if a borrow occurs. Cleared otherwise.

Notes:
1. Neither the <Sea> nor <Dea> is changed.
2. CMPA is used when the <Dea> is an Address Register.
3. CMPI is used when the <Sea> is immediate data.
4. CMPM is used for memory-to-memory compares.

TABLE 6-22

CMPA Instruction

CMPA   Compare Address

Assembler Syntax: CMPA.W <Sea>, An
                        .L

<Sea> Source Effective Address	<Dea> Destination Effective Address An
Dn	*
An	22
(An)	*
(An) +	*
− (An)	*
$w (An)	*
$b (An, Xn)	*
$s	*
$l	*
$w(PC)	*
$b (PC, Xn)	*
#$q	*
SR	
CCR	

22 denotes the mode used in Example 6.22.

Action: The <Sea> is subtracted from the <Dea>, and the condition codes are changed accordingly

Status Register Condition Codes: X   N   Z   V   C
                                      -   *   *   *   *

X is not affected.
N is set if the result is negative. Cleared otherwise.
Z is set if the result is zero. Cleared otherwise.
V is set if an overflow occurs. Cleared otherwise.
C is set if a borrow occurs. Cleared otherwise.

Notes:
1. Neither the <Sea> nor <Dea> is changed.
2. Word-length source operands are sign-extended to longwords before the compare.

TABLE 6-23

CMPI Instruction

CMPI   Compare Immediate

Assembler Syntax: CMPI.B #$q,<Dea>
                        .W
                        .L

<Sea> Source Effective Address	<Dea> Destination Effective Address													
	Dn	An	(An)	(An) +	− (An)	$w (An)	$b (An, Xn)	$s	$l	$w (PC)	$b (PC, Xn)	#$q	SR	CCR
#$q	22		*	*	*	*	*	*	*					

TABLE 6-23 (continued)

CMPI Instruction

---

22 denotes the mode used in Example 6.22.

Action: The <Sea> is subtracted from the <Dea>, and the condition codes are changed accordingly

Status Register Condition Codes: X N Z V C
                                       - * * * *

X is not affected.
N is set if the result is negative. Cleared otherwise.
Z is set if the result is zero. Cleared otherwise.
V is set if an overflow occurs. Cleared otherwise.
C is set if a borrow occurs. Cleared otherwise.

Notes:
1. Neither the <Sea> nor <Dea> is changed.
2. The size of the <Sea> matches the operation size.

TABLE 6-24

Compare Memory Instructions

CMPM   **Compare Memory**

Assembler Syntax: **CMPM.B <Sea>,<Dea>**
                                .W
                                .L

<Sea> Source Effective Address	<Dea> Destination Effective Address													
	Dn	An	(An)	(An) +	− (An)	$w (An)	$b (An, Xn)	$s	$l	$w (PC)	$b (PC, Xn)	#$q	SR	CCR
(An) +				23										

23 denotes the mode used for Example 6.23.

Action: The <Sea> is subtracted from the <Dea>, and the condition codes are changed accordingly

Status Register Condition Codes: X N Z V C
                                         - * * * *

X is not affected.
N is set if the result is negative. Cleared otherwise.
Z is set if the result is zero. Cleared otherwise.
V is set if an overflow occurs. Cleared otherwise.
C is set if a borrow occurs. Cleared otherwise.

Note: Neither the <Sea> nor <Dea> is changed.

Now our assignment.

EXAMPLE 6.22

If memory locations 7100 to 710F contain an array of 16 unsorted byte-sized numbers, create a program to sort them in ascending order. To avoid the need to *preload* memory with an array of numbers, we will tack onto the top of our program a few steps that will load the array with a sequence of descending numbers. (This will present a *worst-case* situation for our bubble sort, by the way.)

Using the flowchart in Figure 6-25, our program looks as follows:

```
; Example 6.22
;
; Sort an array of numbers (bytes) into ascending order
;
          .ORG $7100
START:    .RS 16              ; reserve space for numbers
NUM:      .DW 16              ; the number of numbers in the array
;
          .ORG $1000
;
; Preload an array with descending numbers
          LEA.L START,A0      ; point to beginning of array
          MOVE.B #$FF,D0      ; start filling array with FF
          MOVE.W NUM,D1       ; load the countdown counter
          SUBQ.W #1,D1        ; take care of DBF quirk
LOOP:     MOVE.B D0,(A0)+     ; store the number
          SUBQ.B #1,D0        ; decrease it by one
          DBF.W D1,LOOP       ; loop until done
; Beginning of bubble sort
AGAIN:    CLR.L D7            ; clear the swap flag
          LEA.L START,A0      ; point to BEGINNING of array
          LEA.L NUM-1,A2      ; point to END of array
READ1:    MOVE.B (A2),D0      ; get pointed to
                              ; (first number) to compare
          SUBQ.L #1,A2        ; decrement pointer
          CMPA.L A2,A0        ; pointing past beginning of array?
          BHI CHECK           ; if so, see if swap flag is set
          MOVE.B (A2),D2      ; if not, get pointed to
                              ; (second number) to compare
          CMP.B D2,D0         ; compare the numbers,
                              ; is the first > second?
          BHI READ1           ; if so, don't swap,
                              ; read another number to compare
          MOVE.B D2,1(A2)     ; if not, swap the numbers
          MOVE.B D0,(A2)      ;      swap the numbers
          MOVEQ.L #1,D7       ; set swap flag
          BRA READ1           ; read another number to compare
CHECK:    CMPI.L #1,D7        ; is swap flag set?
          BEQ AGAIN           ; if so, do sort again
HERE:     BRA HERE            ; if not, sorting is complete
          .END
```

See Figure 6-26 for execution of this example.

Addressing Modes Used

For the CMP instruction, the **Data Address Direct** is used. To compare two Address Registers that are being used as pointers, the CMPA instruction uses the **Address Direct** addressing mode.

For the MOVE.B D2,1(A2) instruction, we have used a new addressing mode called **Address Indirect with Displacement.** Since we wanted to point to one location past that being pointed to by Address Register A2, we can add a displacement [the 1 in front of (A2)] and that value will be added to the current value in A2 to get the effective <Dea> in which to put the number contained in Data Register D2. The displacement value would be preceded by a $ if a larger hex number was desired.

```
7100  FF   FE   7101        7100  F0   F1  7101
7102  FD   FC   7103        7102  F2   F3  7103
7104  FB   FA   7105        7104  F4   F5  7105
7106  F9   F8   7107        7106  F6   F7  7107
7108  F7   F6   7109        7108  F8   F9  7109
710A  F5   F4   710B        710A  FA   FB  710B
710C  F3   F2   710D        710C  FC   FD  710D
710E  F1   F0   710F        710E  FE   FF  710F
```

Before Sort   After Sort

Figure 6-26 Execution of Example 6.22.

A new instruction used, **BHI,** is in the **Branch Conditionally Bcc** group. Refer to Table 6-12 for details. Table 6-13 gives information on the choice of these instructions for various comparisons and checks. Basically, the BHI will branch if BOTH the Z bit and the C bit are 0 or cleared. If a compare is done and the result is NOT zero, and a carry/borrow did not occur, the comparison indicates that the <Sea> was **greater than or equal to** the <Dea>.

We also used some assembler *tricks* to simplify the pointing process. If you wish to change the number of numbers to be sorted, change the .RS 16 and the value for NUM to the desired value (in decimal for this example). The label NUM is used for two purposes: to point to the memory location PAST the end of the reserved 16 locations, and to contain the number of numbers in the array.

The MOVE.W NUM,D1 instruction will load the CONTENTS of the location labeled NUM (0010 hex for this example) into D1, and the LEA.L NUM-1,A2 instruction is used to point to the LAST of the 16 numbers, located in location 710F (not 7110). The assembler does the arithmetic for us.

The most important characteristic of the COMPARE instructions: NOTHING is changed as a result of a COMPARE except for the applicable bits in the CCR.

Hands-On Execution of This Example

Refer to Section 10.1.9, 10.2.9, or 10.3.9 for execution on the ECB, assembler/ simulator, or MAX 68000.

EXAMPLE 6.23

A useful utility program is one that compares two ASCII character strings, looking for an identical match. Write a program that compares two 10-character-long ASCII text strings for an identical match.

Assume that locations $7000 to 7003 contain the address of the first string and locations $7004 to 7007 contain the starting address of the area to be searched. Upon completion, location $7008 to $700B should point to an ASCII string saying "MATCH" or "NO-Match," depending on the results.

```
; Example 6.23
;
; Compare two ASCII strings (10 characters long),
; look for a complete match
; The FIRST string address is contained in locations 7000–7003
; The SECOND string address is contained in locations 7004–7007
; The MESSAGE address is contained in
; locations 7008–700B (and Address Register A6)
; FOR THIS SPECIFIC EXAMPLE
; Location 7100 contains the first string
; Location 7120 contains the second string
```

```
            ;
                        .org $7000
                        .DL FIRST
                        .DL SECOND
                        .DL MESSAGE
            ;
                        .org $7100
            FIRST:      .DB "String 1"
                        .org $7120
            SECOND:     .DB "String 2"
                        ; To get a MATCH, change the 2 above to a 1
                        ;
                        .org $1000
                        LEA.L FIRST,A0          ; The address of first string
                        LEA.L SECOND,A1         ;                    second
            STRING:     MOVE.W #10,D0           ; init 10 character countdown counter
                        SUBQ.L #1,D0            ; take care of the DBF quirk
            LOOP:       CMPM.B (A0)+,(A1)+      ; check for character match
                        BNE NOPE                ; mismatch found
                        DBF D0,LOOP             ; not done, keep checking for a match
            YEP:        LEA.L MATCH,A6
                        MOVEA.L A6,MESSAGE
                        BRA DONE
            NOPE:       LEA.L NOMATCH,A6
                        MOVEA.L A6,MESSAGE
            DONE:       BRA DONE
            MATCH:      .DB "MATCH"
            NOMATCH:    .DB "NOMATCH"
                        .END
```

New Instructions

Refer to Table 6-24 for information on the CMPM instruction. It allows us to compare the memory contents being pointed to by two Address Registers without altering them, only setting the appropriate bits in the CCR.

Another new assembler technique, the .DB "String 1", allows us to *define a byte*, and to enclose ASCII text with '', letting the assembler convert and store them as the appropriate hex numbers.

Also, the program is made more versatile by letting addresses 7000 to 7007 contain the possibly variable starting addresses of the strings to be compared. The pseudo-op .DL lets us store the address of a label in a specific address.

An Exercise for the Student: Write the program above without using the CMPM instruction, and compare program size and execution speed.

6.7 DOING LOGICAL OPERATIONS

Refer to Section 5.4.4 for the **Logical Instructions** and to Section 4.1.4 for fundamentals on logical AND, OR, exclusive OR, and NOT operations. These instructions are handy for bitwise comparisons of data stored in memory. If a memory location represented eight individual switches, which could be open or closed and a situation required an action **if and only if** TWO of the switches were closed (logic 0), a MASK can be used to mask out the bits of no concern.

For example, checking for bits 1 and 2:

INPUT MEMORY (switch condition)	xxxxx**00**x	xxxxx**01**x
	both closed	one closed
AND'd with MASK	00000110	00000110
gives	0000000**0**	0000001**0**

which gives a ZERO result regardless of the other bits when BOTH bits 1 and 2 are 0. A nonzero result occurs if only one or none of the switches are closed.

A simple graphical example may help. To find the condition of bits 1 and 2 of a memory location, we can move the memory contents into a Data Register, say D0, and then we AND the register with the binary number 00000110, which represents a number with a "1" in the positions we want to test (Figure 6.27). D0 will then be zero if bits 1 and 2 are both a zero or will be nonzero if either of them is a "1". The Z bit of the CCR will be 0 or 1, respectively, therefore mirroring the condition of bit 2. We then use this result to perform the desired action.

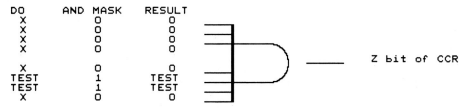

Figure 6-27   Bit testing with the AND instruction.

Many other applications can be found for this class of instructions. They allow comparisons and decisions of multiple bits in a single instruction.

Some general rules for consideration when using the logical instructions:

ANDing a bit with a 0 CLEARS it.
ANDing a bit with a 1 leaves it unchanged.
ORing a bit with a 0 leaves it unchanged.
ORing a bit with a 1 SETS it.
XORing a bit with a 0 leaves it unchanged.
XORing a bit with a 1 complements it (changes to a 1 if 0, to a 0 if 1).

Practical applications and discussion of these instructions are covered in Chapter 7.

6.8   USING THE SHIFT, ROTATE, AND BIT MANIPULATION INSTRUCTIONS

Refer to Section 5.4.5 for the **Shift and Rotate Instructions** and to Section 5.4.6 for the **Bit Manipulation Instructions.** The Bit Manipulation Instructions are extremely useful in manipulation of individual bits in a memory location. They are ideal for input/output applications where each bit represents a different external condition that needs to be monitored or controlled. These instructions will be put to good use in Chapter 7. Use of the shift and rotate instructions was required in earlier microprocessors not having instructions for testing individual bits. They are also handy for simple multiplication or division applications.

The following example shows how a binary number is multiplied by 2 if it is shifted one bit to the left. Similarly, if you take a decimal number, such as 3, and shift it left one digit, you get 30, or a multiplication by 10.

EXAMPLE 6.24

Create a program that multiplies a longword positive integer by a positive power of 2 (i.e., 2, 4, 8, 16, etc.). Allow for a 64-bit result and detection if the result is greater than 64 bits.

```
; Example 6.24
;
;
; Multiply a longword positive integer by 2 raised to a positive
; power
; (i.e., multiply the longword by 2, 4, 8, 16, etc.)
; Allow for a 64-bit result and detection if the result is > 64 bits
;
; Result is left in D1 (most significant 32 bits)
;           and D0 (least significant 32 bits)
;
              .ORG $1000
              MOVE.L NUMBER,D0    ; get the number to multiply
              MOVE.B POWER,D3     ; get the power to be used
              MOVEQ.L #1,D6       ; set an error flag
              CLR.L D1            ; to be used to get the result
TOP:          LSL.L #1,D0         ; shift left one bit (multiply by 2)
              ROXL.L #1,D1        ; shift possible carry bit
                                  ; from D0 to D1
              BCS EXIT            ; exit with error if carry bit is set
              SUBQ.B #1,D3        ; decrement counter
              BNE TOP             ; loop until done
              CLR.B D6            ; normal exit, no error,
                                  ; clear error flag
EXIT:         BRA EXIT            ; loop forever
NUMBER:       .RS 4               ; store number to be multiplied here
POWER:        .RS 2               ; store power here
              .END
```

The new instructions introduced are the **LSL** (Table 6-25) and **ROXL** (Table 6-26). Their action is definitely best demonstrated by execution of the program on the simulator. The LSL shifts the number left one bit, multiplying it by 2. A 0 is shifted in as the least significant bit, and the most significant bit is shifted into the C (carry) bit (and the X bit as well).

Look at the following binary number:

0001 0000   which is equal to 8 decimal. Shifted left one bit gives

0010 0000   which is equal to 16 decimal, which is twice the original value

Repeated shifting would multiply by 4, 8, 16, and so on.

The ROXL instruction is similar, except that the previous C bit is shifted in as the least significant bit. This allows us to have a 64-bit result (D1 and D0 together). If a carry is shifted out by the ROXL, the next instruction BCS exits to tell us we have exceeded a 64-bit answer.

Otherwise, we decrement our counter and continue multiplying.

TABLE 6-25

LSL Instruction

LSL   Logical shift left

Assembler Syntax: **LSL.B <Sea>,Dn**
                         .W
                         .L

<Sea> Source Effective Address	Dn	An	(An)	(An) +	– (An)	$w (An)	$b (An, Xn)	$s	$l	$w (PC)	$b (PC, Xn)	#$q	SR	CCR
Dn	*													
#$q	24													

24 denotes the mode used in Example 6.24.

Action: Left Shift <Sea> by <count stored in Dn> → <Dea>

Assembler Syntax: **LSL.W <ea>**

	Dn	An	(An)	(An) +	– (An)	$w (An)	$b (An, Xn)	$s	$l	$w (PC)	$b (PC, Xn)	#$q	SR	CCR
			*	*	*	*	*	*	*					

Action: Left Shift <ea> by one bit → <ea>

Status Register Condition Codes: X N Z V C
                                      * * * 0 *

X is set according to the last bit shifted out.
    Unaffected for a shift count of zero.
N is set if the result is negative. Cleared otherwise.
Z is set if the result is zero. Cleared otherwise.
V is cleared.
C is set according to the last bit shifted out.

Notes:
1. The carry and extend bits receive the last bit shifted out.
2. The shift count is contained in Dn for Data Register Direct mode.
3. The shift count is between 1 and 8 for Immediate Data mode.

TABLE 6-26

ROXL Instruction

ROXL   Rotate Left with Extend

Assembler Syntax: ROXL.B <Sea>,Dn
                           .W
                           .L

<Sea> Source Effective Address	Dn	An	(An)	(An)+	−(An)	$w (An)	$b (An, Xn)	$s	$l	$w (PC)	$b (PC, Xn)	#$q	SR	CCR
						<Dea> Destination Effective Address								
Dn	*													
#$q	24													

24 denotes the mode used in Example 6.24.

Assembler Syntax: **ROXL.W <Dea>**

	Dn	An	(An)	(An)+	−(An)	$w (An)	$b (An, Xn)	$s	$l	$w (PC)	$b (PC, Xn)	#$q	SR	CCR
						<Dea> Destination Effective Address								
			*	*	*	*	*	*	*					

Action: Rotate Left <Sea> by <count> → <Dea>

Status Register Condition Codes: X N Z V C
                                                    * * * 0 *

X is set according to the last bit shifted out.
    Unaffected for a shift count of zero.
N is set if the most significant bit of the result is negative.
    Cleared otherwise.
Z is set if the result is zero. Cleared otherwise.
V is cleared.
C is set according to the last bit shifted out.
    Cleared for a shift count of zero.

Notes:
1. The carry and extend bits receive the last bit shifted out, and the extend bit is shifted to the least significant bit of the operand.
2. The shift count is contained in Dn for Data Register Direct mode.
3. The shift count is between 1 and 8 for Immediate Data mode.

6.9 COMPUTATIONS WITH SIGNED NUMBERS

Most work with a microprocessor used as a controller involves manipulation down at the bit level, and work with unsigned integers. Occasionally, we need to work with numbers that can be either positive or negative. The 68000 has several instructions to deal with these operations.

Previous examples have used UNSIGNED INTEGERS. To perform a longword-sized add using word-sized integers, allowing for a possible 32-bit longword sum, we simply clear out the 32-bit sum, load the words as longwords, and use longword addition.

The upper 16 bits of the registers containing the words will be zeros, so the sum comes out correct.

Using SIGNED INTEGERS, we have to keep track of the most significant bit of the words carefully. If it is a 0, the number is positive; if 1, it is negative. For example, the positive decimal number 16 is represented as 0000 0000 0001 0000 in a 16-bit binary fashion. The most significant bit is automatically a 0, denoting a positive integer if we want the 16 to be a **positive signed** integer. The representation for this number is the same if we want the number to be an **unsigned** integer. Decimal numbers up to 32,767 (0111 1111 1111 1111 binary) would be represented similarly, whether they are to be used as signed or unsigned integers. Numbers greater than this, on up to 65,535 (1111 1111 1111 1111 binary) would be used similarly as **unsigned** integers, but they are considered differently if they are used as **signed** integers.

Let's look at the *first* negative signed integer, 1000 0000 0000 0000. At first glance, it looks like a *minus zero* (-0). We will see shortly that it is really equal to -32,768. A minus zero is the same as a plus zero! We will find -1 to be 1111 1111 1111 1111.

The most significant bit denotes a negative signed integer. Unfortunately, it is not that simple. Refer to Chapter 4 for the binary subtraction examples. Negative numbers were represented by taking the TWO'S COMPLEMENT of the number.

Looking at the negative decimal number 16, first as a byte-sized signed integer, we take the two's complement of the 16 (0001 0000). This gives

$$1110 \ 1111 + 1 = 1111 \ 0000$$

Doing the same process as a word-sized signed integer, we get 1111 1111 1111 0000 with the sign bit being EXTENDED up through the remaining bits of the word. The most significant bit is a 1, denoting a negative number. We have SIGN-EXTENDED the byte to word size.

When an 8-bit number is to be used as a 16-bit number, or when a 16-bit number is to be used as a 32-bit number, there is a possibility of losing its sign if it is a signed number. Copying FF (which represents the signed number -1) to a 16-bit location would give 00FF, which would represent the signed number +255. To eliminate this problem, some of the 68000's instructions copy the sign bit into all of the upper 8 bits of the new 16-bit number, thus preserving the sign of the original 8-bit number. Sign extension of the FF above would give FFFF, which represents a word-sized -1 signed integer. Note that the UNSIGNED value of the number, which we are not using, is changed drastically. Table 6-27 shows how the signed numbers fall into place.

TABLE 6-27

Unsigned and Signed Numbers

BINARY INTEGER	HEX INTEGER	UNSIGNED DECIMAL	SIGNED DECIMAL
0000 0000 0000 0000	0000	0	0
0000 0000 0000 0001	0001	1	1
0000 0000 0000 0010	0002	2	2
0000 0000 0000 0011	0003	3	3
....			
0111 1111 1111 1110	7FFE	32,766	32,766
0111 1111 1111 1111	7FFF	32,767	32,767
1000 0000 0000 0000	8000	32,768	-32,768
1000 0000 0000 0001	8001	32,769	-32,767
1000 0000 0000 0010	8002	32,770	-32,766
....			
1111 1111 1111 1110	FFFE	65,534	-2
1111 1111 1111 1111	FFFF	65,535	-1

Let's try our first example.

EXAMPLE 6.25

Find the average value of an array of five word-sized SIGNED INTEGERS stored in locations starting at $7000. Store the result in location $7100.

Recall that a 16-bit signed integer is a decimal number between -32,768 and +32,767. They will, of course, be represented in hex as $0000 to $FFFF, with the most significant bit being used to represent the sign of the integer. To show a new instruction, the word-sized numbers will be chosen between $80–0–$7F (1000 0000–0000 0000– 0111 1111), equivalent to -128–0–+127 decimal.

An algorithm and perhaps a flowchart are in order. Without checking, you would guess that we do not have an *averaging* instruction at our disposal. The average of a series of numbers is simply the sum of the numbers divided by the NUMBER of numbers we are averaging. The *computer approach* is fairly straightforward and is shown in the flowchart in Figure 6-28. The special treatment needed for signed integers was omitted from the flowchart and will be discussed after we look at the program.

Our program would look as follows:

```
; Example 6–25
; Find the average of an array of five word-sized
; signed integers (starting at 7000),
; store the result in memory location 7100
;
.ORG $1000
        .EQU DATA,$7000         ; beginning of stored data
        .EQU AVERAGE,$7100      ; where the average is stored
        .EQU NUMBERS,5          ; five (decimal) numbers
        CLR.L D0               ; used to get the number from memory
        CLR.L D1               ; used to total the numbers (= SUM)
        LEA.L DATA,A0          ; point to the beginning of data
        MOVE.W #NUMBERS-1,D2   ; initialize counter to 1
                               ; less than number of numbers
LOOP:   MOVE.W (A0)+,D0        ; get first number
        EXT.L D0              ; sign-extend number to 32 bits
        ADD.L D0,D1           ; add to the sum
        DBF D2,LOOP          ; loop until done
        DIVS #NUMBERS,D1      ; divide sum by number of numbers
        MOVE.W D1,AVERAGE     ; store answer in AVERAGE
DONE:   BRA DONE
        .END
```

Recall that when using the DBF instruction we have to start it with one less than the number of times we wish to loop. In this example we used the power of the assembler to do it, using arithmetic in our instruction as MOVE.W #NUMBERS-1,D2. The ASSEMBLER takes the value of NUMBERS (5, for this example) and subtracts one from it, putting this value in the object file that is created.

To deal with SIGNED numbers, two new instructions were introduced: **EXT** (Table 6-28) and **DIVS** (Table 6-29). The **EXT.L D0** sign extends our WORD in D0 to a LONG-WORD, thus preserving the sign of the number. We do this because our divide instruction always uses 32 bits for the numerator (<Sea>). At the completion of the DBF loop, D1 will contain our possible 32-bit sum. The **DIVS #NUMBERS,D1** will divide the contents of D1 by the number of numbers (five), leaving the quotient in the lower word of D1 and any possible remainder in the upper word. In our example we ignore the possible remainder.

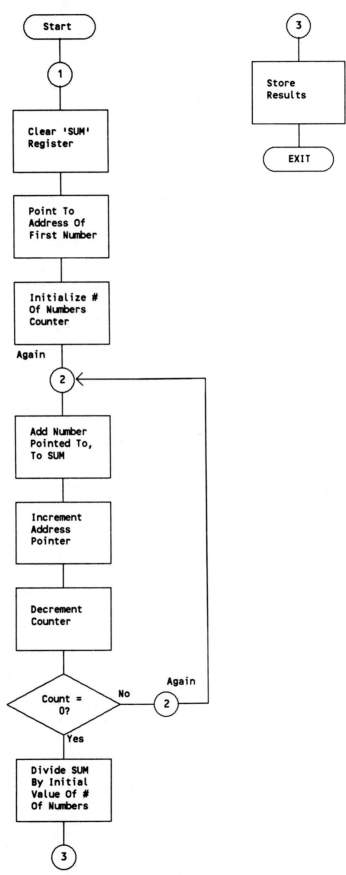

Figure 6-28 Average value example flowchart.

Before executing the program, we need to load up the array starting at 7000 with some signed numbers (i.e., -2, 5, 4, -6, 9). They would be entered as 00FE 0005 0004 00FA 0009. -2 is represented by FE, -6 is FA.

TABLE 6-28

EXT Instruction

___

EXT   **Sign Extend**

Assembler Syntax: **EXT.W Dn**
                              .L

							<Dea> Destination Effective Address							
	Dn	An	(An)	(An)+	−(An)	$w(An)	$b(An, Xn)	$s	$l	$w(PC)	$b(PC, Xn)	#$q	SR	CCR
	25													

25 denotes the mode used for Example 6.25.

Action: The sign bit is extended from a byte to a word or from a word to a longword, depending on the size selected.

Status Register Condition Codes: X   N   Z   V   C
                                                                  -   *   *   0   0

X is not affected.
N is set if the result is negative. Cleared otherwise.
Z is set if the result is zero. Cleared otherwise.
V is cleared.
C is cleared.

___

Notes:
1. If the operation is word sized, bit 7 of the Data Register is copied to bits 8 to 15.
2. If the operation is longword sized, bit 15 of the Data Register is copied to bits 16 to 31.

TABLE 6-29

DIVS Instruction

___

DIVS   **Signed Divide**

Assembler Syntax: **DIVS.W <Sea>,Dn**

<Sea> Source Effective Address	<Dea> Destination Effective Address Dn		
Dn	*		
An			
(An)	*		
(An) +	*		
− (An)	*		
$w (An)	*		
$b (An, Xn)	*		
$s	*		
$l	*		
$w(PC)	*		
$b (PC, Xn)	*		
#$q	25		
SR			
CCR			

TABLE 6-29 (continued)

DIVS Instruction

___

25 denotes the mode used in Example 6.25.

Action: <Dea>/<Sea> → <Dea>

Status Register Condition Codes: X   N   Z   V   C
                                      -   *   *   *   0

X is not affected.
N is set if the quotient is negative. Undefined if overflow. Cleared otherwise.
Z is set if the quotient is zero. Undefined if overflow. Cleared otherwise.
V is set if division overflow occurs. Cleared otherwise.
C is cleared.

___

Notes:
1. The longword-sized <Dea> is divided by a word-sized <Sea>, and the result is stored in the <Dea>.
2. The operation is performed using signed arithmetic.
3. The quotient is stored in the lower word of the <Dea>.
4. The remainder is stored in the upper word of the <Dea>.
5. The sign of the remainder is the same as the dividend unless it is equal to zero.
6. Division by zero causes a TRAP.
7. Overflow may occur before completion of the instruction. If this occurs, the condition is flagged and the operands are unaffected.
8. Overflow occurs if the quotient is larger than a 16-bit signed integer.

6.10   SOME MISCELLANEOUS EXAMPLES

Following are a series of examples that use instructions and techniques covered earlier.

EXAMPLE 6.26

A useful utility program is one that moves the contents of a block of memory from one area of memory to another. Assuming that $7000 and 7001 contain the number of bytes to be moved, $7002 to $7005 contain the SOURCE address of the data to be moved, and $7006 to $7009 contain the DESTINATION address for the data, write a program that moves the data from the source to the destination.

In simple terms it looks like we need to set a couple of pointers, initialize a countdown counter, and then move/loop until the counter is zero. A flowchart is hardly necessary. Our program would look as follows:

```
; Example 6.26
;
; Move a block of data from one memory area to another
; 7000–7001 contains the number of bytes to be moved
; 7002–7005 contains the SOURCE address for the data
; 7006–7009 contains the DESTINATION address for the data
;
          CLR.L D0            ; clear out all of D0
          MOVE.W $7000,D0     ; load the countdown counter
          SUBQ.W #1,D0        ; take care of the DBF quirk
          LEA.L $7002,A0      ; set the SOURCE pointer
          LEA.L $7006,A1      ; set the DESTINATION pointer
LOOP:     MOVE.B (A0)+,(A1)+  ; move a byte, bump pointers
          DBF.W D0,LOOP       ; countdown until done
DONE:BRA DONE                 ; done, loop forever
```

No new tricks were used here. The astute student may ask why we moved the data a byte at a time instead of a word or longword at a time. Example 6.26 will work for any number of bytes up to 65,535 (the maximum size of our countdown counter) regardless of the source and destination addresses, which may be even or odd.

EXAMPLE 6.27

Finding the maximum value contained in a block of memory is often required. Write a program that will find the largest **unsigned integer word** contained in a block of memory. Assume that $7002 to $7005 contain the starting address of the block, and $7006 to $7009 contain the ending address of the block. Put the maximum value found in locations $7000 and $7001.

Our program would look as follows:

```
; Example 6.27
;
; Find the maximum value contained in a block of memory
; 7000–7001 will contain the maximum value found
; 7002–7005 contains the STARTING address for the search
; 7006–7009 contains the ENDING address for the search
;
            LEA $7002,A0        ; point to start of data block
            LEA $7006,A1        ; point to the end of data block
            CLR.W $7000         ; make the maximum found so far = 0
LOOP:       MOVE.W (A0)+,D0     ; get the first number and bump
                                ; pointer up one
            CMP.W $7000,D0      ; is the old max > the number
                                ; just obtained?
            BCC.S NOPE          ; nope, the old max was larger
            MOVE.W D0, $7000    ; yep, store new maximum value in D0
NOPE:       CMPA.L A1,A0        ; are we done?
            BNE LOOP            ; loop until done
HERE:       BRA HERE            ; loop forever
```

Note that we used the CMPA instruction to compare the start and ending addresses rather than counting down a counter with the DBF instruction. The DBF is limited to word-sized countdowns and could not handle cases where more than 65,536 word (= 131,072 bytes) locations were to be checked. A variation, left for the student, would be a modification to use the DBF instruction. (Subtract the contents of A0 from A1, put it in a Data Register, and so on.)

EXERCISES

Exercises for the student are consolidated at the end of the book.

chapter 7

a look at some industrial control applications

Roughly categorizing the applications of the 68000, we have:

1. Desktop computers
2. Minicomputers
3. Stand-alone devices (i.e., laser printers)
4. Industrial controllers

A 68000 system can easily take on the task of monitoring and controlling a large number of devices. With a little care, a computer neophyte can take a readily available economical 68000 development board, such as the Motorola ECB or MAX 68000, and add interfacing projects to perform industrial control operations. With minor modifications, the system can be modified to act as a completely stand-alone controller. In this chapter we cover some very elementary interface examples to show how industrial control projects could be undertaken, and also to expand on our coverage of the 68000's instruction set. Both the hardware and software are discussed, providing challenging laboratory experiments.

Those that become *addicted* to programming at the assembly level often become hooked because their efforts involve both hardware and software. Going from the pencil to the hot soldering iron and back provides an interesting change of scenery during programming assignments. This serious programmer has been known to pick up the wrong tool, luckily by the correct end, during late-night work.

This book is intended as an introductory text, without the details needed for circuit design. However, to provide some excitement or possibility for laboratory experimentation, a little hardware interfacing will be introduced. In other words, if you use only the simulator program or one of the development boards without added hardware, the most you will see are some characters on the screen, and keyboard input will be your only method of providing data to a program while it is executing. However, by simple addition of a little hardware, even the most inexperienced person can add to the development boards enough to make them do meaningful tasks.

7.1 INTERFACING FUNDAMENTALS

The basic concept of *reading* and *writing* to mailboxes still holds, but now we must be concerned with the electrical details of interfacing input and output devices to our computer system. Realizing that one of the main goals of this book is coverage from the ground up for beginners, we should discuss the electrical characteristics of the output circuitry of our computer. A basic understanding of AC/DC voltages and currents is assumed. The primary interfacing connections to our computer system will be to a class of integrated circuits referred to as transistor-transistor logic (TTL).

Referring to Figure 7-1, we see the various dc voltage levels that represent logic 1 and logic 0. These 1's and 0's are the same as those we use to construct nibbles, bytes, words, and longwords. To *turn on* a single bit and turn off all others bits at a single memory address, such as $7500, we would use an instruction such as **MOVE.B #01,$7500.** The least significant bit (bit 0) would then output a dc voltage in the range of 2.4 to 5.0 V to any I/O device addressed to that location. The other 7 bits would be in the range of 0.0 to 0.4 V. Getting the desired output **voltage** levels is straightforward. However, when we interface the TTL IC to the outside world we have to consider the available **current** drive. A TTL logic 1 is only capable of providing 400 μA of current and **still maintain the logic 1 voltage level.** The lines that are logic 0 must *sink,* or absorb, up to 16 mA of current. The input/output considerations are also shown in Figures 7-1 and 7-2.

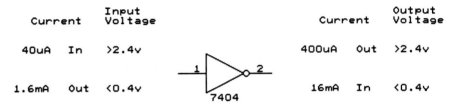

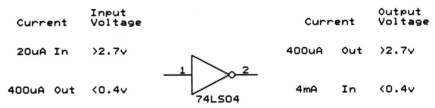

Figure 7-2   TTL and LSTTL input/output current capabilities.

Looking at Figure 7-3, we have directly interfaced a light-emitting diode (LED) to the output of a TTL IC. An LED typically needs about 20 mA of current to light brightly. An inverter IC is shown, so a logic 0 will light the LED and a logic 1 will extinguish it. In the preceding paragraph we said the IC could only provide 400 μA of current and **still maintain the logic 1 voltage level.** If you measure the voltage across the lit LED, you will observe between 1.6 and 2.0 V, which is below that of a logic 1 for an input. Our main concerns are to light the LED when we give the IC a logic 0 and to prevent destruction of the IC. The IC has an internal resistor that protects it from excessive current, and the LED works just fine. The only thing we cannot do is connect across the LED and use that voltage level as the input to another TTL IC that expects the normal logic 1 voltage levels. (*Note:* A LED does not have an internal resistor, and connecting one directly across 5 V will destroy it!)

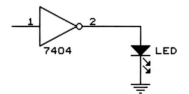

Figure 7-3   Typical TTL LED connection.

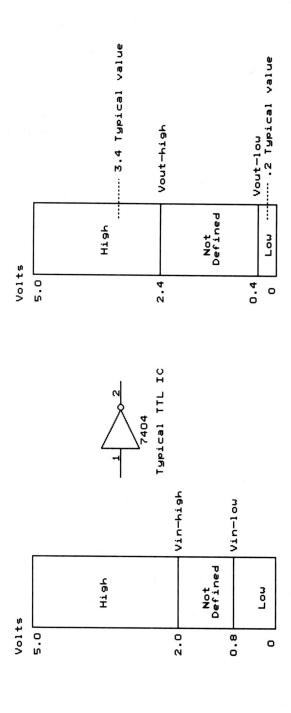

Figure 7-1 TTL input/output voltage ranges.

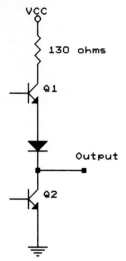

Figure 7-4 Partial output circuitry for a TTL IC.

Figure 7-4 shows the typical internal circuitry of a TTL IC. Without going into the details, since mastery of transistor circuits is not required for this book, let's assume that transistor Q1 is an open circuit and Q2 is a short circuit. The output voltage, that across Q2, would be almost zero. On the other hand, if Q1 is short-circuited and Q2 an open circuit, the output voltage (with nothing connected to the output) would be almost 5 V. If we connect a **short circuit** (worse case) to ground across the output, approximately $(5 - 0.3 - 0.7)/130 = 30$ milliamperes (mA) of current flows. The approximate voltage drop across a diode is 0.7 V, and a turned-on transistor has a voltage drop of about 0.3 V. If we connect an LED instead, which typically drops between 1.6 and 2.0 V at a normal brilliance of 20 mA, we have $(5 - 0.3 - 0.7 - 1.6)/130 = $ approximately 18 mA, which is about right. The IC circuitry is not damaged, but the output voltage is not a logic 1 and cannot be connected to other ICs.

An alternative connection, shown in Figure 7-5, is for purists who want to maintain the logic levels and do not know about the limiting resistor in the IC. A logic 1 is used to turn the LED on. The low output voltage of the IC allows current flow from the power supply through the resistor and LED, then through the IC to ground. The IC must be capable of sinking the 16 mA current and the resistor chosen accordingly, 180 ohms (Ω) being a typical value. If we are using LS TTL ICs, this approach is not recommended since the maximum sinking current is 4 mA instead of 16 mA.

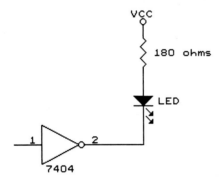

Figure 7-5 Alternative TTL LED connection.

Considering an external input to a TTL IC, we ensure that the IC input sees either a logic 1 or a logic 0 voltage level. A *pull-up* resistor is usually installed to ensure a true logic 1 when the switch is opened. Figure 7-6 shows a typical application. A current of about 1 mA flows through the switch when it is closed. When open, about 20 to 40 μA leakage current flows into the IC. The input voltage is pulled up to a solid 5 V to ensure a valid 1 on the input.

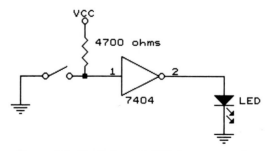

Figure 7-6 Typical switch TTL input connection.

Using a TTL IC in our interfacing serves another valuable purpose. It serves as protection for the more expensive peripheral chip, such as the Motorola 68230 or 6821. It can be omitted if careful consideration of the drive specifications of the peripheral chip is given. Initially, our interfacing will be to low-level dc voltages (0 to 5 V). When we use higher voltages and perhaps 115 V ac, more stringent considerations must be used to avoid **total destruction** of our computer system.

In summary, it looks like a TTL IC can do a better job at **sinking** current than it can **source** current. It can be used to source current, such as the LED example, in some cases. We must provide proper current limiting for the outputs when they are set up to sink current. IC data sheets indicate current INTO (sinking) a pin as POSITIVE and current OUT (source) of a pin as NEGATIVE. LS TTL ICs consume less power but also cannot source as much current as standard TTL ICs.

The examples in this chapter show not only the actual connections and hardware to perform an industrial control task but also builds on our knowledge of programming and coverage of the 68000's instruction set. If the hardware cannot be built, the programs should still be done on the simulator, resulting in less spectacular visible effects, of course.

7.2 SIMPLE INPUT AND OUTPUT

Previous examples have indicated that memory locations are used to store both the input data or information, and the output data. When dealing with external interfacing, we will refer to memory locations used to input data as an INPUT **PORT,** and those used to output data as an OUTPUT **PORT.** Some microprocessors have separate instructions and circuitry when dealing with I/O ports. The Motorola family of microprocessors allows **all** instructions to be applied to **all** memory locations and does not use separate port decoding or instructions. Again, we will refer to the input/output addresses as ports, simply to save some typing.

For our first example interfacing with the outside world, let's consider a single LED as discussed above.

EXAMPLE 7.1

```
; Example 7.1 - Assume that bit 0 of memory location $7500 is attached to
; a LED
; Create a program that blinks the LED approximately once per second
; Assume a computer clock speed of 4 megahertz
;
; Blink LED connected to bit 0 of location $7500 at a rate of one per
; second
;
                .ORG   $1000
                .EQU   OUTPUT,$7500  ; assign an address to a label
TOP:            MOVE.B  #0,OUTPUT    ; turn LED OFF
                MOVE.W  #$FFFF,D0     ; initialize countdown counter to
                                      ; 65536
OFFLOOP:  NOP                         ; do nothing, burns up some time
          NOP                         ;
          NOP                         ;
          NOP                         ;
          DBF   D0,OFFLOOP            ; decrement counter, burn up time
          MOVE.B  #1,OUTPUT          ; turn LED on
          MOVE.W  #$FFFF,D0          ; reinitialize countdown counter
ONLOOP:   NOP
          NOP                         ;
          NOP                         ;
          NOP                         ;
          DBF   D0,ONLOOP             ; decrement counter, burn up time
          BRA   TOP                   ; keep looping indefinitely
          .END
```

Execution Speed of an Instruction

In many cases it is of no concern how long it takes an instruction to execute. In some cases, where it is desired to burn up some time, *software timing loops* are written. The instructions and how long it takes them to execute, along with the operating speed of the 68000 itself, are used to come up with a desired time interval.

Recall that the postmaster uses its *brain* (instruction decoder) to decode and execute an instruction. What actually takes place is a series of steps being performed on a vast array of digital logic ICs. Lots of gates, flip-flops, register, and so on, are *stepped* through their paces to perform the instruction execution. To synchronize all this, a clock circuit is used. It is simply a square-wave generator, producing a series of pulses. If it has a frequency of 10 megahertz (MHz), a pulse is produced every 1/10,000,000 second (s) or every 0.1 μs. Complete execution of a single instruction takes several of these clock pulses.

Referring to the instruction execution time chart in Appendix A, the MOVE instruction above indicates that 20 (4/1) clock periods are needed. The 20 represents the total number of periods needed if no additional waiting periods are introduced by the hardware. Four read and one write cycles are used (the numbers in the parentheses) with each requiring four clock periods (4 times 4 plus 4 times 1 = 20). If slow memory chips were used, and one period of delay was used for each read or write cycle, we would add 4 times 1 + 1 times 1 = 5 to the 20 periods needed for a nondelayed system, giving a total of 25 clock periods. To find the time needed to execute this instruction, we divide the number of periods by the clock frequency used by the 68000. For a 4-MHz system, and a 20-period-long instruction, we have 20/4,000,000 = 5 μs.

The 68000 uses a technique called *pre-fetch* to enhance its operation, done completely transparent to the user. When an instruction is read from memory (fetched), it actually reads two, saving the second one just in case it is needed next. If the first instruction is a branch instruction, the second one is discarded if the branch is executed. Otherwise, the computer progresses normally down through memory, and the processor already has the next instruction and is ready to execute it. The execution time chart reflects this in its values.

To blink our LED at one blink per second, we need to turn it on, kill ½ s, turn it off, kill ½ s, indefinitely. Again, the ***elapsed time* of an instruction is the number of clock cycles it requires divided by the clock frequency.** To *kill* ½ s, we would need 0.5 × 4 MHz = 2,000,000 cycles.

For our program we simply turn the LED off, burn up some time with a *timing loop,* turn the LED on, burn up some time with a timing loop, and then branch back to the beginning of our program. To create a timing loop, we load a countdown counter, insert a series of "do-nothing" instructions, and decrement the counter until it exits the loop. For the DBF decrement/branch instruction, the largest value we can use is $FFFF, which will give us 65,536 times through the loop. Note that the DBF requires 14 cycles if it branches and only 12 during its final count/exit.

Looking at the program above, we need to calculate the time delay between the LED on instruction and the LED off instruction, and vice versa.

			#cycles	#times executed	
TOP:	MOVE.B	#0,OUTPUT	; 16	1	OFF (start)
	MOVE.W	#$FFFF,D0	; 8	1	TIME
OFFLOOP:	NOP		; 4	65536	
	NOP		; 4	65536	
	NOP		; 4	65536	
	NOP		; 4	65536	
	DBF	D0,OFFLOOP	; 14 (12)	65535 (1) (see below)	
	MOVE.B	#1,OUTPUT	; 6	1	OFF (end)
	MOVE.W	#$FFFF,D0	; 8	1	TIME

```
ONLOOP:   NOP                    ; 4       65536
          NOP                    ; 4       65536
          NOP                    ; 4       65536
          NOP                    ; 4       65536
          DBF  D0,ONLOOP         ; 14 (12) 65536
          BRA  TOP               ; 10      1
```

To simplify matters, we will assume that we have no memory waiting periods. In other words, the number of cycles is as shown in bold in the instruction execution time table of Appendix A. If you are using the ECB, note that it does add an additional wait state.

The off duration will be from the completion of the top: MOVE.B #0,OUTPUT instruction through the MOVE.B #1,OUTPUT instruction. The on duration will be from the second MOVE.W #$FFFF,D0 instruction up through the first instruction. Doing our arithmetic:

Off: $8 \times 1 + 4 \times (4 \times 65{,}536) + 14 \times 65{,}535 + 12 \times 1 + 6 \times 1 = 1{,}966{,}102$ machine cycles

On: $8 \times 1 + 4 \times (4 \times 65{,}536) + 14 \times 65{,}535 + 12 \times 1 + 10 \times 1 + 16 \times 1 = 1{,}966{,}112$ machine cycles

The off time would be the number of machine cycles times the period of one cycle, which is the reciprocal of the clock frequency. We would have 1,966,102/4 MHz = 0.49 s. Similarly, the on time would be 1,966,112/4 MHz = 0.49 s.

The purists will note that the ON/OFF times are not identical and that the timing is not exactly 0.5 s. On a 4-MHz computer with no wait states, the minimum time increment we can create would be 4/4 MHz = 1 μs, and we can adjust the timing to this resolution if desired. (The NOP is one of the fastest instructions, taking four cycles.)

To adapt the preceding for the creation of other time periods, replace the 65,536 with a variable called COUNT, the 65,535 with COUNT-1, and the number of total cycles (1,966,102) with ON × CLK, where ON is the desired on time, and CLK is the computer clock frequency (i.e., 4 MHz for the ECB). By inserting the desired ON value and solving for COUNT, we can then use this value in the preceding program. (Leave the $ off and the assembler will convert it to the proper hex value.)

Hands-On Execution of This Example

Refer to Section 10.**1.10** for execution of this example for the **ECB,** to Section 10.**2.10** for execution on the Pseudo Sam/Pseudo Max **assembler/simulator,** or to Section 10.**3.10** for the **MAX 68000.**

EXAMPLE 7.2

This is another input/output (I/O) problem that needs no flowchart.

```
; Example 7.2 — Assume that bit 0 of memory location $7500 is connected
; to a switch and that bit 7 of the same address is connected to a LED
; Create a program that turns off the LED when the switch is closed
;
;
          .ORG $1000
          .EQU  INPUT,$7500      ; assign an address to a label
          .EQU  OUTPUT,$7500     ; assign an address to a label
```

```
LOOP:   MOVE.B   INPUT,D0      ; get the switch conditions
        BCHG    #0,D0          ; invert switch condition
                               ; (makes closed = on)
        ASL.B   #7,D0          ; shift it left to bit 7
        MOVE.B  D0,OUTPUT      ; turn LED on or off
        BRA     LOOP           ; keep on checking switch
        .END
```

Two new instructions, **ASL** and **BCHG,** are introduced. Refer to Tables 7-1 and 7-2 for information on their use. The **BCHG** instruction is used here to convert the switch

TABLE 7-1

ASL Instruction

ASL   Arithmetic shift left

Assembler syntax: ASL.B <Sea>,<Dea>
                          .W
                          .L

<Sea> Source effective address	<Dea> Destination effective address														
	Dn	An	(An)	(An) +	– (An)	$w (An)	$b (An, Xn)	$s	$l	$w (PC)	$b (PC, Xn)	#$q	SR	CCR
Dn #$q	* 2													

2 denotes the mode used in Example 7.2.

Assembler syntax: ASL.W <ea>

		<ea> Effective address													
	Dn	An	(An)	(An) +	– (An)	$w (An)	$b (An, Xn)	$s	$l	$w (PC)	$b (PC, Xn)	#$q	SR	CCR
			*	*	*	*	*	*	*					

Action: Left shift <Sea> by <Dea> → <Dea>
   or   Left shift <ea> by one bit → <ea>

Status register condition codes: X   N   Z   V   C
                                       *   *   *   *   *

X is set according to the last bit shifted out. Unaffected for a shift count of zero.
N is set if the most significant bit of the result is set.
   Cleared otherwise.
Z is set if the result is zero. Cleared otherwise.
V is set if the most significant bit is changed at any time during the shift operation.
   Cleared otherwise.
C is set according to the last bit shifted out. Cleared for a shift count of zero.

Notes:
1. The <Sea> is shifted left by the count contained in <Dea>, and the result is stored in the <Dea>.
2. The carry and extend bits receive the last bit shifted out.
3. A zero is shifted into the least significant bit of the operand.
4. The shift count is contained in Dn for data register direct mode.
5. The shift count is between 1 and 8 for immediate data mode.
6. The ASL <ea> instruction shifts the memory word left by one bit only.

TABLE 7-2

BCHG Instruction

BCHG Bit test and change

Assembler syntax: BCHG.B <Sea>,<Dea>
.L

<Sea> Source effective address	<Dea> Destination effective address													
	Dn	An	(An)	(An) +	– (An)	$w (An)	$b (An, Xn)	$s	$l	$w (PC)	$b (PC, Xn)	#$q	SR	CCR
Dn	*		*	*	*	*	*	*	*					
#$q	2		*	*	*	*	*	3	*					

2 denotes the mode used for Example 7.2. 3 denotes the mode used for Example 7.3.

Action: The bit indicated by the <Sea> is tested, the Z bit is set or cleared accordingly, and the corresponding bit in the <Dea> is inverted.

Status register condition codes: X N Z V C
                                    - - * - -

X is not affected.
N is not affected.
Z is set if the bit tested was zero before inversion. Cleared otherwise.
V is not affected.
C is not affected.

Notes:
1. When <Dea> is a data register, only longword operations are allowed.
2. When <Dea> is a memory location, only byte operations are allowed.
3. Bits are numbered from 0, 0 being the least significant.

closure, which is a 0, to a 1 to turn the LED on when the switch is closed. If the switch is open, the 1 is converted to a 0 to extinguish the LED. The BCHG instruction is covered in more depth in Example 7.3.

We then use the **ASL** instruction to shift the inverted input switch condition over to bit 7 to the output bit. Note that the remaining bits of $7500 are *don't cares* and that we will be outputting a 0 to the input bit 0 position. **Output**ting to an **input** location has no effect.

===

✓**EXAMPLE 7.3**

Figure 7-7 shows a possible industrial control situation. Our task is to *read* the switches and *write* to the LEDs. We need to:

(a) Turn on LED-3 when switch-4 is closed.
(b) Turn on LED-7 when switch-1 **and** switch-2 are closed.
(c) Toggle LED-2 for each momentary closing of switch-0. In other words, when switch-0 is closed and then opened, LED-2 is turned off if it is on, or on if it is off.
(d) No other actions are taken. (Do not disturb any of the other outputs.)
(e) Continually loop through the program.

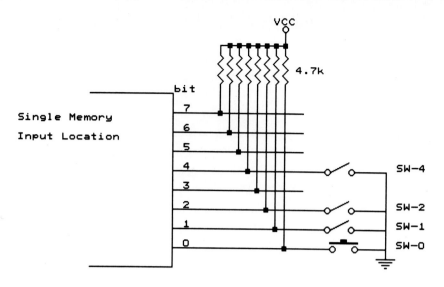

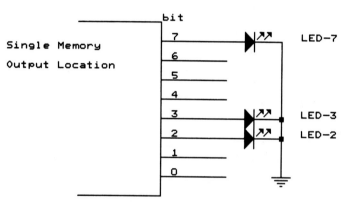

Figure 7-7   Simple industrial control schematic diagram.

The *algorithm* for this task is rather straightforward, and the flowchart in Figure 7-8 shows the steps to be taken. We need to cover a few new instructions to accomplish the job, however. Referring to Figure 7-8, we see that closing a switch results in a 0 for that particular bit of the input memory location. The 0 will remain as long as the switch is closed. For the LEDs, writing a 1 to a bit in the output memory location will result in the LED coming on and staying on. We will find one pitfall with our setup and program, however. Also, we purposely chose an awkward configuration for the switches and LEDs—perhaps to complicate the task.

To see what *tools* are at our disposal to **test,** to **set,** or to **clear** a specific bit, refer to Section 5.4.6 for the **bit manipulation instructions.** They are detailed in Tables 7-3 to 7-5.

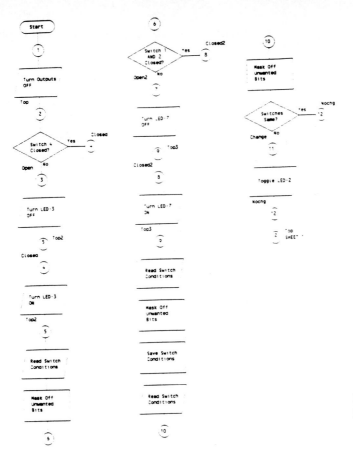

Figure 7-8 Flowchart for Example 7.3.

TABLE 7-3

BCLR Instruction

BCLR **Bit test and clear**

Assembler syntax: BCLR.B <Sea>,<Dea>
.L

<Sea> Source effective address	<Dea> Destination effective address													
	Dn	An	(An)	(An) +	– (An)	$w (An)	$b (An, Xn)	$s	$l	$w (PC)	$b (PC, Xn)	#$q	SR	CCR
Dn	*		*	*	*	*	*	*	*					
#$q	*		*	*	*	*	*	3	*					

3 denotes the mode used for Example 7.3.

Action: The bit indicated by the <Sea> is tested, the Z bit is set or cleared accordingly, and the corresponding bit in the <Dea> is cleared

Status register condition codes: X N Z V C
                                     - - * - -

X is not affected.
N is not affected.
Z is set if the bit tested was zero before being cleared. Cleared otherwise.
V is not affected.
C is not affected.

Notes:
1. When <Dea> is a data register, only longword operations are allowed.
2. When <Dea> is a memory location, only byte operations are allowed.
3. Bits are numbered from 0, 0 being the least significant.

129

TABLE 7-4

BSET Instruction

BSET Test a bit and set

Assembler syntax: **BSET.B <Sea>,<Dea>**
.L

<Sea> Source effective address	*<Dea> Destination effective address* Dn	*An*	*(An)*	*(An) +*	*− (An)*	*$w (An)*	*$b (An, Xn)*	*$s*	*$l*	*$w (PC)*	*$b (PC, Xn)*	*#$q*	*SR*	*CCR*
Dn	*		*	*	*	*	*	*	*					
#$q	*		*	*	*	*	*	3	*					

3 denotes the mode used for Example 7.3.

Action: The bit indicated by the <Sea> is tested, the Z bit is set or cleared accordingly, and the corresponding bit in the <Dea> is set

Status register condition codes: X N Z V C
- - * - -

X is not affected.
N is not affected.
Z is set if the bit tested was zero before being set. Cleared otherwise.
V is not affected.
C is not affected.

Notes:
1. When <Dea> is a data register, only longword operations are allowed.
2. When <Dea> is a memory location, only byte operations are allowed.
3. Bits are numbered from 0, 0 being the least significant.

We have four different BIT testing instructions from which to choose. Here are the tasks at hand for Example 7.3:

	INPUT LOCATION Bit 7 6 5 4 3 2 1 0	**OUTPUT LOCATION** Bit 7	6	5	4	3	2	1	0		
(a)	× × × **0** × × × ×					**1**				Switch-4 closed	LED-3 on
	× × × **1** × × × ×					**0**				Switch-4 open	LED-3 off
(b)	× × × × × **0 0** ×	1								Switches closed	LED-7 on
	× × × × × **1 1** ×	0								Switches open	LED-7 off
	× × × × × **0 1** ×	0								One switch open	LED-7 off
	× × × × × **1 0** ×	0								One switch open	LED-7 off
(c)>	× × × × × × × **1**						0			Top of loop	LED-2 off
|	× × × × × × × **0**						0			No change yet	LED-2 off
|	× × × × × × × **1**						1			Switch back off	LED-2 on
|	× × × × × × × **0**						1			No change yet	LED-2 on
|	←—— loop										

Note: The x's represent "don't cares."

In (a) and (b) we want to test a bit (or bits) in our input location, and set (or clear) a bit in the output location. So the **BCHG, BCLR,** and **BSET** appear, at first, to be good candidates to use.

TABLE 7-5

BTST Instruction

_____

BTST   Bit test

Assembler syntax: **BTST. B <Sea>,<Dea>**
                            .L

<Sea> Source effective address	<Dea> Destination effective address													
	Dn	An	(An)	(An) +	– (An)	$w (An)	$b (An, Xn)	$s	$l	$w (PC)	$b (PC, Xn)	#$q	SR	CCR
Dn	*		*	*	*	*	*	*	*					
#$q	*		*	*	*	*	*	3	*					

3 denotes the mode used for Example 7.3.

Action: The bit indicated by the <Sea> is tested, and the Z bit is set or cleared accordingly

Status register condition codes: X   N   Z   V   C
                                                             -    -   *   -   -

X is not affected.
N is not affected.
Z is set if the bit tested was zero. Cleared otherwise.
V is not affected.
C is not affected.

_____

Notes:
1. When <Dea> is a data register, only longword operations are allowed.
2. When <Dea> is a memory location, only byte operations are allowed.
3. Bits are numbered from 0, 0 being the least significant.

The <Sea> indicates which bit is to be tested, changed, set, or cleared. **The specific bit** must be contained in either a data register or as immediate data. The two forms of this instruction:

```
MOVE.L  #04,D0
BCLR.B  D0,OUTPUT       or      BCLR.B  #4,OUTPUT
```

Either form tests bit 4 (remember, they are numbered starting at 0) of the output memory location, copies the bit to the Z bit of the CCR, and then clears (changes to zero) bit 3 of the output location. The **BSET** instruction operates similarly, setting bit 4. The **BCHG** instruction changes the condition of bit 4.

Note in Table 7-4 that **<Sea>** cannot be of the form **$s,** representing a short address (less than $8000) or **$l** for a long address (greater than $7FFF). In other words, we cannot test the bit of one location and change another bit of another address with a single instruction. We will be able to use these instructions to turn our LEDs on and off, however.

The first part of our program can be handled like this:

```
.EQU  INPUT,$xxxx      ; assign an address to a label
.EQU  OUTPUT,$yyyy  ; assign an address to a label
; We will use $10013 for the input and $10011 for the output with
; the Motorola ECB
; or any convenient addresses for the simulator,
; such at $7000 and $7010
```

```
TOP:        BTST.B  #4,INPUT        ; test bit 4 of the input location
            BEQ   CLOSED            ; branch if Z bit is set,
                                    ; indicating switch is closed
OPEN:       BCLR  #3,OUTPUT         ; turn LED-3 off,
                                    ; since switch is open
            BRA   TOP               ; keep checking
CLOSED:     BSET  #3,OUTPUT         ; turn LED-3 on,
                                    ; since switch is closed
            BRA   TOP               ; keep checking
```

Recall that if switch-4 is closed, this represents a logic 0 (0 V), and if the switch is open, bit 4 will be a logic 1 (5 V). For the output locations, the LED is turned off by a logic 0 and on by a logic 1.

The first instruction checks bit 4 of the input, setting the Z bit if the switch is closed. If the switch is open, we skip over the instruction that turns LED-3 on.

To do part (b) of our task, we need to check two switches for a closed condition. Using what we just covered, a typical solution would be:

```
TOP2:       BTST.B  #2,INPUT        ; test bit 2 of the input location
            BNE   OPEN2             ; branch if Z bit is not set
                                      ; (switch open)

            BTST.B  #1,INPUT        ; test bit 1 of the input location
            BNE   OPEN2             ; branch if Z bit is not set
                                      ; (switch open)
CLOSED2:    BSET  #7,OUTPUT         ; turn LED on,
                                      ; BOTH switches are closed
            BRA   TOP2              ; keep checking
OPEN2:      BCLR  #7,OUTPUT         ; turn LED off,
                                      ; BOTH switches are not closed
            BRA   TOP2              ; keep checking
```

This looks straightforward enough. If we had several more bits to check, the number of instructions and time to execute would of course increase. Looking at another approach, discussed briefly in Section 6.7, we can develop a mask, do a logical operation, and act accordingly. This allows several bits to be checked simultaneously.

Choice of the proper mask takes a little thought concerning how AND and OR logic functions work. Refer to Figure 6-27 to see how the AND mask works. The truth tables for the AND and OR operation are as follows:

	AND			OR	
Inputs		Output	Inputs		Output
0	0	0	0	0	0
0	1	0	0	1	1
1	0	0	1	0	1
1	1	1	1	1	1

We need an output that is unique when BOTH switches are closed (the same). At first glance it looks like either the AND mask of 1 1 or the OR mask of 0 0 could be used. However, the inputs for the AND and OR tables above represent a single input bit and the applicable mask bit. The AND mask for the 8-bit input would be 00000110, the OR mask would be 11111001. Since we want to mask OUT undesired bits, the AND mask is the proper choice. When the mask **AND** the input bits match, we will have an output. Using the AND mask, a replacement program for the one above would be:

```
TOP2:      MOVE.B  INPUT,D0      ; get the switch conditions
           AND.B   #00000110B,D0 ; mask off unwanted bits
           BEQ  CLOSED2          ; branch if Z bit set, both closed
OPEN2:     BCLR  #7,OUTPUT        ; turn LED off
                                  ; since BOTH are not closed
           BRA  TOP2             ; keep checking
CLOSED2:   BEST  #7,OUTPUT        ; turn LED on
           BRA  TOP2             ; keep checking
```

Looking at a few of the possible input conditions:

	1110<u>1</u>111	11111<u>111</u>	11111<u>0</u>11	11111<u>001</u>	(switch inputs)
AND	00000110	00000110	00000110	00000110	(AND mask)
gives	000001<u>10</u>	00000<u>110</u>	00000<u>0</u>10	<u>00000000</u>	(result)

The first one represents when switch 4 is closed, and the result is NONZERO. The second one represents all switches open, and the result is NONZERO. The third one represents only switch-2 closed, and the result is NONZERO. The fourth represents both switches 1 and 2 closed, and the result is ZERO.

To do part (c) takes a little more thought [assuming (a) and (b) did not take much thought!]. The LED-2 can be assumed to be intially off and switch-0 open (logic 1). We want the LED to come on after switch-0 is closed and returned to open. The LED remains on until switch-0 is again closed and returned to open. Turning the LED off and on will be simple with the **BCHG** instruction, which changes a specific bit. Unfortunately, there is no instruction that tells us when a specific input bit is changed by something external!

One approach would be to *read* switch-0 and store its condition in a temporary location. We again read the switch and compare with the previous reading. If there has been a change, we change the LED. Let's try this program:

```
TOP3:      MOVE.B  INPUT,D0      ; get the switch conditions
           AND.B   #01,D0        ; mask out unwanted bits
                                  ; (= 00000001)
           MOVE.B  D0,D1         ; save the switch condition
           MOVE.B  INPUT,D0      ; read the switches again
           AND.B   #01,D0        ; mask out unwanted bits
           CMP.B   D0,D1         ; compare now with previous
           BEQ  TOP3             ; loop back since
                                  ; there was no change
CHANGE:    BCHG  #2,OUTPUT        ; change LED-2
           BRA  TOP3             ; loop back
```

What looks like a perfectly good program may have problems. Without doing the math, it looks like switch-0 gets checked every few microseconds. Assuming a *perfect* switch, which is not a good assumption, the odds are still pretty good that a closing/opening operation will not go undetected. Depending on the speed of our program and the erratic bouncing of the switch, we may or may not get the desired action.

In reality, the activation of a switch looks like Figure 7-9.

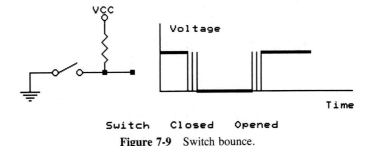

Figure 7-9 Switch bounce.

Two approaches are taken to ensure accurate sensing of switches. One is done with software, called switch *debouncing*, and the other involves simple use of an IC to get reliable operation. The circuit used is shown in Figure 7-10. Note that this circuit requires a single pole double throw (SPDT) switch. When it is in the *set* position, the output is a 1. When moved to the *reset* position, the output goes to a 0 cleanly with no erratic results. As the switch goes from the set to reset position, it will be *in between* both contacts for a period of time, will then contact the reset contact on the switch, and then perhaps bounce off the reset contact several times before eventually resting on the reset contact. It does not bounce high enough to hit the set contact, however. The switch will be at rest in a few milliseconds, by the way. The truth table shows the resulting action.

Back to our total task: putting the program(s) together to take care of all three parts of the problem. If we have problems with the switch bouncing, one of the debouncing approaches will have to be used.

Our finished program could look like this:

```
; Example 7.3
; Turn ON LED-3 when switch-4 is closed
; Turn ON LED-7 when switch-1 AND switch-2 are closed
; Toggle LED-2 for each momentary closing of switch-0
;
                .ORG   $1000
                .EQU   INPUT,$7000       ; assign an address to a label
                .EQU   OUTPUT,$7010      ; assign an address to a label
; We will use $10013 for the input and $10011 for the output
; with the Motorola ECB
; or any convenient addresses such as $7000 and $7010
; for the simulator
; Test first situation
                MOVE.B  #0,OUTPUT         ; turn off all LEDs
TOP:            BTST.B  #4,INPUT          ; test bit 4 of the input location
                BEQ    CLOSED             ; branch if Z bit is set
                                          ; (switch closed)
OPEN:           BCLR   #3,OUTPUT          ; turn LED-3 off (switch is open)
                BRA    TOP2               ; check next situation
CLOSED:         BSET   #3,OUTPUT          ; turn LED-3 on (switch is closed)
; Test second situation
TOP2:           MOVE.B  INPUT,D0          ; get the switch conditions
                AND.B   #00000110B,D0     ; mask off unwanted bits
                BEQ    CLOSED2            ; branch if Z bit set
                                          ; (if both are closed)
OPEN2:          BCLR   #7,OUTPUT          ; turn LED off
                                          ; since both are not closed
                BRA    TOP3               ; check next situation
CLOSED2:        BSET   #7,OUTPUT          ; turn LED on
                                          ; (both switches are closed)

; Test third situation
;                                         Switch 0 must make its changes between
TOP3:           MOVE.B  INPUT,D0          ; get the switch conditions _____ here __
                AND.B   #01,D0            ; mask out unwanted bits
                                          ; (= 00000001)
                MOVE.B  D0,D1             ; save the switch condition
                MOVE.B  INPUT,D0          ; read the switches again _____  here __
                AND.B   #01,D0            ; mask out unwanted bits
                CMP.B   D0,D1             ; compare now with previous
                BEQ    NOCHG              ; skip over since
                                          ; there was no change
CHANGE:         BCHG   #2,OUTPUT          ; change LED-2
NOCHG:          BRA    TOP                ; loop back to top of program
                .END
```

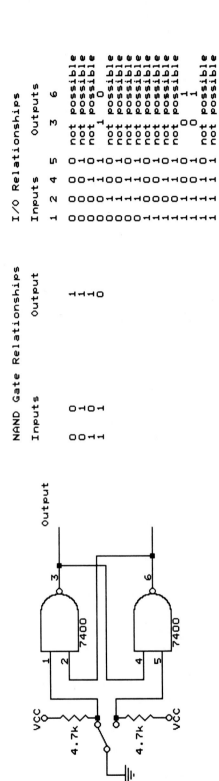

Figure 7-10 Switch debounce circuit.

Note that switch 0 had to change state during a very brief time. IF the switch had sufficient contact bouncing, the LED would have changed during a single change of the switch position, even before the switch got to the opposite state. It is unlikely that the switch could be physically moved closed and back open during that time period. To get a reliable check for a closing/opening of the switch, we would have to put a time delay in the program between the checks. It is left as an exercise for the student to *fine-tune* this program.

New Instructions

The AND instruction (Tables 7-6 and 7-7), discussed in Chapter 6 was used in this example. Remember, it does the AND operation on a bit-by-bit basis.

TABLE 7-6

AND Instructions

AND   Bitwise logical AND

Assembler syntax: AND.B Dn,<Dea>
                           .W
                           .L

<Sea> Source effective address	<Dea> Destination effective address												
	Dn	An	(An)	(An) +	− (An)	$w (An)	$b (An, Xn)	$s	$l	$w (PC)	$b (PC, Xn)	#$q	SR	CCR
Dn			*	*	*	*	*	*	*					

Assembler syntax: AND.B<SEA>,Dn
                           .W
                           .L

<Sea> Source effective address	<Dea> Destination effective address
	Dn
Dn	*
An	
(An)	*
(An) +	*
− (An)	*
$w (An)	*
$b (An, Xn)	*
$s	*
$l	*
$w(PC)	*
$b (PC, Xn)	*
#$q	3
SR	
CCR	

3 denotes the addressing mode used in Example 7.3.

Action: Logical-AND each bit of <Sea> with <Dea> → <Dea>

Status Register Condition Codes:
X   N   Z   V   C
-   *   *   0   0

TABLE 7-6 (continued)

AND Instructions

X is not affected.
N is set if the result is negative. Cleared otherwise.
Z is set if the result is zero. Cleared otherwise.
V is cleared.
C is cleared.

Notes:
1. Each bit of <Sea> is AND'd with the <Dea>, and the result is stored in the <Dea>.
2. ANDI is used when the source is immediate data, and is selected automatically by some assemblers.

TABLE 7-7

ANDI Instructions

ANDI   Bitwise logical AND (immediate data)

Assembler syntax: ANDI.B <Sea>,<Dea>
                                      .W
                                      .L

<Sea> Source effective address	Dn	An	(An)	(An)+	–(An)	$w (An)	$b (An, Xn)	$s	$l	$w (PC)	$b (PC, Xn)	#$q	SR	CCR
							<Dea> Destination effective address							
#$q	*		*	*	*	*	*	*	*			*	*	*

Action: Logical-AND each bit of <Sea> with <Dea> → <Dea>

Status register condition codes: X   N   Z   V   C
                                                         –   *   *   0   0

X is not affected.
N set if the result is negative. Cleared otherwise.
Z set if the result is zero. Cleared otherwise.
V is cleared.
C is cleared.

Notes:
1. Each bit of the immediate data is AND'd with the <Dea>, and the result is stored in the <Dea>.
2. ANDI #$q,SR is a privileged instruction (supervisor mode only).
3. ANDI #$q,SR and ANDI #$q,CCR instructions affect the CCR bits as follows: Bits are cleared if the corresponding bits are low, unaffected otherwise.

<Sea> data bit:	4	3	2	1	0
CCR flags:	X	N	Z	V	C

Hands-On Execution of This Example

Refer to Section 10.**1.11** for execution of this example for the **ECB,** to Section 10.**2.11** for execution on the **Pseudo Sam/Pseudo Max assembler/simulator,** or to Section 10.**3.11** for the **MAX 68000.**

EXAMPLE 7.4

Checking with the list of available arithmetic instructions, we see missing a square root instruction. The following program shows how we can calculate the square root of an unsigned integer number.

If the longword memory location at $7000 contains a 32-bit unsigned integer, representing a decimal number between 0 and 4,294,967,295, write a program that calculates the square root of the number and stores the 16-bit result in $7004.

This type of problem is one whose plan of attack is not immediately obvious. In other words, if we do not have an algorithm, tackling this one is almost impossible. Perhaps you have seen the following approach to finding the square root.

Here's our algorithm, in words: Take the number, subtract from it the odd integers (1, 3, 5, 7, 9, 11, etc.) until the sum is zero. The square root will be the number of odd integers it takes to reduce the number to zero. Two examples:

$$
\begin{array}{rr}
16 & 18 \\
-\ 1 & -\ 1 \\
\hline
15 & 17 \\
-\ 3 & -\ 3 \\
\hline
12 & 14 \\
-\ 5 & -\ 5 \\
\hline
7 & 9 \\
-\ 7 & -\ 7 \\
\hline
0 & 2 \\
-\ 9 & -\ 9 \\
\hline
-\ 9 & -\ 7 \\
\end{array}
$$

In both cases it took four odd integers to reduce the number to zero, indicating that the square root of both the numbers is equal to 4. (This procedure rounds off our answer.) We took the subtractions one step further until the sum was negative to show that a shortcut to be used in our program. Note that if we divide the last odd integer used (the one that causes the sum to go negative) by 2, we also get the correct answer. Try a few more examples to convince yourself.

Therefore, our refined algorithm is to take the number, subtract the odd integers from it until the sum is negative, divide the last odd integer used by 2, and store that number as the result. An alternative approach would be to subtract until the sum was equal to zero, keeping count of the number of numbers used, and store that as the result. We will show the results for both methods.

Our program looks as follows:

```
; Example 7.4
; Compute the integer portion of the square root of a 32-bit number
;    stored in $7000
; Store the 16-bit result obtained by counting the number of
;    odd numbers in $7010
; Store the 16-bit result obtained by dividing the last odd
;    integer by 2 in $7020
;
        .ORG    $1000
        .EQU    INPUT,$7000
        .EQU    OUTPUT1,$7010
        .EQU    OUTPUT2,$7020
```

```
        ;
                CLR.L   D2              ; used to store the number of subtracts
                CLR.L   D1              ; used for the odd integers
                MOVE.L  INPUT,D0        ; get the number
                BEQ.S   EXIT            ; if it is zero, we are done
                MOVE.L  #1,D1           ; load the first odd integer
        LOOP2:  SUB.L   D1,D0           ; subtract the odd integer
                                        ; from the number
                BCS     DONE            ; is the number negative yet?
                ADDQ.L  #1,D2           ; increment the number counter
                ADDQ.L  #2,D1           ; add two to get the next odd integer
                BRA     LOOP2           ; subtract again
        DONE:   ASR.L   #1,D1           ; divide last odd integer by 2
                MOVE.W  D1,OUTPUT1      ; store the answer (method 1)
                MOVE.W  D2,OUTPUT2      ; store the answer (method 2)
        EXIT:   BRA     EXIT            ; done, keep looping
                .END
```

No new addressing modes were used, but a new instruction, **ASR,** was used (see Table 7-8). Use of this instruction to divide by 2 is similar to the previous LSL used to multiply by 2.

Look at the following binary number:

0001 0000   which is equal to 8 decimal. Shifted right one bit gives

0000 1000   which is equal to 4 decimal, which is one-half our original value

Repeated shifting would divide by 4, 8, 16, and so on.

Although not discussed yet, the **DIVU** (unsigned divide) instruction could have been used.

TABLE 7-8

ASR Instruction

ASR   Arithmetic shift right

Assembler syntax: ASR.B <Sea>,<Dea>
                         .W
                         .L

<Sea> Source effective address	Dn	An	(An)	(An) +	– (An)	$w (An)	$b (An, Xn)	$s	$l	$w (PC)	$b (PC, Xn)	#$q	SR	CCR
Dn #$q	* 4													

<Dea> Destination effective address

4 denotes the addressing mode used in Example 7.4.

Action: Right Shift <Sea> by <Dea> → <Dea>

Status Register Condition Codes: X N Z V C
                                      * * * * *

X is set according to the last bit shifted out. Unaffected for a shift count of zero.
N is set if the most significant bit of the result is set. Cleared otherwise.
Z is set if the result is zero. Cleared otherwise.
V is set if the most significant bit is changed at any time during the shift operation. Cleared otherwise.
C is set according to the last bit shifted out. Cleared for a shift count of zero.

TABLE 7-8 (continued)

ASR Instruction

Notes:
1. The <Sea> is shifted right by the count contained in the <Dea>, and the result is stored in the <Dea>.
2. The carry and extend bits receive the last bit shifted out.
3. The sign bit is replicated into the high-order bit.
4. The shift count is contained in Dn for Data Register Direct mode.
5. The shift count is between 1 and 8 for Immediate Data mode.
6. The ASR <ea> instruction shifts the memory word right by one bit only.

Note that this example was done using a unique algorithm. It involved typical arithmetic and comparison instructions. More sophisticated problems would involve more complicated algorithms and perhaps a wider variety of the normal instructions.

Suppose that this example only required the square roots of a small number of numbers to be computed, numbers within a specific range. Another algorithm, called a **lookup table,** could be used. Lookup tables are often used for problems involving conversion of data from one form to another. One that comes to mind is a temperature conversion from Fahrenheit to Celsius. These are related mathematically, so a mathematical algorithm could be used as well. An example of a table lookup is used in Example 7.6.

EXAMPLE 7.5

Let's set up a simple but typical application for a computer. We need to design a warning buzzer system for an aircraft. The buzzer is a hazard indicator, warning the pilot of a potential hazardous condition. Four aircraft functions are to be monitored:

	(abbreviation used)
STALL indicator	(S)
Gear-DOWN indicator	(G)
HIGH-airspeed indicator	(A)
LOW-rpm indicator	(R)

The buzzer (B) should sound only for either of the two following conditions:

SITUATION 1

STALL indicator on	$(S = 1)$
Gear-DOWN indicator off	$(G = 0)$
LOW-rpm indicator on	$(R = 1)$
(HIGH-airspeed is a "don't care")	

or

SITUATION 2

Gear-DOWN indicator on   (G = 1)

HIGH-airspeed indicator on   (A = 1)

(STALL and LOW-rpm are "don't cares")

Situation 1 would be when the airplane is trying to climb with too low an RPM setting or trying to land with its landing gear UP. Situation 2 would be when the airplane is traveling fast enough possibly to damage the landing gear, which is down.

Lacking details of the exact hardware available, we will assume for now that an on condition can be represented by a logic 1, or a fixed voltage, when it comes to building the hardware. An off condition is considered as a logic 0, or a zero voltage. The basic operations of the logical AND and OR functions have been covered. We can represent the **AND** function with **&** and the **OR** function with a *.

To come up with a circuit configuration for a hardware solution, or before we can come up with a computer program to tackle it with our software–hardware approach, we first develop an "equation" representing the situation. Denoting an ON condition for gear DOWN with a **G** and an OFF condition for gear UP with a **G̲**:

Situation 1: S **&** G̲ **&** R
Situation 2: G **&** A

Since we want a buzzer to sound if either of the situations occur, we simply **OR** the two situations: (S **&** G̲ **&** R) * (G **&** A). Using the logic symbols for the AND and OR gate, we construct the logic diagram (Figure 7-11). Note that we need an input condition for when the gear is DOWN or OFF, so an inverter gate is used to perform this function.

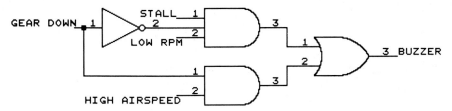

Figure 7-11   Logic diagram for Example 7.5.

To *build* this circuit we select appropriate ICs and voltage levels. In addition, we have to consider the current sourcing (output) and sinking (input) capabilities of the specific ICs used. Using readily available TTL ICs, Figure 7-12 shows a schematic that would solve this situation. Note that a transistor was used between the OR gate and the buzzer, since most buzzers require current beyond the capability of the logic gate.

If we wanted to seek a software solution using a μP, we would hook up our inputs to proper I/O chips that would appear as a single memory address to the μP. When we READ from the location, we would obtain the input conditions. While the same memory location could be used for the output, it is simpler to use a different address. When we WRITE to the output location, we would turn the buzzer (LED) on or off. Figure 7-13 shows a typical setup for testing purposes, with a LED substituted for the buzzer. Another alternative would be to connect a small speaker to the output bit. However, when a logic 1 is output to a speaker, all that is heard is a faint click. A routine to create an audible sound can be created with a little ingenuity.

We have several techniques or approaches for creating the necessary software program. We can load a data register (i.e., D0) with a mask for situation 1 and a register (i.e., D1) with the mask for situation 2:

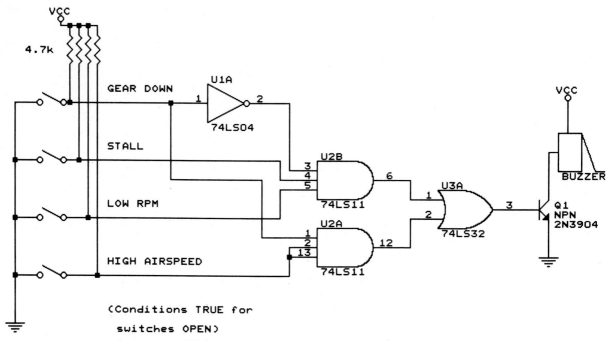

Figure 7-12 Schematic diagram for Example 7.5.

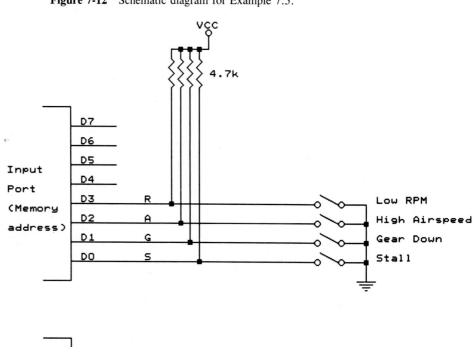

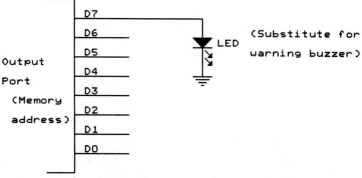

142

Figure 7-13 Input/output bit arrangement for Example 7.5.

```
MOVE.B   #00001011B,D0      (putting 0's for the "don't cares")
MOVE.B   #00000110B,D1
```

The first mask needs explanation. We put 1's for the bits we want to check, even though one of them is supposed to be a 0 for situation 1 to occur (gear DOWN = OFF). A flowchart is hardly needed, and the COMPARE and AND instructions have been covered in a previous example. Our simple program would look as follows:

```
; Example 7.5
; Turn a BUZZER ON if the STALL indicator is ON, the GEAR is DOWN,
; and the LOW-RPM indicator is ON
; or if
; The GEAR is UP AND the HIGH-AIRSPEED indicator is ON
;
          .ORG   $1000
          .EQU   INPUT,$7000      ; assign an address to a label
          .EQU   OUTPUT,$7010     ; assign an address to a label
          BCLR.B #7,OUTPUT        ; turn the buzzer off initially
          MOVE.B #00000110B,D1    ; situation 2 mask
TOP:      MOVE.B #00001011B,D0    ; situation 1 mask
          MOVE.B INPUT,D2         ; get the input condition
          AND.B  D0,D2            ; mask out undesired bits
          BCHG   #1,D0            ; invert the gear-down bit
                                  ; to reflect desired test
          CMP.B  D0,D2            ; look for a match
          BEQ    BUZZ             ; turn on buzzer for situation 1
          MOVE.B INPUT,D2         ; read switches again
          AND.B  D1,D2            ; mask out undesired bits
          CMP.B  D1,D2            ; look for a match
          BEQ    BUZZ             ; turn on buzzer for situation 2
          BCLR.B #7,OUTPUT        ; turn off buzzer
                                  ; since no situation
          BRA    TOP              ; keep looping
BUZZ:     BSET   #7,OUTPUT        ; turn on buzzer
          BRA    TOP              ; keep looping
          .END
```

Note that we alter the situation 1 mask to use it for our desired test condition and reinitialize it at the beginning of the loop each time. Yet another way to visualize the possible conditions for these four inputs is to construct an overall truth table. We have 16 possible input conditions:

RAGS	Buzzer	RAGS	Buzzer
0000	0	1000	0
0001	0	1001	1
0010	0	1010	0
0011	0	1011	0
0100	0	1100	0
0101	0	1101	1
0110	1	1110	1
0111	1	1111	1

We see that there are actually six situations that will cause a buzzer output.

Hands-On Execution of This Example

Refer to Section 10.1.12 for execution of this example for the **ECB,** to Section 10.2.12 for execution on the **Pseudo Sam/Pseudo Max assembler/simulator,** or to Section 10.3.12 for the **MAX 68000.**

EXAMPLE 7.6

This example utilizes a concept known as TABLE LOOKUP. It shows how a lot of programming or calculating steps can be eliminated by using the memory address itself as an index to look up in a table (memory contents) some type of conversion data. In our example we need to look up for each of the decimal digits, the hex pattern needed to turn on specific LEDs on a seven-segment display. A display with a common cathode could be used, with each of the segment LEDs being connected to the bits of an output location as outlined below.

Suppose that a single seven-segment display is hooked up to an output address as shown in Figure 7-14. To display a single decimal digit, the following hex code would be output to the memory location:

Digit	Hex Code	Digit	Hex Code
0	3F	5	6D
1	06	6	7D
2	5B	7	07
3	4F	8	7F
4	66	9	6F

Write a program that takes the 4-bit BCD character stored in an input location and displays it on the seven-segment display located at the output location.

This program is so simple that a flowchart is trivial. The indexing concept is the feature we need to concentrate on. Before progressing, take a look at how the program and data are stored in memory in Table 7-9. For clarity, we have started the data table at an address having 00's for the lower byte. Note that address 1100 contains a 3F, 1101 contains a 06, 1102 contains a 5B, and so on. The lower byte of the address represents the decimal digit between 0 and 9 that we need to *convert* or *lookup* to find the proper hex or binary output to light the proper LEDs.

TABLE 7.9

Lookup Table Stored in Memory

Memory location	Contents
1100	3F
1101	06
1102	5B
1103	4F
1104	66
1105	6D
1106	7D
1107	07
1108	7F
1109	6F

```
Memory Output Location    Display
Bit Number                Segment
------------------------------------------
        7                    none
        6                     g
        5                     f
        4                     e
        3                     d
        2                     c
        1                     b
        0                     a
```

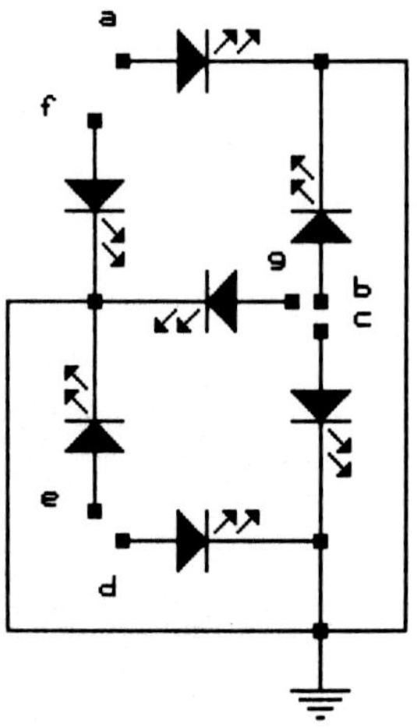

Figure 7-14   Seven segment display interfacing.

The program is:

```
; Example 7.6
; Read a single 4-bit BCD character from an input memory location and
; display the corresponding digit on a seven-segment display
; connected to an output memory location
;
            .ORG    $1000
            .EQU    INPUT,$7000         ; assign an address to a label
            .EQU    OUTPUT,$7010        ; assign an address to a label
            LEA.L   TABLE,A0            ; point to conversion table
            CLR.L   D0                  ; clear out the register,
                                        ;  used as index
TOP:        MOVE.B  INPUT,D0            ; get digit
            CMP.B   #9,D0               ; ensure that it is not > 9
            BHI.S   DARK                ; turn off all segments
                                        ;  if digit > 9
            MOVE.B  0(A0,D0),OUTPUT     ; output the seven-segment code
                                        ;  from table
            BRA     TOP                 ; keep looping
DARK:       CLR.B   OUTPUT              ; darken all segments
                                        ;  since digit was > 9
            BRA     TOP                 ; keep looping
            .ORG    $1100               ; the data table starting address
TABLE:      .DB     $3F,06,$5B,$4F,$66,$6D,$7D,07,$7F,$6F
            .END
```

The **DB** tells the assembler to **define byte**(S) or simply store a series of bytes in the memory at the next available addresses. If this assembler directive was not available, and the operating system or monitor program on our 68000 development computer would not allow simple, direct storage of numbers into memory, it would take a lot of programming steps to make our program itself do this task.

The **BHI** is used again in this example. Refer to Table 6-12 for details. Table 6-13 gives information on the choice of these instructions for various comparisons and checks. Basically, the BHI will branch if BOTH the Z bit and the C bit are 0 or cleared. If a compare is done and the result is NOT zero and a carry/borrow did not occur, the comparison indicates that the <Sea> was **greater than or equal to** the <Dea>. In our example we checked to ensure that a decimal value greater than 9 was not used. The 9 is greater than or equal to the value stored in Data Register D0.

Addressing Modes Used

The strange MOVE instruction is the powerful one used for the table lookup. It uses the **Address Register with index and displacement** mode. To come up with a Source effective address (<Sea>), which is the desired memory location for our output data, three things are added up: The <Sea> is equal to the sum of the displacement (which is zero in this case), the value stored in the Address Register (which is 1100 in this case), and the value stored in the Data Register (which is the actual decimal digit for which we need the output data). For example, if we need the hex code to turn on a decimal 4 on the seven-segment display, the number is first transferred from the input location into Data Register D0. Then the **MOVE.B 0(A0,D0),OUTPUT** will add 0 + 1100 + 4 to get 1104, take the contents of memory location 1104 (= 66), and then write the 66 to the output memory location. The LEDs corresponding to a decimal digit 4 will be turned on.

A refinement to the program was a check to see if the input value exceeded a decimal digit 9. If so, all segments of the LED are turned off. With a little ingenuity, the program could be modified to provide the hex digits A, b, C, d, E, and F.

This technique of table lookup is powerful and has a lot of applications.

EXAMPLE 7.7

Suppose that we have a data acquisition system (Figure 7-15) hooked to a 68000 system. It consists of 50 sensors, capable of providing an 8-bit (binary) number from 0 to 255 (decimal), representing pressures measured throughout an industrial environment. Each sensor will be *connected* to a different address in our system, starting at 7000H.

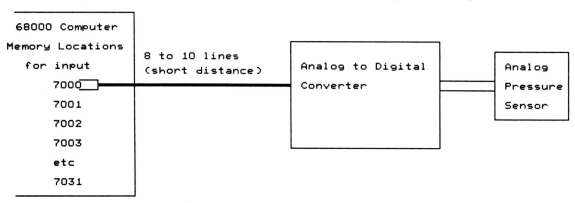

Figure 7-15 Data acquisition system block diagram.

Note that special considerations must be taken when I/O devices are located more than a few feet from the computer itself. A later example will show typical hardware.

The task is to monitor the incoming data continually, which can be changing, and provide the following information every 10 seconds:

1. The number of locations reporting a pressure of 0
2. The number of locations reporting a pressure of 255
3. The addresses of the locations reporting a pressure of 255
4. The average pressure for the 50 sensors
5. The maximum pressure reported
6. The minimum pressure reported

Note that if a pressure of 255 is reported, it would of course be the maximum; similarly, for a pressure of 0, it would be the minimum. We record the addresses of locations reporting the 255 pressure, so perhaps corrective action can be taken promptly.

A primitive version of this program, without display capability, would simply leave the results in an area of memory or registers. Provisions for the video display would depend on the specific hardware used. Both the Motorola ECB and the Micro Designs MAX 68000 provide routines for video display (and keyboard input into a program). A typical display would be as follows:

"Pressure Monitoring System"
Pressure   **minimum =**   **maximum =**
 # stations reporting   **0 =**   **255 =**
Addresses of 'danger'   **(255) pressures =**
Average pressure =

and see the flowchart (Figure 7-16).

We are going to initialize the minimum to 255 and the maximum to 0, a rather strange way to start things. Comparisons with these numbers will be easier as we search for the actual minimum and maximum. Let's discuss the results, or what this program does, before looking at the program. Figure 7-17 shows the affected registers and memory locations just *before* execution of the program, after execution of the *initialization* portion, immediately after execution of the program through *one* of its cycles, and upon completion of the program. We have provided some dummy data, and the result using this data after complete execution of the program would be:

1. The sum of the numbers: 765 (hex)
2. The average of the numbers: 25 (hex)
3. The minimum: 01
4. The maximum: FF
5. The number of 0's: 00
6. The number of 255's: 03
7. The addresses of the 255's: 0000703D, 0000703F, 00007041

We purposely avoided any 0's, so we could check out the portion of the program that finds the minimum and totals the number of 0's. The program should be tested further by introducing some zeros and reducing the 255's to smaller numbers to check remaining portions of the program. Several new instructions and techniques are introduced.

Next, look at our program:

```
; Example 7.7
; Continually monitor incoming data, which can be changing,
; and provide every 10 s the following information:
;1) The number of locations reporting a pressure of 0
;2) The number of locations reporting a pressure of 255
;3) The addresses of the locations reporting a pressure of 255
;4) The average pressure for the 50 sensors
;5) The maximum pressure reported
;6) The minimum pressure reported
;
            .ORG   $1000
            LEA.L  $7FFE,A7      ; Load the stack pointer
START:      CLR.L  D0            ; Init the maximum to zero
            CLR.L  D3            ; Clear upper portion of D3
            MOVE.B #255,D3       ; Init the minimum to 255
;
            CLR.L  D1            ; Clear the zero counter
            CLR.L  D2            ; Clear the 255 counter
            CLR.L  D4            ; Clear the sum totalizer
```

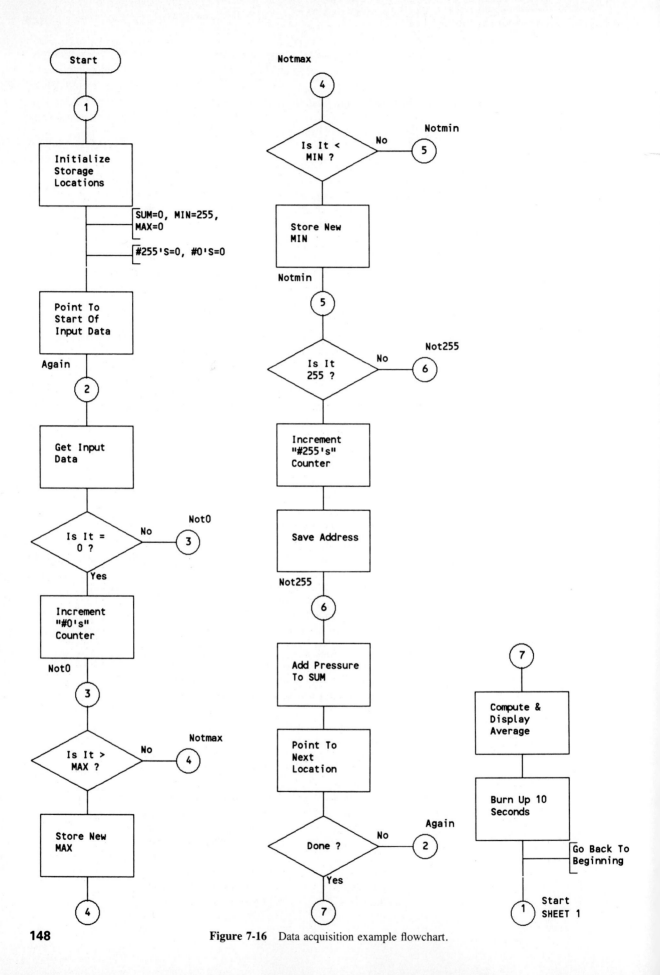

148   **Figure 7-16**   Data acquisition example flowchart.

Usage:

Register		Prior To Initialization	After Initialization	After Evaluation Of First Data Value (down to CONT: line)	After Evaluation Of All Data (down to BSR DISPLAY line)
MAX	D0	XXXXXXXX	00000000	00000001	000000FF
ZERO	D1	XXXXXXXX	00000000	00000000	00000000
255	D2	XXXXXXXX	00000000	00000000	00000003
MIN	D3	XXXXXXXX	000000FF	00000001	00000001
SUM	D4	XXXXXXXX	00000000	00000001	002B0025
255 CHECK	D6	XXXXXXXX	000000FF	000000FF	000000FF
0 CHECK	D7	XXXXXXXX	00000000	00000000	00000000

Register		Prior To Initialization	After Initialization	After Evaluation Of First Data Value	After Evaluation Of All Data
DATA POINTER	A0	XXXXXXXX	00007010	00007011	00007042
255 ADR STOR.	A1	XXXXXXXX	00007044	00007044	00007044
STACK POINTER	A7	XXXXXXXX	00007FFE	00007FFE	00007FFE

Memory — Prior To Initialization:

Label	Addr		
MAX	7000	XX	XX	ZEROs
255s	7002	XX	XX	MIN.
AVER.	7004	XX	XX	
		etc		
	7010	01	02	
	7012	03	04	
		etc		
	7040	2F	FF	
	7044	XX	XX	
		XX	XX	
	7FFE	XX	XX	stack

Memory — After Initialization:

Addr		
7000	XX	XX
7002	XX	XX
7004	XX	XX
etc		
7010	01	02
7012	03	04
etc		
7040	2F	FF
7044	XX	XX
	XX	XX
7FFE	XX	XX

Memory — After Evaluation Of First Data Value (down to CONT: line):

Addr		
7000	XX	XX
7002	XX	XX
7004	XX	XX
etc		
7010	01	02
7012	03	04
etc		
7040	2F	FF
7044	XX	XX
	XX	XX
7FFE	XX	XX

Memory — After Evaluation Of All Data (down to BSR DISPLAY line):

Addr		
7000	FF	00
7002	03	01
7004	00	25
etc		
7010	01	02
7012	03	04
etc		
7040	2F	FF
7044	00	00
	70	3D
etc.		
7FFE	XX	XX	stack

X's are unknowns or don't cares

Figure 7-17 Execution of Example 7.7

149

```
;
                CLR.L   D7              ; Init the zero checker
                CLR.L   D6              ; Clear upper portion of D6
                MOVE.B  #255,D6         ; Init the 255 checker
;
                LEA.L   DATA,A0         ; Point to beginning of data
                LEA.L   HIADRS,A1       ; Point to beginning
                                        ;  of '255' addresses
; End of initialization
; Check for a zero
LOOP:           CMP.B   (A0),D7         ; Is data = 0?
                BNE.S   NOTZERO         ; Nope
                ADDQ.B  #1,D1           ; Yes, increment zero counter
; Check for a new maximum
NOTZERO:        CMP.B   (A0),D0         ; Is it greater than old maximum?
                BCC.B   NOTMAX          ; Nope
                MOVE.B  (A0),D0         ; Yes, store new maximum
; Check for a new minimum
NOTMAX:         CMP.B   (A0),D3         ; Is it less than old minimum?
                BCS.B   NOTMIN          ; Nope
                MOVE.B  (A0),D3         ; Yes, store new minimum
; Check for a 255
NOTMIN:         CMP.B   (A0),D6         ; Is it 255?
                BNE.S   NOT255          ; Nope
                ADDQ.B  #1,D2           ; Yes, increment 255 storage
                MOVE.L  A0,(A1)+        ; Save address of 255 pressure
; Add number to SUM
NOT255:         ADD.B   (A0),D4         ; Not = 255, add to sum
                BCC     CONT            ;
                ADD.L   #256,D4         ; Byte sum was > 255,
                                        ;  add carry to D4
CONT:           ADDA.L  #1,A0           ; Point to next address
                CMPA.L  #DATAEND,A0     ; Done with 50 addresses?
                BNE.S   LOOP            ; Nope, loop again
; Find the average
                DIVU.W  #50,D4          ; Divide by 50, result in D4
; Store the results in memory locations (for use by DISPLAY routines)
                MOVE.W  D4,AVERAGE      ; Store average
                MOVE.B  D0,MAXIMUM
                MOVE.B  D1,ZEROS
                MOVE.B  D2,TWO55
                MOVE.B  D3, MINIMUM
; Call subroutine to display results
                BSR     DISPLAY         ; (1) Stack operation explanation
                                        ;   see Figure 7-2.11

; Call subroutine to kill 10 seconds
                BSR     BURNTIME
EXIT:           BRA     START           ; Start cycle all over again
;
DISPLAY:        NOP                     ; Display routine goes here (2)
                NOP                     ; (see comments below)
                RTS                     ; Return to main program (3)
BURNTIME:       NOP                     ; Burn up 10 seconds here
                NOP                     ; (see comments below)
                RTS
                .ORG    $7000           ; Storage locations for the results
```

```
MAXIMUM:   .RS   1
ZEROS:     .RS   1
TWO55:     .RS   1
MINIMUM:   .RS   1
AVERAGE:   .RS   2
           .ORG  $7010              ; The starting address
                                    ; of the input data
DATA:      .DB   01,02,03,04,05,06,07,08,09,10,11,12,13,14
           .DB   15,16,17,18,19,20,21,22,23,24,25,26,27,28
           .DB   29,30,31,32,33,34,35,36,37,38,39,40,41,42
           .DB   43,44,45,255,46,255,47,255
DATAEND:   .RS   2
; End of data
; Storage area for addresses having 255's starts here
HIADRS:    .RS   200
           .END
```

Hands-On Execution of This Example

Refer to Section 10.**1.13** for execution of this example for the **ECB,** to Section 10.**2.13** for execution on the **PseudoSam/PseudoMax assembler/simulator,** or to Section 10.**3.13** for the **MAX 68000.**

Addressing Modes Used

No new addressing modes were introduced, but a new concept called a **subroutine** was introduced and several very important considerations need to be discussed as well. Note the **MOVE.L A0,(A1)+** instruction. It moves the current longword contents of Address Register A0 to the **longword** address pointed to by A1. The value of A1 is incremented by 4 after execution of the instruction. A0 will contain the address (even or odd) of a location containing a 255. A1 is pointing to an area of memory. You will recall from an earlier discussion of byte, word, and longword organization, **all longword references must be to an even address.** Note that initially, A1 is set to the value of HIADRS. If this address is not set to an even address, the program will not run properly!

Note the **CMPA.L #DATAEND,A0** instruction. The same restriction as above applies. DATAEND must be an even address. Note also the # sign, which indicates the value of DATAEND, not the CONTENTS of the location labeled DATAEND.

How do you ensure **even** addresses for **longword** address references? We normally ORG (originate) our program at an even address, all of the instructions occupy an even number of locations, and the DW (define word) and DL (define longword) assembler directives use an even number of locations. The only way we land up on an odd address is with an odd number of DB (define bytes) or when a RS (reserve storage) reserves an odd number of locations.

In our example above, we ORG'd the storage area for the results at an even address. Also, the total number of locations reserved with RS's is an even number. We could have had an odd number, since we re-ORG'd the data area at $7010 below the RS statements. Next, we have our 50 data points, followed by DATAEND with a dummy RS 2. This ensures that HIADRS will also be on an even address. An alternative approach, a lot simpler, is to use the assembler directive **.ALIGN** on a line prior to HIADRS to force assembly to continue on an even address.

Next, note the choices made concerning the .B, .W, and .L extensions. They are chosen to match the expected values for the specific program. For our program having 50 BYTE-sized data values, the following quantities are known to be BYTE sized: the DATA, the number of ZEROS, the number of 255's, the MAXIMUM, and the MINIMUM. The SUM could be LONGWORD in size, and we will reserve a WORD-sized location for the AVERAGE, even though we will obtain a byte-sized (rounded) value.

Accordingly, the moves and compares will use .B extensions. However, when we add the byte-sized DATA to the longword-sized SUM we need to take into consideration a possible overflow (carry) when the byte-sized addition is done. We cannot ADD.L (A0),D4 to add the pointed to data to the sum. This would take four data values together and add them to the sum! To provide the correct sum, we simply check for a carry after the addition and add 256 to the sum each time there is a carry.

Note the two compares that find the maximum and minimum:

```
CMP.B   (A0),D0
BCC.B   NOTMAX
```

The compare is done by subtracting the data value (in A0) from the old maximum value (in D0). The operation is: D0 − (A0). Neither the contents of D0 nor the memory location pointed to by A0 are changed. The carry (borrow) bit will be set IF the data value is larger than the old maximum. It will be clear (= 0) if the data value is smaller than the old maximum. The BCC instruction will skip over the maximum replacement instruction [MOVE.B   (A0),D0] for the latter case.

```
CMP.B   (A0),D3
BCS.B   NOTMIN
```

The compare here is: D3-(A0), which is THE OLD MINIMUM–THE DATA VAL-UE. The carry (borrow) will be set if the data value is larger than the old maximum. For this case, the BCS instruction will skip over the minimum replacement instruction [MOVE.B (A0),D3]. Remember that **the compare instructions do not change any values; they only set/clear the appropriate Status Register bits.**

Note the **DIVU.W #50,D4** instruction. This takes the **longword-**sized SUM (in D4), divides it by the **word-**sized value 50, and leaves the **word-**sized AVERAGE in the **lower** word of D4. The value obtained will be rounded off. The **remainder** is left in the **upper** word of D6. For the example above, with the dummy data, we obtain an average of 25 hex (37 decimal) and a remainder of 2B hex (43 decimal). If you take this remainder and divide it by 50 (the number of data values), you will get 0.86, which is the fractional part for the average. A good programmer would take this into consideration, rounding the average up when the fractional part exceeds 0.5.

Introduction to Subroutines

Subroutines serve two basic purposes. (1) They allow creation of *modular programs*—programs that are easier to understand, easier to create, and easier to debug. Each subroutine performs a small dedicated function. (2) They also save time and computer memory for functions that need to be done repeatedly during a program. In other words, they save duplication of parts of the program.

How do they work? Recalling that program execution is "straight down" through a program, with deviations caused by the compare and branching instructions, subroutines cause *temporary exits* from the normal flow to perform a particular, or self-contained task, and then resumption in the main program at the location following the call to the subroutine.

How does the *postmaster* keep track of where it has been and where to resume? Before execution of the BSR (branch to subroutine) instruction, the address of the NEXT instruction that would have normally been executed is saved in an area of memory known as **THE STACK. The subroutine always has a RTS (return from subroutine) instruction as its last instruction.** This causes restoration of the *resumption* address into the Program Counter.

The stack is just another portion of memory used as a temporary holding area for addresses and data. In the example above, we have loaded the Stack Pointer (A7) to a convenient EVEN address near the top of available memory, above any portion of the main program. *We normally load the Stack Pointer **once** at the beginning of a program and do not reload it during the program.*

The new instructions introduced, BSR and RTS, allow us to call a subroutine and to return to the main program. Refer to Tables 7-10 and 7-11.

TABLE 7-10

BSR Instruction

BSR   Branch to subroutine

Assembler syntax: BSR.B \<label\>   or BSR.B displacement
                         .W   .W

Action: $PC \rightarrow -(SP); PC + displacement \rightarrow PC$

Status register condition codes: X   N   Z   V   C

                                       -   -   -   -   -

No bits are affected.

Notes:
1. The address of the next instruction to be executed is pushed on top of the stack. The displacement is added to the program counter contents, and execution continues from that location.
2. The displacement is a two's-complement integer denoting the relative distance in bytes between the beginning of the instruction following the BSR instruction and the next instruction to be executed.
3. Byte displacements give a branch range of -128 to +126 bytes away from the BSR instruction.
4. Word displacements give a branch range of -32,768 to +32,766 bytes away from the BSR instruction.
5. The displacement must always be even, since instructions must begin on an even address.
6. Some assemblers use .S notation to denote .B (byte) displacements. Some do not require the .B or .W displacement size, calculating the proper size.

TABLE 7-11

RTS Instruction

RTS   Return from subroutine

Assembler syntax: RTS

Action: $(SP)+ \rightarrow PC$

Status register condition codes: X   N   Z   V   C

                                       -   -   -   -   -

No bits are affected.

Note: The program counter is pulled from the stack and program execution resumes at that address.

Note that we have left creation of the actual subroutines as an exercise for the student. We consider the display routine and the time-delay routine as *subprograms,* and could perhaps write them in such a fashion that they could be used in other programs as well. Figure 7-18 shows how the return value needed for the Program Counter is saved on the stack and how the Stack Pointer is altered.

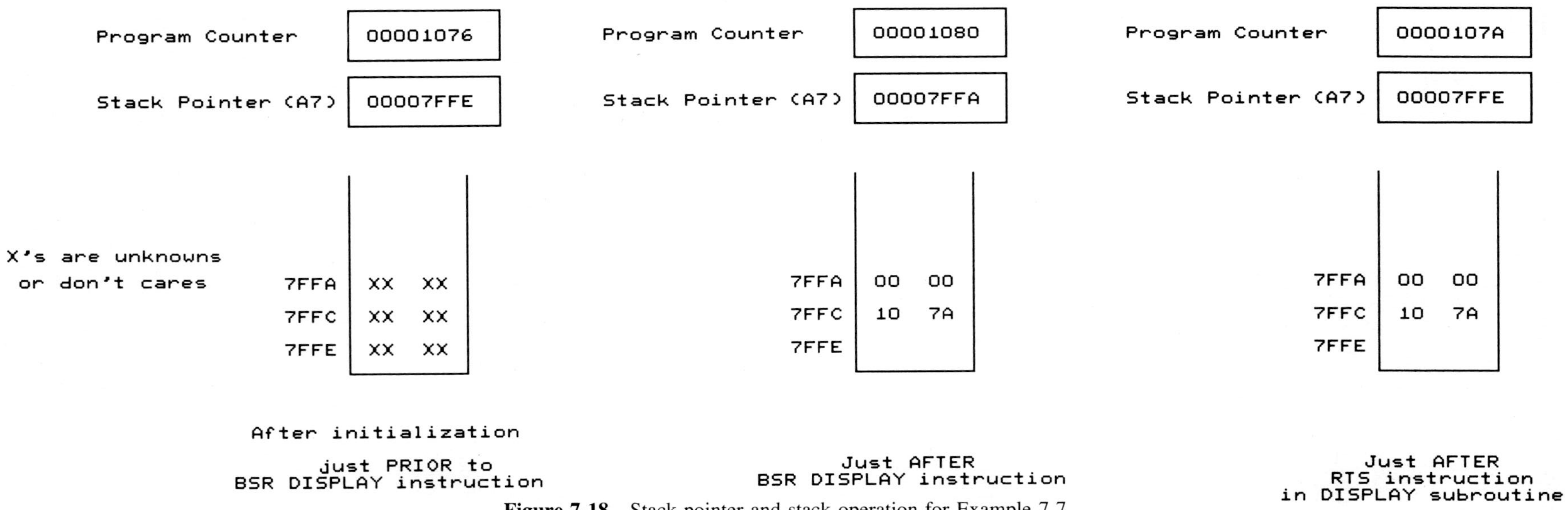

Program Counter 00001076

Stack Pointer (A7) 00007FFE

7FFA XX XX
7FFC XX XX
7FFE XX XX

X's are unknowns
or don't cares

After initialization
Just PRIOR to
BSR DISPLAY instruction

Program Counter 00001080

Stack Pointer (A7) 00007FFA

7FFA 00 00
7FFC 10 7A
7FFE

Just AFTER
BSR DISPLAY instruction

Program Counter 0000107A

Stack Pointer (A7) 00007FFE

7FFA 00 00
7FFC 10 7A
7FFE

Just AFTER
RTS instruction
in DISPLAY subroutine

Figure 7-18 Stack pointer and stack operation for Example 7.7.

154

7.3 A CONTROL APPLICATION: DISPLAYING DATA

Realizing that the primary application of the microprocessor is for control applications, we continue our coverage of input and output interfacing examples. Sufficient detail will be given to the hardware connections to allow construction of some useful experiments. No control application is complete without a useful (or impressive) display of the data being gathered and the status of processes being controlled.

7.3.1 Displaying Decimal Digits

The next step beyond the simple display of LEDs for the presentation of information is to use some type of numeric display. Presented in this section is a wide variety of display techniques. The trade-offs will be:

How many bits of an output location are needed for a digit?
How much hardware and power consumption are needed for each digit?
How much do the displays cost?
How much of the processor's time is used up in the display process?

The Easiest Way

Figure 7-19 shows perhaps the easiest, but most expensive way of displaying a decimal digit. The display contains all of the logic and hardware to display a single hex digit in response to a binary input of 4 bits. These displays will display values from 0 to F. Each digit requires 4 bits. If we only have an 8-bit output, two digits could be displayed.

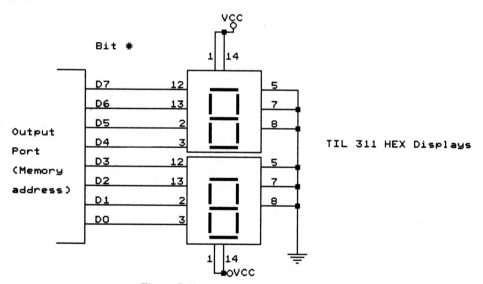

Figure 7-19 Interfacing two hex displays.

EXAMPLE 7.8

Write a program that displays the byte located at location $7000 if the two-digit display shown in Figure 7-19 is located at memory location $7002.
Our program:

MOVE.B $7000,$7002

This looks too simple! Remember that memory locations hold only binary values. These displays will display the equivalent hex value. Refer to previous examples concerning BCD (binary-coded-decimal) numbers if it is desired to represent and display the decimal numbers 00 to 99. A simple exercise for the student: Assume that the memory location contains a value (representing a sum of some numbers, for example) that is between 0 and 63 hex. Write the program to convert and display this sum as two BCD digits on the display as shown above.

Another Alternative

Figure 7-20 shows a much cheaper but identical way of displaying two digits. Each digit requires a seven-segment LED display, seven resistors, and an IC. For this application we use common-anode seven-segment displays. The 4-bit binary equivalent of the numbers 0 to 9 are sent to the 74LS47 ICs. These decoder ICs will not display values greater than 9. It acts as a simple table lookup, lighting the correct segments for each decimal digit.

EXAMPLE 7.9

Write a program that displays the byte located at location $7000 if the two-digit display shown in Figure 7-20 is located at memory location $7002.

Our program:

MOVE.B $7000,$7002

Getting More Digits per Bit

If several digits are needed, and 4 bits of an output location for each display is not practical, another approach is to multiplex the displays. The display program lights a SINGLE digit at a time, advancing from one to the next, faster than the response of the human eye, which is about $1/20$ s. The digits are not latched: in other words, will not be displayed if the multiplexing program is not running.

Figure 7-21 shows how eight digits can be displayed by using two 8-bit output memory locations and proper hardware. This gives us one digit for 2 bits. The price we pay is in the programming effort and in the amount of processor time that is required for the displays. Writing the program to use this display technique is left as an exercise for the student. To display a "7" in the left display, the binary value 11111000 is output to the lower output port address (to take segments a, b, and c low), and the binary value 10000000 is output to the upper output port address to provide voltage for just the left display. The process is continued for the remaining displays, completing the cycle at least once every $1/20$ s.

7.3.2 Displaying Four Alphabetical Characters

To obtain letters as well as numbers, another type of display is needed. Example 7.10 shows how to interface an "intelligent" four-character LED display to a computer system. The following description and program was done as a class project by Abderrahim Rhazi.

First, a few words from the instructor. This example was done as an optional project at the end of the semester by an ambitious (but very busy) student. Time expired before we had a chance to polish the example completely. It had been done previously, interfaced to a Z-80 computer, providing an interesting challenge and is included here as the base for further experimentation. The instructor is grateful for the student's effort.

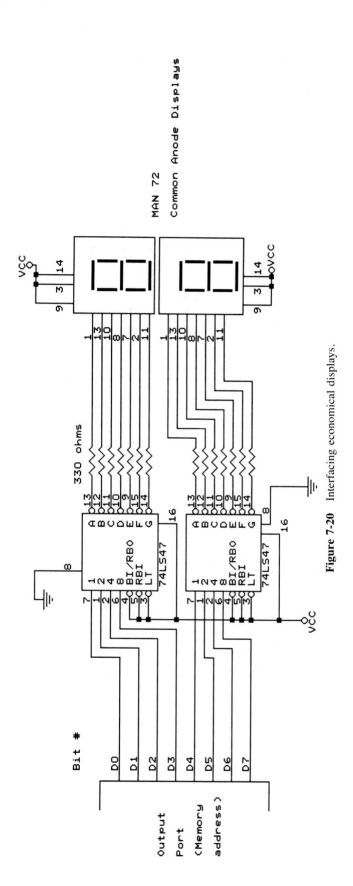

Figure 7-20 Interfacing economical displays.

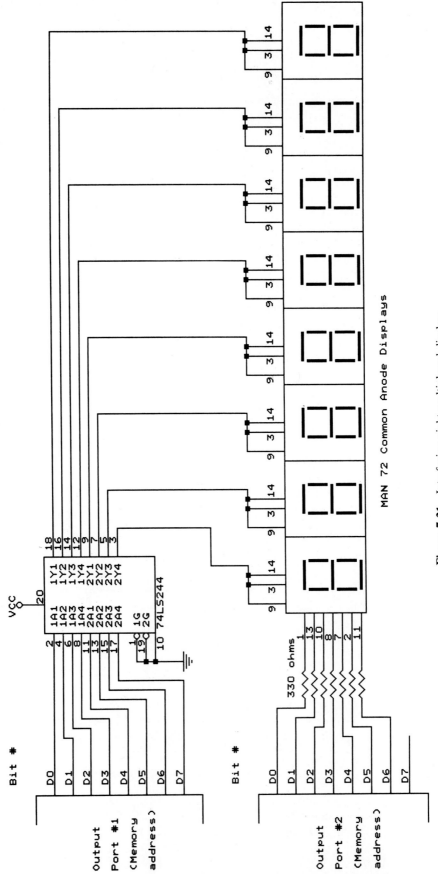

Figure 7-21 Interfacing eight multiplexed displays.

Second, this display, and several of the devices in the following examples, were designed for interfacing DIRECTLY to a microprocessor's data and address bus. Although not difficult, this interfacing requires design considerations not covered in this introductory book. However, these devices can be put to use fairly quickly and simply by interfacing them to an existing I/O port on our computer system.

The devices usually require up to eight lines for the output data, two or three output lines for addressing the device's internal registers, and one or two more output lines to select the device, and to distinguish between a READ and A WRITE operation. The displays may have a *status* line which could be connected to a computer input address and monitored. We will ignore these status or busy lines, provide a time delay between character display, and assume that the display is ready for more characters. When connected directly to a microprocessor, the device's internal registers are addressed as several memory locations. In our case we will manipulate the proper bits of a single output memory address to access the proper registers of the display.

Referring to the schematic shown in Figure 7-22, here are the fundamentals of its operation:

1. The chip enable (CE) line is tied low. The write enable line (WE) is tied low. (We always want the display enabled and will only be writing to it.)
2. The seven data lines used to display the various alphabetical and numerical characters are connected to one memory location used as an output.
3. The cursor update (CU) is connected to bit 6 of another memory output location, and is taken high to set the cursor for data display.
4. Bits 0 and 1 of the second memory output location are connected to the display's A0 and A1 address select lines. Outputting 00, 01, 10, or 11 on bits 0 and 1 will select the individual displays.
5. To display a character, the desired data byte is output to the first memory location, and then the desired character and cursor update bits are output to the second location. The information is latched (stored) in the LED device and remains until replaced with a new character.

While this setup worked (with a few problems), the timing description below shows a more normal setup.

1. The chip enable (CE) line is tied low since there is a single display. The cursor update (CU) should be tied high, or connected to an output bit and taken low momentarily for cursor movement (without updating the displayed character).
2. The seven data bits are connected to a memory output location as above.
3. The write enable (WE) line is connected to the second memory output location and is programmed normally to rest high.
4. To display a character, the data are output to the first output location, the desired bits are output to the second output location to select the desired display, and THEN the write enable (WE) is taken low momentarily and then back high. This more closely resembles the operation done when the device is connected directly to a microprocessor. In other words, the DATA and ADDRESS lines are *stable* before actually doing the write operation. Alternatively, for the schematic shown, the CU line should have been taken high and back low while both the data and address were stable. (The program took it high at the same time as the address lines were set up.)

The schematic shows the connections that would be made to the Motorola ECB computer, interfaced to the complex 68230 PI/T I/O device. Programming the 68230 was easy for this example, and the four instruction lines used for its initialization are commented in the program listing. Programming the simpler 6821 PIA I/O device is covered in Chapter 8.

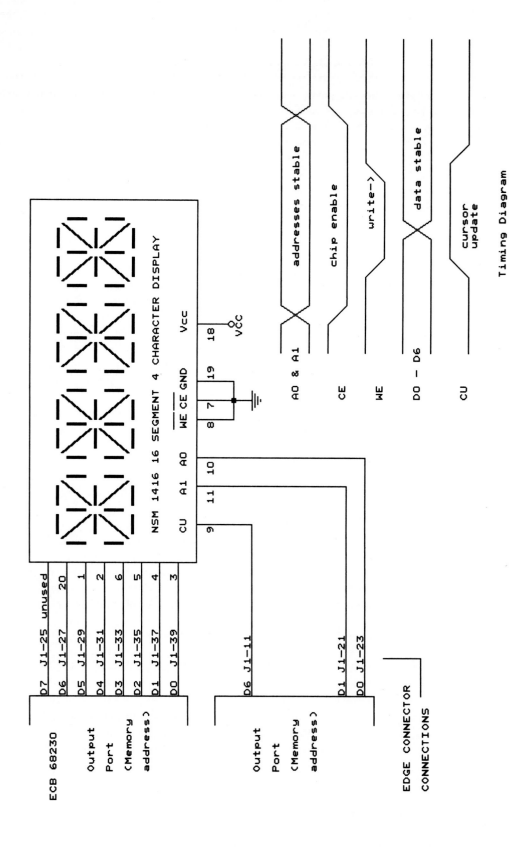

Figure 7-22 Four-character LED display interface schematic.

The program contains some instructions specific to the Motorola ECB that will not be discussed here. Their use is covered in the ECB manual for those using this system. The flowchart is shown in Figure 7-23.

EXAMPLE 7.10

Write a program that displays built-in (canned) messages and messages input from the keyboard.

Introduction

Intelligent displays are an integral part of many microprocessor-based systems. Many displays use a 16-segment configuration such as the Litronix NSM1416 to produce the decimal characters 0 to 9 and alphabetic characters A to Z. Each segment is made up of a material that emits light when current is passed through it. This report presents a functional description of the NSM1416 and a detailed description of the software. A possible replacement for the NSM1416 display is the Siemens DLR 2416.

Discussion

The NSM1416 four-digit display can be directly connected to the output ports of the PIA because it already contains a character decoder, buffers capable of sinking or sourcing the LED current, internal memory, and a multiplexer. The NSM1416 is an interesting device because (1) it does not require external interfacing hardware design, which would have taken a significant amount of time; (2) it consumes very little power (from the computer's output IC) because of MOS technology (IiL = 35 μA); (3) we do not have to worry about current requirements when we are driving all the LEDs high; and (4) it can be driven by very simple software.

As we see in Figure 7-22, the internal memory can be written through the 7-bit data bus (D0 to D6), and the digit locations are addressed by the 2-bit address bus (A0 and A1). The CE lines are used to select the desired four-digit group when more than one four-digit display is used. In this particular project the CE line is tied too low because only one four-digit group is used. The NSM1416 offers an independent and an asynchronous digit access mode. The asynchronous data entry mode is used for two reasons:

1. It allows us to store one of its 64 characters into the digit specified by A0 and A1.
2. It allows us to maintain the data in the digits until new data is entered. To enable data entry the CU line is held high and WE low. The output PIAB is used to select one of the 64 characters and PIAA to address a specific digit. For example, if 41H is written to output PIAB, 40H to PIAA, A0 and A1 and WE low and CU high, a digit 0 will display an A. The A is stored in digit 0 until new data is entered.

Description of the Program

The program, which was constructed from a simple flowchart, is menu driven. The software allows the user to select one of four available operations:

1. Display a message
2. Enter word from keyboard
3. Scroll a message
4. Exit the program

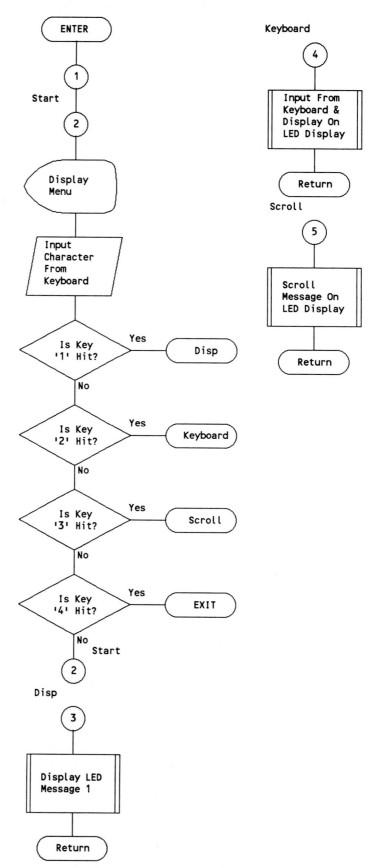

Figure 7-23 Four-character LED display program flowchart.

162

When a number is pressed (1, 2, 3, or 4) the program calls and runs the corresponding subroutine. Every time that the control returns to the main loop, the menu is displayed and another selection can be made.

New Instructions

A new instruction, **MOVEM,** is used in this example (see Table 7-12). It is simply a shortcut instruction, doing several moves with a single instruction. It is used to save the contents of Address Register A5 and Data Register D7 on the stack during a subroutine [the MOVEM.L A5/D7,-(A7) instruction]. See Example 7.7 for an introduction to subroutines. Recall that A7 is our Stack Pointer, pointing to the region of memory that is used to save the value of the Program Counter during subroutines. The stack is also used to save other register contents that are used and altered during the subroutine. The original values are restored at the end of the subroutine with the MOVEM.L (A7)+,A5/D7 instruction.

TABLE 7-12

MOVEM Instruction

MOVEM   Move multiple registers

Assembler syntax: **MOVEM.W <register list>,<Dea>**
                              .L

								<Dea> Destination effective address					
	Dn	An	(An)	(An) +	− (An)	$w (An)	$b (An, Xn)	$s	$l	$w (PC)	$b (PC, Xn)	#$q	SR	CCR
			*		10	*	*	*	*					

10 denotes the addressing mode used in Example 7.10.

Action: Selected registers → consecutive memory locations starting at <Dea>

Assembler syntax: **MOVEM.W <Sea>,<register list>**
                              .L

<Sea> Source effective address	<Dea> Destination effective address		
	Dn	An	
Dn			
An			
(An)		*	
(An) +		10	
− (An)			
$w (An)		*	
$b (An, Xn)		*	
$s		*	
$l		*	
$w (PC)		*	
$b (Pc,Xn)		*	
#$q			
SR			
CCR			

10 denotes the addressing mode used in Example 7.10.

Action: Selected registers are loaded from consecutive memory locations starting at <Sea>

TABLE 7-12 (continued)

MOVEM Instruction

Status register condition codes: X   N   Z   V   C
                                      -   -   -   -   -

Notes:
1. Only word and longword data sizes are allowed.
2. All 32 bits of the <Dea> are affected.
3. Word-sized data is sign-extended to 32 bits.
4. <register list> format examples: D0–Dn   is   D0 to Dn
                                       A2–An   is   A2 to An
                                       D1–D3/A3   is   D1 to D3, and A3

```
; Example 7.10
; ****************************************************************************
; *          MOTOROLA 68000          : FOUR-CHARACTER LED DISPLAY           *
; *  DESIGNED BY ABDERRAHIM RHAZI WITH HELP OF PROFESSOR BARRY              *
; *  Prepared for use on the Motorola ECB Computer                          *
; ****************************************************************************
;
                 .ORG   $1000
                 .COMMAND  +AM2        ; required for proper hex file format
                 .EQU   LF,$0A
                 .EQU   STRING,227
                 .EQU   CR,$0D
                 .EQU   TUTOR,$8146     ; return to TUTOR monitor address
; ------------------------------------------------------------------------------
; Input/Output Port Initialization for 68230 PI/T I/O IC
;
                 .EQU   OUTA,$10013    ; PORTA (USED FOR ADDRESS SELECTION)
                 .EQU   OUTB,$10011    ; PORTB (USED FOR DATA OUTPUT)
                 MOVE.B  #$FF,$10005    ; MAKE PORTA AN OUTPUT PORT
; use             #$00,$10005    ; TO MAKE PORTA AN INPUT PORT
                 MOVE.B  #$FF,$10007    ; MAKE PORTB AN OUTPUT PORT
; use             #$00,$10007    ; TO MAKE PORTB AN INPUT PORT
                 MOVE.B  #$80,$1000F    ; SET SUBMODE FOR PORTA
                 MOVE.B  #$80,$1000D    ; SET SUBMODE FOR PORTB
; end of initialization
; ------------------------------------------------------------------------------
                 LEA.L  $7FFE,SP        ; SET STACK POINTER
MESS:            MOVE.L  #MSG1,A5       ; START OF MESSAGE 1
                 MOVE.L  #MSG1E,A6      ;
                 MOVE.L  #STRING,D7     ; DISPLAY MENU
                 TRAP   #14             ;
                 BSR  AGAIN
                 BRA  MESS
; ------------------------------------------------------------------------------
;
AGAIN:           CLR.B  D1              ;
                 JSR  BOUT              ; GET CHARACTER FROM KEYBOARD
                 MOVE.B  KEY,D1         ; LOAD D1 WITH AN ASCII FROM KEYBOARD
                 CMPI.B  #'1',D1        ; CHECK IF KEY IS 1
                 BNE  TWO
                 BSR  DISP              ; IF SO, GO TO DISP SUBROUTINE
                 BRA  EXITIT
TWO:             CMPI.B  #'2',D1        ; CHECK IF KEY IS 2
                 BNE  THREE
                 BSR  IFK               ; IF SO, GO TO IFK SUBROUTINE
                 BRA  EXITIT
```

```
THREE:    CMPI.B  #'3',D1      ; CHECK IF KEY IS 3
          BNE   FOUR
          BSR   ROLL           ; IF SO, GO TO ROLL SUBROUTINE
          BRA   EXITIT
FOUR:     CMPI.B  #'4',D1      ; CHECK IF KEY IS 4
          BNE   EXITIT
          JMP   TUTOR          ; IF SO, RETURN TO TUTOR MONITOR
EXITIT:   RTS
```

; ---
;

```
DISP:     MOVE.L  #MSG3,A5     ; START OF MESSAGE 3
          MOVE.L  #MSG3E,A6
          MOVE.L  #STRING,D7   ; DISPLAY MESSAGE 3 on screen
          TRAP  #14
          MOVE.B  #$43,OUTA    ; SELECT DIGIT #3 of LED display
          MOVE.B  #$54,OUTB    ; DISPLAY A T
          MOVE.B  #$42,OUTA    ; SELECT DIGIT #2
          MOVE.B  #$48,OUTB    ; DISPLAY AN H
          MOVE.B  #$41,OUTA    ; SELECT DIGIT #1
          MOVE.B  #$49,OUTB    ; DISPLAY AN I
          MOVE.B  #$40,OUTA    ; SELECT DIGIT #0
          MOVE.B  #$53,OUTB    ; DISPLAY AN S
          JSR   TIME           ; CALL TIME
          JSR   TIME
          MOVE.B  #$43,OUTA    ; SELECT DIGIT #3
          MOVE.B  #$49,OUTB    ; DISPLAY AN I
          MOVE.B  #$42,OUTA    ; SELECT DIGIT #2
          MOVE.B  #$53,OUTB    ; DISPLAY AN S
          MOVE.B  #$41,OUTA
          MOVE.B  #$20,OUTB    ; CLEAR DIGIT #1
          MOVE.B  #$40,OUTA
          MOVE.B  #$20,OUTB    ; CLEAR DIGIT #0
          JSR   TIME           ; CALL TIME
          JSR   TIME           ; CALL TIME
          MOVE.B  #$43,OUTA    ; SELECT DIGIT #3
          MOVE.B  #$46,OUTB    ; DISPLAY AN F
          MOVE.B  #$42,OUTA    ; SELECT DIGIT #2
          MOVE.B  #$55,OUTB    ; DISPLAY A U
          MOVE.B  #$41,OUTA    ; SELECT DIGIT #1
          MOVE.B  #$4E,OUTB    ; DISPLAY AN N
          MOVE.B  #$40,OUTA
          MOVE.B  #$20,OUTB    ; CLEAR DIGIT #0
          JSR   TIME           ; CALL TIME
          JSR   TIME
          RTS                  ; RETURN TO MAIN LOOP
```

; ---
; Input from keyboard routine
;

```
IFK:      MOVE.B  #$43,D7      ; SELECT DIGIT #3
          MOVE.B  D7,OUTA      ;
          MOVE.B  #247,D7      ; GET ASCII FROM KEYBOARD
                               ; AND PUT IT IN D7
          TRAP  #14
          MOVE.B  D7,OUTB      ; SEND THE ASCII
                               ; TO THE CHARACTER DISPLAY
          MOVE.B  #$42,D7      ; SELECT DIGIT #2
          MOVE.B  D7,OUTA
          MOVE.B  #247,D7
```

```
                TRAP   #14
                MOVE.B  D7,OUTB
                MOVE.B  #$41,D7        ; SELECT DIGIT #1
                MOVE.B  D7,OUTA
                MOVE.B  #247,D7
                TRAP   #14
                MOVE.B  D7,OUTB
                MOVE.B  #$40,D7        ; SELECT DIGIT #0
                MOVE.B  D7,OUTA
                MOVE.B  #247,D7
                TRAP   #14
                MOVE.B  D7,OUTB
                RTS                    ; RETURN TO MAIN LOOP
; -------------------------------------------------------------------------------
; Scrolling message routine
;
ROLL:           MOVE.L  #MSG4,A5       ; START OF MESSAGE 4
                MOVE.L  #MSG4E,A6
                MOVE.L  #STRING,D7     ; DISPLAY OF MESSAGE 4 on screen
                TRAP   #14
                LEA.L   MSGX,A0        ; PUT LED MESSAGE LOCATION IN A0
                TRAP   #14
                MOVE.B  #$20,D3        ; PUT A BLANK IN D3
                MOVE.B  D3,D4          ; PUT A BLANK IN D4
                MOVE.B  D3,D5          ; PUT A BLANK IN D5
LOOP3:          MOVE.B  #$43,D0        ; SELECT DIGIT #3
                MOVE.B  D0,OUTA
                MOVE.B  D3,D0
                MOVE.B  D0,OUTB        ; PUT A BLANK IN DIGIT #3
                MOVE.B  #$42,D0        ; SELECT DIGIT #2
                MOVE.B  D0,OUTA
                MOVE.B  D4,D0
                MOVE.B  D0,OUTB        ; PUT A BLANK IN DIGIT #2
                MOVE.B  #$41,D0        ; SELECT DIGIT #1
                MOVE.B  D0,OUTA
                MOVE.B  D5,D0
                MOVE.B  D0,OUTB        ; PUT A BLANK IN DIGIT #1
                MOVE.B  #$40,D0        ; SELECT DIGIT #0
                MOVE.B  D0,OUTA
                MOVE.B  (A0),D0        ; PUT MESSAGE IN D0
                MOVE.B  D0,OUTB
                MOVE.B  D4,D3
                MOVE.B  D5,D4
                MOVE.B  D0,D5          ; SCROLL THE MESSAGE
                JSR    TIME
                MOVE.B  (A0),D0
                CMPI   #$4,D0          ; IS END OF MESSAGE?
                BEQ    EXMESS          ; IF SO DISPLAY MENU
                ADDQ.L  #1,A0
                BRA    LOOP3           ; LOOP UNTIL DONE
EXMESS:         RTS                    ; RETURN TO MAIN LOOP
; -------------------------------------------------------------------------------
;
BOUT:           MOVEM.L  A5/D7,-(A7)
                LEA.L   KEY,A5
                MOVE.L  A5,A6
                MOVE.B  #241,D7
                TRAP   #14
                MOVE.B  #248,D7
```

```
                    TRAP   #14
                    MOVEM.L  (A7)+,A5/D7
                    RTS
;  ----------------------------------------------------------------------------
;
TIME:       CLR.L D4              ; SUBROUTINE FOR TIME DELAY
            CLR.L  D5
            MOVE.L  #$FFFF,D5
TIME1:      MOVE.B  #$01,D4
TTTT:       SUBQ.B  #$01,D4
            BNE  TTTT
            SUBQ.L  #$01,D5
            BNE  TIME1
            RTS
;  ----------------------------------------------------------------------------
;
MSG1:       .DB CR, " =================================== " ,CR,LF
            .DB "==                 MENU                  ==",CR,LF
            .DB "==   1.DISPLAY MESSAGE                   ==",CR,LF
            .DB "==   2.ENTER FROM KEYBOARD               ==",CR,LF
            .DB "==   3.SCROLLING MESSAGE                 ==",CR,LF
            .DB "==   4.EXIT                              ==",CR,LF
            .DB " =================================== ",CR,LF
MSG1E:      NOP
MSG3:       .DB CR, "THIS IS FUN",CR,LF
MSG3E:      NOP
MSG4:       .DB CR, "UNIVERSITY OF NORTH CAROLINA AT CHARLOTTE",CR,LF
MSG4E:      NOP
MSGX:       .DB "UNIVERSITY OF NORTH CAROLINA AT CHARLOTTE",CR,LF,$04
KEY:        .RS 4
            .END
```

7.3.3 Displaying Even More Characters

Several economical multicharacter and multiline liquid-crystal displays (LCD) displays are available, making them ideal candidates for information display when more than a few characters are needed. This example, prepared by Howard Silinski, shows an 80-character, two-line display connected to a MAX 68000 computer board. Sixteen-character one- and two-line displays are also readily available. These units are attractively priced from discount suppliers, and their only drawback is the fact that ambient lighting is needed in order to see the digits.

In terms of characters displayed per output memory bit, we get 80 characters using a total of 11 bits (one 8-bit output location and 3 additional bits of another). We shift most of the display effort to the display itself, which has *intelligence* of its own.

This brings up an interesting trend. In the beginning we relied on the microprocessor to do *all* of the work. Displays were dumb, keyboards were dumb, and so on. Software was used to perform the entire task. As peripheral ICs became more common (and economical), they were used to perform more of the input/output task. Keyboards today, for example, contain a microprocessor within itself, allowing the main processor more time to perform its tasks. Video displays, disk drives, and other peripheral devices that need to process or move large amounts of information will usually contain at least a microprocessor or some other sophisticated IC to relieve the main processor of such mundane tasks.

Back to our LCD display example. First, a few words from the instructor. This example was done as an optional project at the end of the semester by an ambitious (but very busy) student. Time expired before he had a chance to complete the "whistles and bells" on it, but it was working pretty well at the due date. It had been done previously, interfaced to a Z-80 computer, providing an interesting challenge, and is included here as the base for further experimentation. The instructor is grateful for the student's effort.

Refer to Example 7.10 for the interfacing principle concerning this type of device. Referring to the schematic in Figure 7-24, here are the fundamentals of its operation:

1. The eight data lines used to display the various alphabetical and numerical characters are connected to one memory location used as an output.
2. The combination of the register select (RS) and the read/write (R/$\overline{\text{W}}$) lines determine which internal register is being read from or written to. These two lines and an enable (E) line are connected to bits of a second memory output location. The enable line is programmed to rest low.
3. To display a character, the desired data byte is output to the first memory location, and then the desired register and read/write operation is chosen by outputting the desired bits to the second memory location.
4. The enable line is then taken high for a brief time, and then brought back low. This latches the desired character into the display.

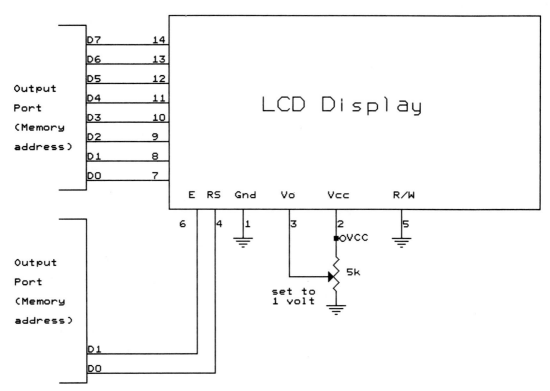

Figure 7-24   Eighty-character LCD display interface schematic.

This device was described in an excellent article by Steven Avritch in the June and July 1990 issues of ***Radio-Electronics***.

The program for Example 7.11 was interfaced to the 6821 PIA on the MAX 68000 computer system. We defer discussion of the PIA initialization to Chapter 8. For this example, simply look at it as two output memory locations. The instructions unique to the MAX 68000 are discussed in its operation manual.

The schematic and the flowchart, and the program are shown in Figures 7-24 and 7-25, respectively.

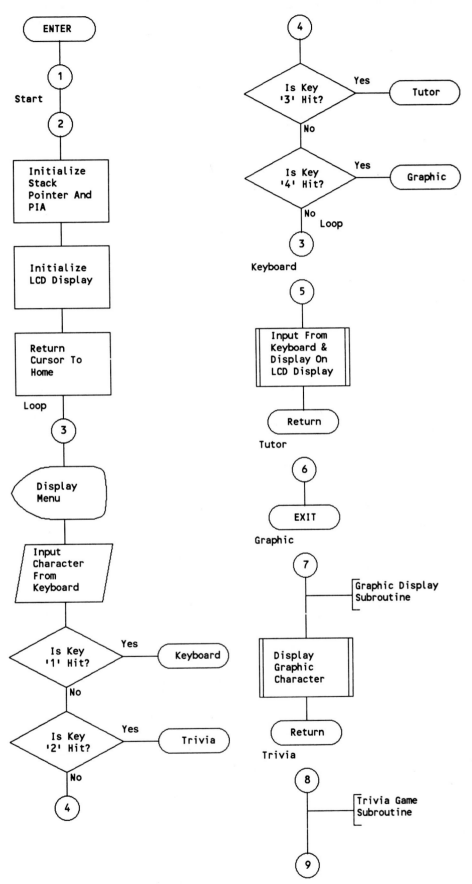

Figure 7-25 Eighty-character LCD display program flowchart.

169

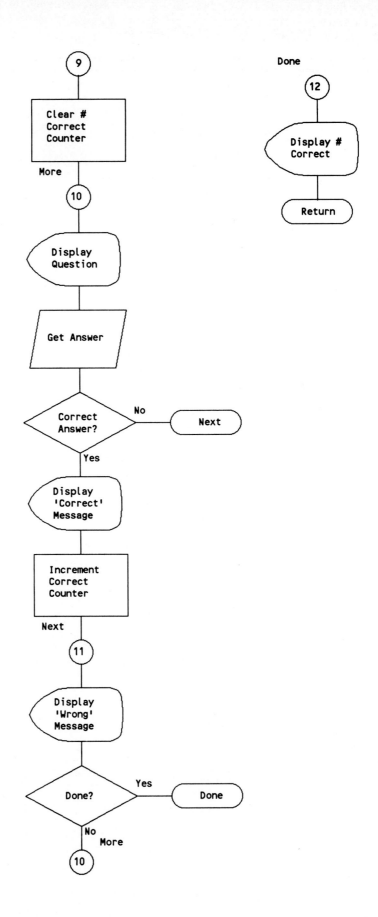

EXAMPLE 7.11

Write a program that displays built-in (canned) **messages and messages input** for the keyboard on the 80-character LCD display.

```
; Example 7.11
; **************************************************************************
;* MAX 68000 Computer/Hitachi HD44780 LCD Display Program
;   By Howard Silinski
; **************************************************************************
;* USES 6821 PIA PORT A FOR D0 → D7
;* USES          PIA PORT B FOR RS, R/W, E
;*                                        BITS         D3   D7   D1
          .COMMAND    +AM2      ; Required for MAX 68000    UPLOAD
          .equ   CR,$0D
          .equ   LF,$0A
          .equ   PIAA,$E8001
          .equ   PIAB,$E8005
          .equ   TIME1,$1FF
          .equ   TIME2,$1FFF
          .equ   TIME3,$4FFF
          .equ   TIME4,$1FFFFE
          .equ   PUTSTR,$0400
          .equ   GETSTR,$0404
          .equ   PUTCHAR,$0408
          .equ   GETCHAR,$040C
          .equ   NEWLINE,$0414
          .equ   PUTSPC,$0480
          .equ   MAKBEEP,$0484
          .equ   WARMST,$0488
          .ORG   $20000
; --------------------------------------------------------------------------
; Input/Output Port Initialization for 6821 PIA I/O IC
;
INIT:     MOVE.B  #0,2+PIAA     ; SELECT PIAA DATA DIR. REG
          MOVE.B  #$FF,PIAA     ; MAKE PIAA OUT
          MOVE.B  #$04,2+PIAA   ; SEL PIAA DATA REG
          MOVE.B  #0,2+PIAB     ; SELECT PIAB DATA DIRECTION REGISTER
          MOVE.B  #$FF,PIAB     ; MAKE PIAB DIRECTION = OUTPUT
          MOVE.B  #$04,2+PIAB   ; SELECT PIAB DATA REGISTER
; --------------------------------------------------------------------------
; end of initialization
;
          LEA.L   $22000,A7     ; SET STACK POINTER
;* LCD INITIALIZATION
          MOVE.B  #$80,PIAB     ; SELECT COMMAND MODE FOR LCD DISPLAY
          MOVE.B  #$30,PIAA     ; FUNCTION SET COMMAND FOR LCD
          BSR   CONTROL         ; TOGGLE ENABLE PULSE
          BSR   DELAY2          ; WAIT 5 MILLISECONDS
          MOVE.B  #$30,PIAA     ; DO IT AGAIN
          BSR   CONTROL         ; TOGGLE ENABLE PULSE
          BSR   DELAY1          ; WAIT 500 MICROSECONDS
          MOVE.B  #$30,PIAA     ; DO IT AGAIN
          BSR   CONTROL         ; TOGGLE ENABLE PULSE
          BSR   DELAY1
          MOVE.B  #$38,PIAA     ; SELECT 2 LINE DISPLAY,
                                ; 5 × 7 DOTS DISPLAY
```

```
                    BSR    CONTROL
                    BSR    DELAY1
                    MOVE.B  #08,PIAA        ; DISPLAY OFF
                    BSR    CONTROL
                    BSR    DELAY1
                    MOVE.B  #01,PIAA        ; CLEAR DISPLAY
                    BSR    CONTROL
                    BSR    DELAY1
                    MOVE.B  #06,PIAA        ; SELECT INCREMENT MODE,
                                            ; SHIFT CURSOR RIGHT
                    BSR    CONTROL
                    BSR    DELAY1
;* END OF INITIALIZATION
;
                    MOVE.B  #$0E,PIAA       ; TURN ON DISPLAY AND CURSOR
                    BSR    CONTROL
                    BSR    DELAY1
                    MOVE.B  #06,PIAA
                    BSR    CONTROL
                    BSR    DELAY1
;
; Main loop
DONE:       BSR    HOME
            BSR    MENU             ; DISPLAY MENU ON MONITOR
            BSR    SELECT           ; GET KEYBD SELECTION
LOOP:       MOVE.B  (A0)+,D0
            CMP.B   #0D,D0
            BEQ    ENDSTR
            MOVE.B  D0,PIAA          ; DISPLAY CHARACTER
            BSR    DATA
            BSR    DELAY3
            DBRA   D6,LOOP
ENDSTR:     BSR    DELAY4
            BRA    DONE
; ------------------------------------------------------------------------------
;
DELAY1:     MOVE.W  #TIME1,D1  ; DELAY LOOPS
DY1:        SUBQ.L     #1,D1
            BNE.S      DY1
            RTS
; ------------------------------------------------------------------------------
;
DELAY2:     MOVE.W  #TIME2,D1
DY2:        SUBQ.L     #1,D1
            BNE.S      DY2
            RTS

; ------------------------------------------------------------------------------
;
DELAY3:     MOVE.W  #TIME3,D1
DY3:        SUBQ.L     #1,D1
            BNE.S      DY3
            RTS

; ------------------------------------------------------------------------------
;
DELAY4:     MOVE.L  #TIME4,D1
DY4:        SUBQ.L     #1,D1
            BNE.S      DY4
            RTS
```

```
;  --------------------------------------------------------------------------------
;
CONTROL: MOVE.B  #0,PIAB        ; CLEAR RS FOR COMMAND MODE
         BSR  DELAY1
         MOVE.B  #02,piab       ; TOGGLE ENABLE LINE
         BSR  DELAY1
         MOVE.B  #00,piab
         BSR  DELAY1
         MOVE.B  #$88,PIAB
         BSR  DELAY1
         RTS
;  --------------------------------------------------------------------------------
;
DATA:    MOVE.B  #08,PIAB       ; SET RS FOR DATA MODE
         BSR  DELAY1
         MOVE.B  #$0A,PIAB      ; TOGGLE ENABLE LINE
         BSR  DELAY1
         MOVE.B  #08,PIAB
         BSR  DELAY1
         MOVE.B  #$88,PIAB
         BSR  DELAY1
         RTS
;  --------------------------------------------------------------------------------
;
CGRAM:   MOVE.B  #40,PIAB       ; SET RS FOR CHARACTER MODE
         BSR  DELAY1
         MOVE.B  #$0A,PIAB      ; TOGGLE ENABLE LINE
         BSR  DELAY1
         MOVE.B  #40,PIAB
         BSR  DELAY1
         MOVE.B  #$44,PIAB
         BSR  DELAY1
         RTS
;  --------------------------------------------------------------------------------
;
MENU:    LEA  MSG1,A0           ; DISPLAY MENU USING MONITOR CALL
         JSR  PUTSTR
         LEA  MSG2,A0
         JSR  PUTSTR
         LEA  MSG3,A0
         JSR  PUTSTR
         LEA  MSG4,A0
         JSR  PUTSTR
         LEA  MSG6,A0
         JSR  PUTSTR
         BSR    TWO             ; DISPLAY WELCOME ON LCD
         BSR    TP
         RTS
;  --------------------------------------------------------------------------------
;
HOME:    MOVE.B #01,PIAA
         BSR CONTROL
         RTS
;  --------------------------------------------------------------------------------
;
SELECT:  JSR  MAKBEEP
         JSR  GETCHAR           ; GET CHARACTER FROM KEYBD
         CMP.B  #'1',D0
         BEQ  ONE
         CMP.B  #'2',D0
```

```
                    BEQ   TRIVIA
                    CMP.B  #'3',D0
                    BEQ   THREE
                    CMP.B  #'4',D0
                    BEQ   FOUR
                    BRA   SELECT
        ;
        ONE:        BSR    HOME            ; LOAD CHARACTER FROM KEYBD TO A0
        UNO:        LEA.L  KEYBD,A0
                    JSR   GETSTR
                    JSR   NEWLINE
                    LEA.L  KEYBD,A0
                    BRA   DONE2
        ;
        TWO:        LEA.L  MSG5,A0         ; WELCOME MESG
                    BSR   DELAY3
                    BRA   DONE2
        ;
        THREE:      JMP   WARMST
        ;
        FOUR:       BSR    CGRAM           ; DISPLAY CGRAM CREATED CHARACTER
                    BSR    DELAY3
                    LEA.L  PLANE,A0
                    MOVE.B  #24,D6
        AGN:        MOVE.B  (A0)+,D0
                    MOVE.B  D0,PIAA
                    DBRA    D6,AGN
                    BSR    DELAY4
                    BRA    DONE
        ;
        TRIVIA:     CLR    D5              ; CLEAR CORRECT ANSWER COUNTER
                    MOVE.B  #1,D4
                    LEA.L  TRIV,A0
                    BSR   DELAY3
                    BSR   TP
                    BSR   DELAY4
        QUES1:      BSR HOME
                    LEA.L Q1,A0            ; QUESTION1
                    BSR DELAY3
                    BSR    TP
                    BSR    DELAY4
                    BSR    HOME
                    LEA.L Q1A,A0
                    BSR DELAY3
                    MOVE.W  #37,D6
                    BSR    TLP
                    BSR    DELAY4
                    BSR    UNO             ; GET ANSWER
                    BSR DELAY4
                    CMP.B #'1',(A0)        ; RIGHT OR WRONG ANSWER
                    BNE    WRG1
                    BSR    HOME
                    LEA.L Y1,A0
                    BSR    LINE1
                    ADD.B D4,D5
                    BRA    QUES2
        WRG1:       LEA.L N1,A0
                    BSR    HOME
                    BSR    LINE1
```

```
QUES2:    BSR     HOME
          LEA.L   Q2,A0
          BSR     DELAY3
          BSR     TP
          BSR     DELAY4
          BSR     UNO
          CMP.B   #'3',(A0)
          BNE     WRG2
          BSR     HOME
          LEA.L   Y2,A0
          BSR     LINE1
          ADD.B   4,D5
          BRA     QUES3
WRG2:     LEA.L   N2,A0
          BSR     HOME
          BSR     LINE1
QUES3:    BSR     HOME
          LEA.L   Q3,A0
          BSR     DELAY3
          BSR     TP
          BSR     DELAY4
          BSR     UNO
          CMP.B   #'2',(A0)
          BNE     WRG3
          BSR     HOME
          LEA.L   Y3,A0
          BSR     LINE1
          ADD.B   D4,D5
          BRA     QUES4
WRG3:     LEA.L   N3,A0
          BSR     HOME
          BSR     LINE1
QUES4:    BSR     HOME
          LEA.L   Q4,A0
          BSR     DELAY3
          BSR     TP
          BSR     DELAY4
          BSR     UNO
          CMP.B   #'3',(A0)
          BNE     WRG4
          BSR     HOME
          LEA.L   Y4,A0
          BSR     LINE1
          ADD.B   D4,D5
          BRA     GRD1
WRG4:     LEA.L   N4,A0
          BSR     HOME
          BSR     LINE1
GRD1:     CMP.B   #4,D5           ; MESSAGES ACCORDING TO # RIGHT ANS.
          BNE     GRD2
          BSR     HOME
          LEA.L   C1,A0
          BSR     DONE3
GRD2:     CMP.B   #3,D5
          BNE     GRD3
          BSR     HOME
          LEA.L   C2,A0
          BSR     DONE3
```

```
GRD3:       CMP.B   #2,D5
            BNE     GRD4
            BSR     HOME
            LEA.L   C3,A0
            BSR     DONE3
GRD4:       CMP.B   #1,D5
            BNE     GRD5
            BSR     HOME
            LEA.L   C4,A0
            BSR     DONE3
GRD5:       BSR     HOME
            LEA.L   C5,A0
DONE3:      BSR     DELAY3
            BSR     TP
            BSR     DELAY4
            BRA     DONE
;   ----------------------------------------------------------------------------------------
;
LINE1:      BSR     DELAY3      ; USE ONLY ONE LINE ON LCD
            MOVE.W  #39,D6
            BSR     TLP
            BSR     DELAY4
            BRA     DONE2
TP:         MOVE.W  #79,D6      ; SAME AS LOOP
TLP:        MOVE.B  (A0)+,D0
            MOVE.B  D0,PIAA
            BSR     DATA
            BSR     DELAY3
            DBRA    D6,TLP
DONE2:      RTS
;   ----------------------------------------------------------------------------------------
;
MSG1:       .DB "*       LCD display menu                          *",CR,LF,00
MSG2:       .DB "*   1) Input from keyboard                        *",CR,LF,00
MSG3:       .DB "*   2) Trivia Game                                *",CR,LF,00
MSG4:       .DB "*   3) Return to MAX 68000 monitor
MSG5:       .DB "Welcome to UNCC ENGINEERING TECHNOLOGY!!"
            .DB "Make selection from menu on monitor              ",CR,LF,00
MSG6:       .DB "*   4) Plane Display                              *",CR,LF,00
TRIV:       .DB "Test your trivia expertise!                       "
            .DB "Type answer and then RET) To Enter answer         "
Q1:         .DB "WHAT CHESS PIECE HAS THE HIGHEST                  "
            .DB "FIGHTING POWER NEXT TO THE QUEEN?                 "
Q1A:        .DB "1)ROOK, 2)BISHOP, 3)KNIGHT, OR 4)KING             "
N1:         .DB "WRONG! IT'S THE ROOK! IT'S POWER = 5              "
Y1:         .DB "CORRECT! ROOK = 5, BISHOP & KNIGHT = 3            "
Q2:         .DB "What composer published 4 sonatas at              "
            .DB "age 7: 1)Bach, 2)Beethoven, or 3)Mozart           "
N2:         .DB "Wrong! It was Amadeus Wolfgang Mozart             "
Y2:         .DB "RIGHT! WHAT A GREAT, YOUNG PRODIGY!               "
Q3:         .DB "The NBA record for most points scored by          "
            .DB "a player in 1 game is 1)50 2)100 3)110            "
N3:         .DB "Wrong! 100 points was the highest scored          "
Y3:         .DB "RIGHT! Wilt Chamberlain did it in 1962            "
Q4:         .DB "Who was the 3rd President of the U.S.?             "
            .DB "1)J. Adams 2)J. Monroe or 3)T. Jefferson          "
N4:         .DB "Wrong! It was Thomas Jefferson!                   "
Y4:         .DB "Correct! Adams was 2nd and Monroe 5th             "
```

```
C1:        .DB "You answered all 4 correctly!!!                                    "
           .DB "Consider yourself a Renaissance Man                                "
C2:        .DB "You answered 3 correctly!                                          "
           .DB "Almost a perfect score!                                           "
C3:        .DB "You answered 2 out of 4 correctly!!!!                              "
           .DB "Not bad for a beginner!!!!!!!!!!!!!!!!                             "
C4:        .DB "Only one correct answer?                                           "
           .DB "Better luck next time!!                                            "
C5:        .DB "No correct answers??????                                           "
           .DB "You should be ashamed of yourself!!                                "
PLANE:     .DB $00,$00,$00,$1C,$1F,$00,$00,$00,$10,$0C
           .DB $06,$1F,$1F,$06,$0C,$10,$18,$18,$18,$1F
           .DB $1F,$00,$00,$00
KEYBD:     .RS 80
           .DB 00
           .END
```

The next step up would be to display an entire video screen of information. Special ICs are available, but construction of a suitable project is beyond the scope of this book. Example 7.11, however, shows how characters can be displayed on the PC video display, using subroutines built into the MAX 68000. Similar capability is available on the Motorola ECB and was used in Example 7.10.

7.4 SOME MORE I/O PROJECTS

The application of a microprocessor is limited only by one's imagination. Following are some more interesting interfacing projects.

7.4.1 Digital Thermometer

The following in not intended to be a complete tutorial on the operation of analog-to-digital circuits. However, this very basic project does show how a versatile analog-to-digital (A/D) IC can be used. It takes an input voltage of 0 to 5 V dc and provides an 8-bit (byte) result between 0000000 11111111 (0 to 255 decimal). Figure 7-26 gives an indication of the input/output relationships of this chip.

The A/D converts the input periodically, providing a new result for each conversion. In Figure 7-25 we see that an analog signal input of 0.160 V would produce a digital output value of 08 hex. Some analog signals vary quite rapidly, requiring the sampling or conversion to be done at a fast rate. For example, a video image contains an enormous amount of *information* to be converted, and very fast converters are required. Needless to say, storage of audio or video binary data in our computer memory occupy large amounts of memory.

To get a voltage that is proportional to a temperature, we use a device known as a thermistor. It is actually a resistor that changes its resistance value as its temperature changes. It is connected as a voltage divider with a variable resistor, allowing a calibration adjustment. Figure 7-27 shows the variation of a thermistor's resistance versus temperature. We try to adjust the calibration resistor to provide as much *voltage swing* as possible, but will not get the full 5-V swing, since we are using a 5-V power supply.

The A/D converter IC is set up to "free-run," providing continuous conversions every 100 μs, after it is reset or started. For the configuration we are using, the input port will always have a valid conversion value. An alternative approach, for purists only, would be to pulse the WR line (an output bit) and then monitor the INTR line (an input bit). It goes low when the conversion is complete.

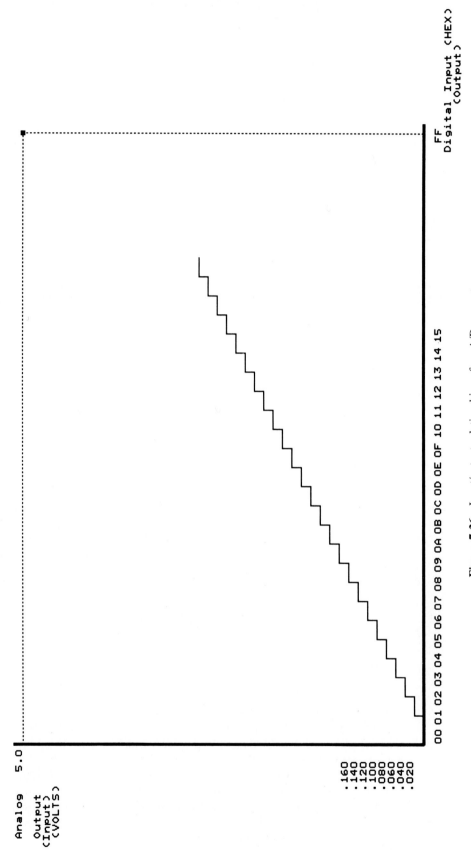

Figure 7-26 Input/output relationships of an A/D converter.

178

THERMISTOR CHARACTERISTIC CURVES
TEMPERATURE VS RESISTANCE

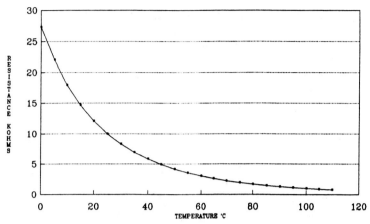

Figure 7-27 Thermistor characteristic curve.

EXAMPLE 7.12

The task is the interfacing of an ADC 0804 A/D IC and a thermistor to your 68000 system. If a thermistor is not available, try an ordinary diode, whose resistance also varies with temperature.

A flowchart is hardly needed, since the function of the program is simply to:

(a) Read the digital input from the A/D.
(b) Convert the digital value to the corresponding temperature value.
(c) Provide a display of the temperature.
(d) Loop to step (a).

Referring to Figure 7-28, we see that a single memory address used as an 8-bit input port is all that is needed. A single bit of an output address can be used to provide the reset instead of the switch if desired. As soon as the circuit is built, some means of calibrating the thermistor–A/D combination must be developed. Assuming that our input memory location is $7000 and that bit 0 of location $7002 is used to start the circuit, a simple program such as

```
        BSET.B  #0,$7002
TOP:    MOVE.B  $7000,D0
        BSR   DISPLAY
        BRA   TOP
```

will allow us to constantly read and display the input location, which represents the conversion made by the A/D chip. There is no need to measure the voltage across the thermistor and then use the A/D conversion chart to get the correct digital value. Instead, we will get a digital value for each measured temperature. This test program would be single-stepped or the display routine tailored for the specific computer used.

Next, you will need to beg, borrow, or steal an accurate thermometer, a hair dryer, and some ice cubes. Putting the thermistor and thermometer as close as possible, apply varying temperatures while single-stepping our calibration program. Create a chart containing the temperature and hexadecimal value observed in register D0. Once we know the relationship between temperature and the A/D output, construction of the computer program is a simple task.

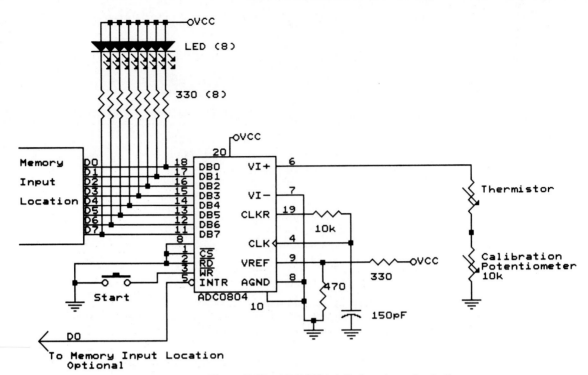

Figure 7-28   ADC 0804 A/D thermistor circuit diagram.

To convert our binary number, which represents a temperature, to an easy-to-read decimal number, we use the table lookup approach. We let some of the least significant digits of the memory address represent the digitized hex values obtained for the various temperatures, and will store in the corresponding memory locations the decimal value of the temperature. Note that no conversions from hex to decimal will be needed. If you do not have the patience to get temperature readings for all possible 255 outputs from the A/D, you can interpolate the values for your table. Table 7-14 shows this approach.

TABLE 7.14

Typical Temperature
Lookup Table

Memory address	Measured temperature
7000	30
7001	30
7002	31
7003	31
.... etc.	
.....	
70FF	99

We will leave the program as an assignment for the student. Refer to Example 7.6 for ideas on the use of a table lookup. The temperature values are stored in the format needed for the display portion of the program. We have stored them as two BCD digits for possible display on two numeric displays connected to a single memory output address. If they were to be displayed in ASCII format on the PC's display, they would have been stored in the proper ASCII format. (A little ingenuity with the lookup table is needed, since *each* ASCII character occupies a byte.) To put the numbers into our program, we use a define byte (DB) statement as

.ORG $7000
TABLE: .DB $30,$30,$31,$31, etc

Note that we include the $ so that the assembler will store them as shown. If we had omitted the **$,** they would have been interpreted as decimal numbers, with their hex equivalents being stored in memory. A bit confusing, since we really want to save and display them as two BCD digits.

7.4.2   Variable-Speed Control of a Small DC Motor

The following in not intended to be a complete tutorial on the operation of digital-to-analog circuits. However, it does show how by going the other way, we now can use binary values to selectively vary the speed of a simple dc motor. We use an 8-bit digital-to-analog converter (D/A) to do this, which can provide 256 different increments or values of voltage. Experimentation showed that the least significant bit, which represents a voltage change of only a few millivolts, had negligible effect on the speed of the motor. Instead, this bit was used to allow selection of the direction of the motor. For speed control we output from our computer output port to the D/A, a value between 00 and 7F allowing it to produce a variable voltage for speed control of the motor. This value is OR'd with 80 or 00 to select the motor direction. The DAC 0807 is an easy-to-use D/A converter but has limited current sourcing capability. Its output is connected to an operational amplifier IC, which converts the current to a voltage. The voltage in turn is used to control the turn-on of a transistor capable of controlling the current required by the motor. It should be capable of dissipating several watts since it will not be fully **on** at slow motor speeds. The direction switching transistors remain cool, since they are either fully **on** or **off.** A surplus 12-V floppy disk drive motor (which does not need a full 12 V) is an excellent motor for the project. A typical schematic of a circuit to be used for this project is shown in Figure 7-29.

EXAMPLE 7.13

(a) Interface the DAC 0807 D/A IC and a dc motor to your 68000 system.
(b) Write a program that controls the speed and direction of the motor.

The program to control the motor is straightforward:

1. Display a message, requesting direction and speed control instructions.
2. If a change in direction is desired, stop the motor, then output the desired speed and direction information; otherwise, just output the speed information.
3. Loop back to the beginning.

As you run the motor, you will not be able to get an idea of its revolutions per minute (rpm), nor will you be able to guarantee a constant speed if the load applied to the motor varies. Our next assignment utilizes a motor that also has a tachometer output, which provides us with an output voltage that is proportional to the motor speed. We can use this *feedback* to obtain rpm information which can in turn be used to increase the motor's applied voltage should the load on the motor increase. This is called a closed-loop feedback system.

The program is left for the student.

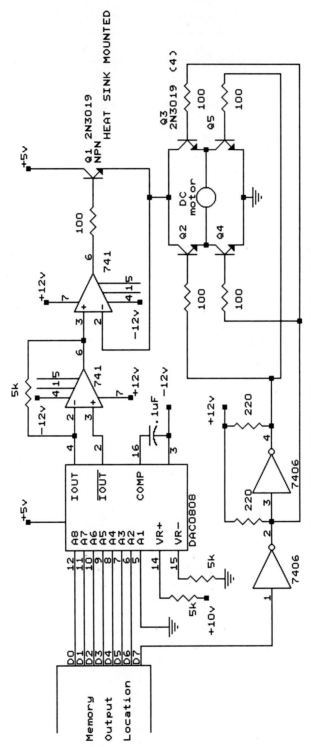

Figure 7-29 DAC 0807 DC motor control schematic diagram.

7.4.3 Variable-Speed Control (with Tach Feedback) of a Small DC Motor

To control the speed and direction of the motor, we proceed as in Example 7.13. This is an *output process* as far as our computer is concerned. Next, we take the voltage available from the motor and use it as input to give us an indication of the motor's speed. Using the A/D circuit similar to the digital thermometer to give us a digital input proportional to the motor speed, we will get a digital value that can be added or subtracted to the value being sent to the D/A that is controlling the motor. The schematic to be used is shown in Figure 7-30.

EXAMPLE 7.14

(a) Interface the DAC 0807 D/A IC and a DC motor to your 68000 system.
(b) Interface the DAC 0804 A/D IC and tach feedback winding to your system.
(c) Write a program that controls the speed and direction of the motor, taking into account variations in the load on the motor. Display the output voltage being sent to the motor, the tach feedback voltage being input, and an estimate of the rpm. Apply variations in the load to see the effect of the closed-loop operation.

The program to control the motor is straightforward:

1. Display a message, requesting direction and speed control instructions.
2. If a change in direction is desired, stop the motor, then output the desired speed and direction information; otherwise, just output the speed information, taking into account the value of the tach feedback.
3. Loop back to the beginning.

Again the program is left for the student.

7.4.4 A Tank Filling/Draining/Heating Process Control Experiment

---

EXAMPLE 7.15

Congratulations! You have landed a big job. Unfortunately, you were assigned to an old noncomputerized Mechanical Engineer who is determined to put you to work. (Don't laugh, it could happen!) This old codger has a task in need of a solution. (He's been doing a particular task manually for 20 years and is getting tired.) He's heard about things called computers and knows you happen to have one in your back pocket.
　　Here is all he has been doing:

1. Opens a valve to fill a tank, after making sure that the drain valve is closed
2. Waits for the tank to fill up and closes the fill valve
3. Turns on a heater to heat the liquid
4. Waits for the liquid to reach 175°F, using a thermometer
5. Turns off the heater
6. Opens the drain valve to empty the tank
7. Goes for coffee break
8. Goes back to step 1 unless it is quitting time

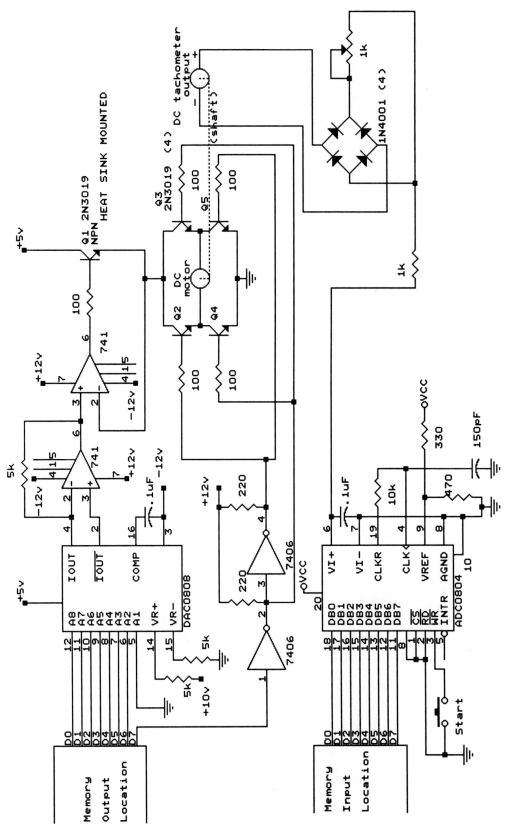

Figure 7-30 DC motor closed-loop speed control schematic diagram.

184

The industrial process he is trying to control simply involves filling up a tank with liquid, heating it to a particular temperature, and emptying it into whatever it is to be used with. Let's make two assumptions: that the drain rate is fast enough so that the liquid temperature stays close enough to 175°F during draining, and that the tank is sealed and the air being expelled from the sealed tank during filling can normally escape back through the liquid inlet valve while it is open, *unless* excessive temperature or pressure is encountered.

A sketch of the present setup is shown in Figure 7-31. The engineer never got around to hooking up the solenoids and sensors; again, he opened/closed the valves manually and used a thermometer. He told you that the sensors provide *dry contacts*. This means an isolated set of *relay* contacts that do not provide a voltage or current, but act as a short circuit when the event occurs. They can handle 100 MA dc easily. The solenoids require 1 A of current to operate, and the heater requires 10 A. An interface circuit will be required between the TTL inputs and outputs of the 68000 computer and the sensors and solenoids. He said that other sensors and safety devices may be procured if deemed necessary.

To prove that you really do have your wits about you, you offer to computerize the entire process, hook up the sensors, maybe even provide a few more that you feel are missing, and for safety, provide audible warnings if the pressure gets too high, or if the temperature gets too high, or if any other dangerous situation you can envision crops up.

In other words, clean up this mess, and provide a modern, professional, safe process control system. (The coffee breaks will not be needed by the computer system.)

For simplicity, we will ASSUME that the sensors provide accurate (debounced) closures and that there is no noise on the interconnecting cables. In reality, one should debounce the inputs. A typical hardware setup you can use for testing purposes is shown in Figure 7-32.

Your job, then, is to:

1. Create a program that controls the task above.
2. Provide a professionally created flowchart for the process, or provide a professionally created flowchart for the controlling program. Provide the one you prefer, the one best describing your work.
3. Provide a brief but complete description of how the process is controlled. (Not individual steps of your program—comments on the source listing do that job.) To convince everyone that your system is "really fast," include the actual (calculated) elapsed times that will occur between activation of one of the inputs and its corresponding output.
4. Provide a commented source listing of your program.
5. Provide a complete report in a simple but professional folder. In addition to the requirements above, the report should include a process control block diagram, a schematic of the hardware setup used for testing simulation, and a schematic showing actual circuitry that would be required to make the setup completely functional, replacing the DIP switches and LEDs used for test simulation. The computer input and output ports can be assumed to have standard TTL drive/sinking capabilities.
6. Do not forget to include some ingenuity and originality in order to impress the ol' engineer. [**Your merit raise (project grade) depends on it!**]

Steps to a successful project:

0. Create a flowchart for the process and for your program. Put your ingenuity to work in the very beginning, thinking of enhancements for the system.
1. Create the source code using an ASCII file editor.
2. Assemble your program using the assembler.
3. Debug and single step using the simulator. The simulator is a valuable aid.
4. Upload and execute your program on the target computer after complete debug on the simulator.

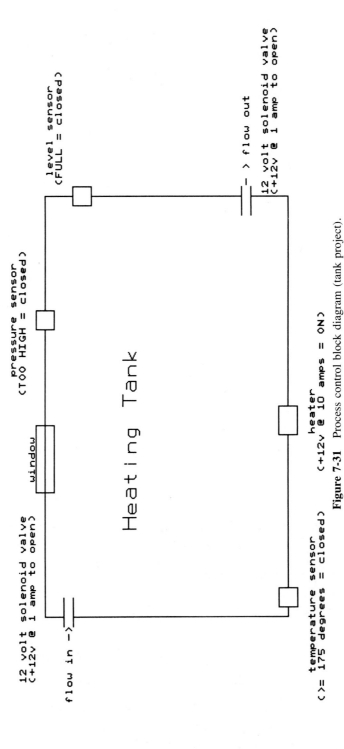

Figure 7-31 Process control block diagram (tank project).

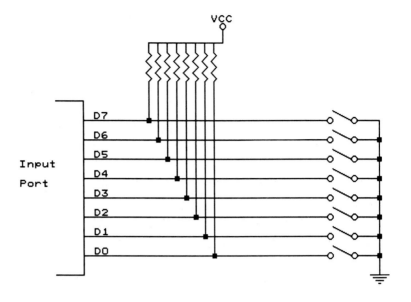

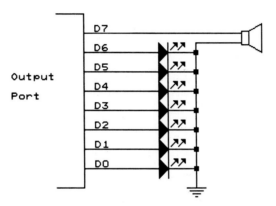

Figure 7-32 Hardware setup for tank project.

Feel free to be innovative in your design, drawing on your knowledge of physics, chemistry, electronics, using common sense if all else fails.

This project has been assigned as a class project by the author for several semesters and always provides a challenge, and uncovers innovative improvements each semester. The innovations and program are therefore left as an exercise for the student in this book.

7.5 THE ULTIMATE I/O PROJECT

The next example provides an excellent challenge for an entire class. It lays the ground-work for a real industrial process control and monitoring application. The main feature is that each of the input and output devices can be located at significant distances from the controlling computer and from each other. Development of this project will further build on your programming skills and should keep your interest in the building or discussion of the hardware aspects of the task.

The idea was conceived from an excellent article appearing in ***Radio-Electronics.*** The article, written by Steven Frickey, outlined the operation of the specialized MC14669 Motorola IC, contained schematics, programming information, and an excellent overall coverage of this unique application project. It is reprinted (with permission) in Chapter 10, together with ordering information for kits useful for laboratory experiments. Figure 7-33 outlines a block diagram of the project:

While the block diagram shows a personal computer (PC) interfaced to a 68000 computer system, which is in turn connected to the I/O devices, a similar system could be set up without the 68000. For a permanent industrial application, this would require dedication of the more expensive PC and a shift of the programming task from the 68000 assembly language to one on the PC, such as BASIC, C, or 8088 assembly language. A more feasible approach would be the development of a system as shown in the diagram, eventually putting the completed program into an EPROM on a more economical stand-alone 68000 system. The PC could then be used for other purposes.

Basically, the project involves several individual efforts in setting up monitor and control points that have either digital or analog inputs and outputs. Each of the individual projects are first completed and tested individually, following guidelines in previous examples. They are merged into a sophisticated process control project whose capability is limited only by the imagination and ingenuity of the students. The heart of the project is the MC14469, which allows selective addressing of each of the *stations* all connected to a single set of wires. Each MC14469 circuit provides a 7-bit output and a 16-bit input. A/D and D/A circuits can be interfaced as shown in previous examples.

Next, the circuits are connected via a single set of lines to a control circuit connected to a serial interface IC on the 68000 computer system. This interface will look like two memory addresses from a programming standpoint, and the selective addressing of each of the monitor and control points will be done with the program.

EXAMPLE 7.16

Develop a data acquisition and control system containing:

1. An 8-bit switch or a numeric keypad input
2. A two-character numeric display output
3. A digital thermometer input
4. A variable-speed dc motor output

Write a program that:

1. Uses the input switch value to control the speed of the dc motor
2. Displays the temperature on the two-digit display
3. Optionally, provides display and control functions on the PC, utilizing the 68000 development system's I/O routines

Setting up the total program is relatively straightforward. While only two of the 68000's memory addresses are used to communicate with the system, each I/O location is addressed in the following manner:

1. A 7-bit I/O *address* value is transmitted to the entire system via the 6850 ACIA serial interface. A 7-bit *data output* value is sent next. The I/O station having a matching address, as determined by a series of DIP switches, receives the data. It is ignored by all other stations.
2. Shortly thereafter, the same station transmits back to the 6850 ACIA 2 bytes of *input data,* corresponding to the conditions on its 16-bit input. This data is ignored by the other I/O stations.

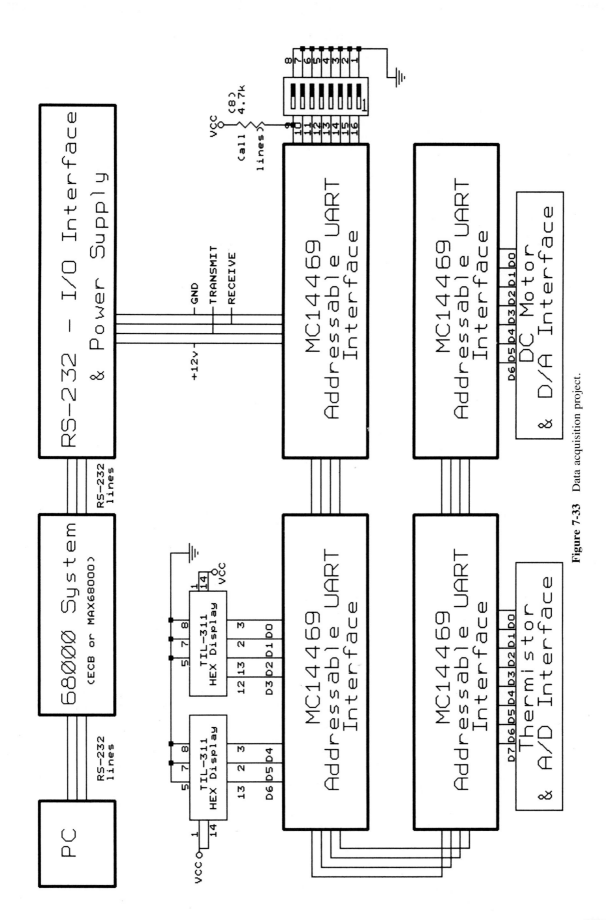

Figure 7-33 Data acquisition project.

The MC14469 has one design shortcoming that adds to the challenge of the project. Only 7 bits of output data are available. The dc motor control interface works fine with 7 bits, providing 128 different voltage levels to the motor. The two-digit display connected as shown will display only the digits from 00 to 79 (ignoring the hex digits). Addition of a few logic gates and latches and the writing of two successive bytes are two approaches that can be used when more than 7 bits are needed.

This challenging group project is left for the student, and details on programming and use of a 6850 ACIA serial RS-232 interface are provided in Chapter 8.

chapter 8

external hardware interfacing utilizing parallel and serial devices

We continue our experimentation into input/output applications in this chapter. There are two basic ways of interfacing to a computer system: in a *parallel* fashion or in a *serial* fashion. A few comparisons/contrasts:

PARALLEL	SERIAL
All the data bits are sent simultaneously.	The data bits are sent one at a time.
At least one wire is needed for each bit.	A minimum of three wires can be used.
Data is transmitted at a very high rate.	Data is transmitted much more slowly.
Cables must be relatively short (10 ft).	Cables can be several hundred feet.
Software/hardware interface is difficult.	Software/hardware interface is easy.

What does all this mean? If we want to control, monitor, and so on, external devices located in the vicinity of our computer system, **parallel** data transmission would usually be chosen. If the devices are remotely located, **serial** data transmission would be the choice. In fact, serial data transmission would allow us to control devices anywhere in the world if telephones were located at both ends! Fortunately, both the **Motorola ECB** and **Micro Designs MAX 68000** have easy-to-use parallel and serial integrated circuits that are available for expansion use.

From a software standpoint, we are still just reading and writing to a few memory locations for either parallel or serial data transfer with external devices. We will have a short initialization routine for the device, a few memory locations that will need to be read to see if new data is ready, or to see if the external device is ready to accept new data, and then the data is read from or written to ordinary memory locations.

8.1 INTERFACING WITH THE 6821 PERIPHERAL INTERFACE ADAPTER (PIA)

This 8-bit IC has long outlasted its 8-bit μP counterpart, the 6800, for several reasons. It was initially the only parallel interface IC when the 68000 was introduced. Most external interfacing involves moving 8-bit data from/to memory, hence the continued use of the 6821. The 6821 is very easy to program compared to its "big brother," the 16-bit 68230. It looks like four memory addresses to the 68000. Keep two things in mind during the following discussion:

1. We will assume that the *design* work is already done for us; all we need to do is program the PIA and hook it up to external devices. The essential electrical considerations and connections are covered.

2. The PIA is a very sophisticated device capable of several modes of parallel data transmission. Only the most elementary nonhandshaking mode will be covered. Handshaking is a technique used to allow automatic transfer of data, with the μP knowing when the PIA is ready, and vice versa. Our examples will write data when we desire, and will read data when we desire, always assuming that the I/O device is ready or that new data is available. This approach works fine for all but the most demanding I/O applications. An advanced text should be used for sophisticated applications of the PIA.

The 6821 is a 40-pin IC containing essentially two 8-bit ports that can be used for either input or output. Figure 8-1 shows the pin assignments for the PIA and Figure 8-3 a software programmer's model.

1	Vss	CA1	40
2	PA0	CA2	39
3	PA1	$\overline{IRQA}$	38
4	PA2	$\overline{IRQB}$	37
5	PA3	RS0	36
6	PA4	RS1	35
7	PA5	Reset	34
8	PA6	D0	33
9	PA7	D1	32
10	PB0	D2	31
11	PB1	D3	30
12	PB2	D4	29
13	PB3	D5	28
14	PB4	D6	27
15	PB5	D7	26
16	PB6	E	25
17	PB7	$\overline{CS1}$	24
18	CB1	$\overline{CS2}$	23
19	CB2	CS0	22
20	Vcc	$\overline{R/W}$	21

Figure 8-1   6821 PIA pin assignment.

A brief description of the PIA pins that are connected to the 68000 computer system:

1: V_{ss} is the **ground** pin.

20: V_{cc} is the 5-V dc **power** pin.

21: **read/write** pin. The 68000 will take it high during read operations and low during writes.

22, 23, 24: used to **select** or activate the PIA when a selected portion of memory is addressed. They, in effect, determine the specific memory address to which the PIA responds.

25: **clock** or timing signal pin.

26 to 33: eight **data** lines used to transfer data between the PIA and the 68000.

34: **reset** line used during power-up. The PIA can also be reset with software instructions.

35, 36: **register select** lines which provide memory decoding for the PIA's six internal registers.

37, 38: **interrupt** lines allowing the PIA to interrupt normal 68000 instruction processing.

Next, the external pins to which we will be connecting:

18, 19, 39, 40: **handshaking** lines used for sophisticated operations.

2 to 9: port A **input/output** lines, referred to as **PA0→PA7.**

10 to 17: port B **input/output** lines, referred to as **PB0→PB7.**

8.1.1   Electrical Characteristics of PA0→PA7 and PB0→PB7

Before interfacing external TTL devices to the PIA, the allowable voltage levels and current capabilities must be covered. Recall the previous discussion concerning TTL characteristics. There are minor differences between port A and port B.

Port A: When a line is programmed as an output line, it is capable of driving two TTL loads. When programmed as an input line, it is equivalent to one TTL load. Figure 8-2 shows the limitations.

Port B: When a line is programmed as an output line, it is TTL compatible but is capable of sourcing (providing) 1.5 mA at 1.5 V. This is adequate for interfacing with an external transistor circuit. The port's contents can be **read** back correctly even when its output is at 1.5 V, which is below the 2.0-V limit for a valid logic 1. **Reading** the contents of an **output** port does not have many applications, by the way.

When a line is programmed as an input line, it has **three-state** capability. Three-state capability means that a line can be (1) a logic 1, (2) a logic 0, or (3) completely disconnected from the circuit. It is similar to putting a switch between the on and off positions. Three-state circuitry is useful when several output circuits need to be connected to a single set of lines. Only one of the outputs is enabled at a time; the remaining are in the three-state condition and do not affect circuit operation. The outputs of two ordinary TTL ICs cannot be tied together. Figure 8-2 shows these characteristics.

From a software programming aspect, the PIA looks like four memory locations to the 68000. A special command sequence is used to allow addressing of the PIA's six internal registers, using only four memory addresses. The programming model shown in Figure 8-3 and initialization information is more than we will need for our simple examples, but will be useful for more advanced applications.

The sequence shown in Figure 8-4 is usually taken to initialize a PIA.

Comparing this with the PIA initialization used in Example 7.11:

```
            .equ   PIAA,$E8001
            .equ   PIAB,$E8005
;  -----------------------------------------------------------------------------
; Input/Output Port Initialization for 6821 PIA I/O IC
;
INIT:   MOVE.B   #0,2+PIAA    ; SELECT PIAA DATA DIR. REG
        MOVE.B   #$FF,PIAA    ; MAKE PIAA OUT
        MOVE.B   #$04,2+PIAA  ; SEL PIAA DATA REG
        MOVE.B   #0,2+PIAB    ; SELECT PIAB DATA DIRECTION REGISTER
        MOVE.B   #$FF,PIAB    ; MAKE PIAB DIRECTION = OUTPUT
        MOVE.B   #$04,2+PIAB  ; SELECT PIAB DATA REGISTER
```

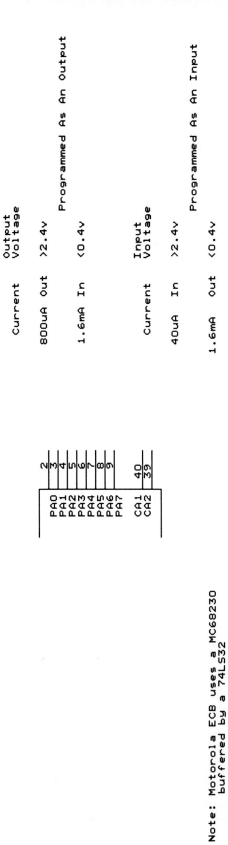

Figure 8-2 6821 PIA port A characteristics.

	Current	Output Voltage	
	800uA Out	>2.4v	Programmed As An Output
	1.6mA In	<0.4v	

	Current	Input Voltage	
	40uA In	>2.4v	Programmed As An Input
	1.6mA Out	<0.4v	

Note: Motorola ECB uses a MC68230
buffered by a 74LS32

Micro Board Designs MAX68000
provides both a MC6821 & MC68230
(unbuffered)

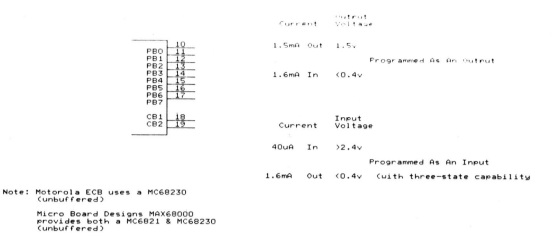

Figure 8-3   6821 PIA port B characteristics.

Memory address	"connection"	Registers	Data in/out
30001		Data Register	A port (DRA)↔PA0 to PA7
		Data Direction Register	A port (DDRA)
30003		Control Register	A port (CRA)
30005		Data Register	B port (DRB)↔PB0 to PB7
		Data Direction Register	B port (DDRB)
30007		Control Register	B port (CRB)

Note: The "switch position" is controlled by bit 2 stored in the Control Register.

Figure 8-4   6821 programming model.

```
CLR.B   $30003                  ; makes $30001 = DDRA (selects Data Direction
                                  Register)
CLR.B   $30007                  ; makes $30005 = DDRB
MOVE.B  #__B,$30001             ; set up INPUT and OUTPUT bits as desired
MOVE.B  #__B,$30005             ; set up INPUT and OUTPUT bits as desired
                                ;           INPUTS = 0, OUTPUTS = 1
MOVE.B  #xxxxx1xxB,$30003       ; makes $30001 = DRA (selects DATA REGISTER)
MOVE.B  #xxxxx1xxB,$30007       ; makes $30005 = DRB (selects DATA REGISTER)
                                ; xxxxx xx bits are set for desired handshaking
                                ; (use HEX values in the following chart)
MOVE.B  D0,$30001               ; output DATA to DRA from D0 DATA REGISTER
MOVE.B  D0,$30005               ; output DATA to DRB from D0 DATA REGISTER
                                ; or
MOVE.B  $30001,D0               ; input DATA from DRA into D0 DATA REGISTER
MOVE.B  $30005,D0               ; input DATA from DRB into D0 DATA REGISTER
```

Figure 8-5   6821 PIA initialization sequence.

Note that we have elected to let the assembler calculate the actual addresses for the Control Registers. They are actually at $E8003 and $E8007. In the example above we have programmed both the A and B output ports to be outputs.

Figure 8-6 is an aid for initialization of the PIA (A output shown). Note:

1. Table shows DRA selected (bit 2 = 1).
2. A data READ from DRA clears bits 6 and 7 of CRA.
3. CA1 is an INPUT ONLY; CA2 can be input or output.
4. The indicated HEX code is stored in the CRA to initialize the port for the desired configuration.

BIT HEX 543210	7	CA1 line	INTERRUPT line (IRQA)	BIT 6	CA2 line	INTERRUPT line (IRQA)
00 000000	(RESET)					
04 000100	1	\	MASKED	1	\	MASKED
05 000101	1	\	GOES LOW	1	\	MASKED
06 000110	1	/	MASKED	1	\	MASKED
07 000111	1	/	GOES LOW	1	\	MASKED
0C 001100	1	\	MASKED	1	\	GOES LOW
0D 001101	1	\	GOES LOW	1	\	GOES LOW
0E 001110	1	/	MASKED	1	\	GOES LOW
0F 001111	1	/	GOES LOW	1	\	GOES LOW
14 010100	1	\	MASKED	1	/	MASKED
15 010101	1	\	GOES LOW	1	/	MASKED
16 010110	1	/	MASKED	1	/	MASKED
17 010111	1	/	GOES LOW	1	/	MASKED
1C 011100	1	\	MASKED	1	/	GOES LOW
1D 011101	1	\	GOES LOW	1	/	GOES LOW
1E 011110	1	/	MASKED	1	/	GOES LOW
1F 011111	1	/	GOES LOW	1	/	GOES LOW
						HANDSHAKE MODE:
24 100100	1	\	MASKED	0		CA2 goes HIGH when CA1 CHANGES
25 100101	1	\	GOES LOW	0		
26 100110	1	/	MASKED	0		CA2 returns LOW after a
27 100111	1	/	GOES LOW	0		DRA DATA READ
						PULSE MODE:
2C 101100	1	\	MASKED	0		CA2 rests HIGH
2D 101101	1	/	GOES LOW	0		
2E 101101	1	\	GOES LOW	0		CA2 goes LOW during DRA DATA READ
2F 101111	1	/	GOES LOW	0		CA2 returns HIGH after DRA DATA READ
						BIT FOLLOWING MODE:
34 110100	1	\	MASKED	0		CA2 GOES LOW
35 110101	1	\	GOES LOW	0		AND
36 110110	1	/	MASKED	0		STAYS
37 110111	1	/	GOES LOW	0		LOW
						BIT FOLLOWING MODE:
3C 111100	1	\	MASKED	0		CA2 GOES HIGH
3D 111101	1	\	GOES LOW	0		AND
3E 111110	1	/	MASKED	0		STAYS
3F 111111	1	/	GOES LOW	0		HIGH

Figure 8-6   6821 PIA Control Register initialization.

Example 8.0

The following application for the 6821 shows how to interface the National MM58167 clock/calendar IC to a 6821 PIA. The program was done as a class project by Ahmed Rhazi. First, a few instructor comments about the following example. It was done as an optional project at the end of the semester by an ambitious (but very busy) student. Time expired before we had a chance to polish the example completely. It had been done previously, successfully interfaced to a Z-80 computer, providing an interesting challenge. It is included here for those interested in a challenging project. The author is grateful to Ahmed for his last-minute efforts! A construction project for a PC add-on clock/calendar circuit, occurring in the June 1989 issue of ***Modern Electronics*** was used for reference, along with the National Semiconductor data sheets for the MM58167.

This example involves another IC that was originally designed for interfacing directly to a microprocessor data and address bus. It can easily be interfaced to an input/output peripheral IC. This approach is much simpler for the beginner, eliminating serious design considerations.

The schematic of the circuit is shown in Figure 8-7, and the flowchart used for creation of the program is shown in Figure 8-8.

Note the structure of the program. A main loop fairly small in size is used, and each individual task to be performed is done in the form of subroutines. This makes the program much easier to follow, easier to debug, and is the preferred way of creating programs. If battery backup is not needed, the diodes, resistor, capacitor, and 3-V battery can be omitted, with pins 23 and 24 connected to 5 V.

One of the 68000 I/O ports is used in a bidirectional manner, programmed as an output for setting (writing to) the clock, then programmed as input for reading from the clock. The second port is programmed as an output port, with some of the bits used to address the clock chip internal registers and two used to toggle the read and write lines low momentarily. An alarm feature of the clock IC was not used.

```
; Example 8.0
; **************************************************************************
; Motorola ECB Computer, MM58167 Clock Program
; Adapted from a program by Jim Herman
; Prepared and demonstrated by Ahmed Rhazi
; (With a little assistance from Prof. Bo Barry)
; **************************************************************************
;* USES 6821 PIA PORT A FOR D0 → D7   (BIDIRECTIONAL)
;* USES                B FOR CLOCK REGISTER SELECT AND CONTROL (OUTPUT)
;*
          .ORG  $1000
          .COMMAND  +AM1        ; Sets proper hex format for .OBJ file
          .EQU  CR,$0D
          .EQU  LF,$0A
; ---------------------------------------------------------------------------
;
          .EQU  PIAA,$30001     ; DATA REGISTER (A)
                                  ; ADDRESS OF ADDED 6821 PIA IC
          .EQU  PIAB,$30005     ; DATA REGISTER (B)
                                  ; ADDRESS OF ADDED 6821 PIA IC
; ---------------------------------------------------------------------------
;
          .EQU  TUTOR,$8146     ; RETURN TO TUTOR MONITOR ADDRESS
          LEA.L  $7FFE,A7       ; SET STACK POINTER
; ---------------------------------------------------------------------------
```

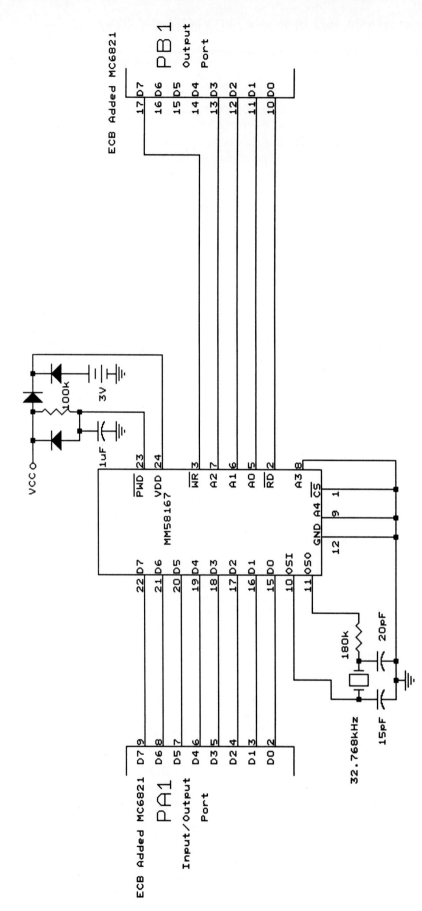

Figure 8-7 MM58167 clock IC: 6821 PIA schematic diagram.

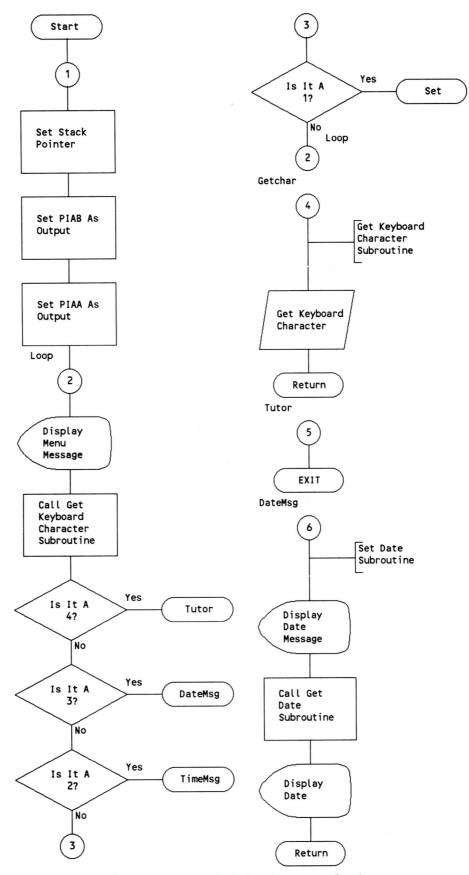

Figure 8-8   MM58167 clock project program flowchart.

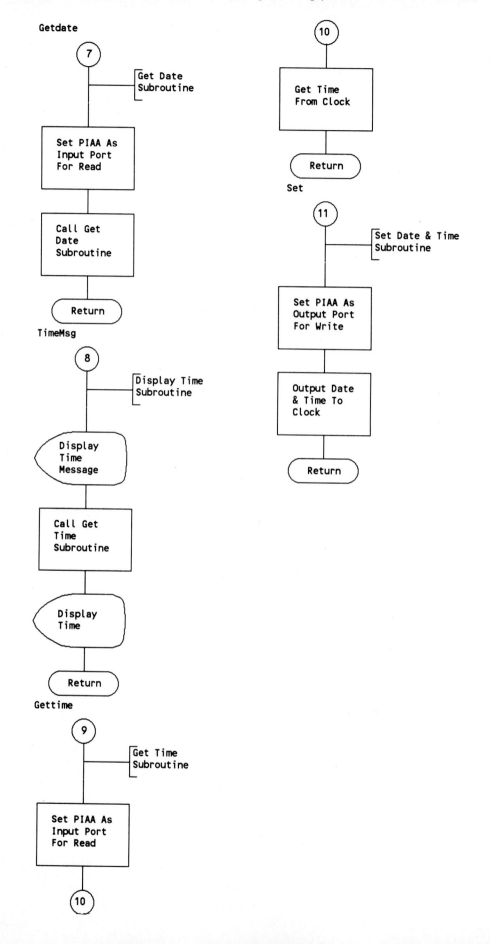

```
; INITIAL initialization of 6821 PIA
              BSR   OUTPT1           ; Sets up PIAA for OUTPUT
              BSR   OUTPT2           ;           PIAB for OUTPUT
;                  NOTE: Use BSR INPT1 to set PIAA for INPUT
; ------------------------------------------------------------------------------------
;
; MAIN  LOOP
SEL:          MOVE.L  #MENU,A5     ; DISPLAY MAIN MESSAGE
              MOVE.L  #MENUE,A6
              MOVE.B  #227,D7
              TRAP  #14
              BSR   GETCHAR
              NOP                   ; D0 HAS ASCII MENU SELECTION CHAR
              CMPI.B  #4,D0         ; RETURN TO TUTOR IF 4
              BNE   NO4
              JMP   TUTOR
NO4:          CMPI.B  #3,D0         ; DISPLAY DATE MESSAGE
              BNE   NO3
              MOVE.L  #DATE,A5
              MOVE.L  #DATEE,A6
              MOVE.B  #227,D7
              TRAP  #14
              BSR   FDIG2           ; GET DATE DIGITS FROM CLOCK
              BRA   SEL
NO3:          CMPI.B  #2,D0         ; DISPLAY TIME
              BNE   NO2
              MOVE.L  #STIME,A5
              MOVE.L  #STEE,A6
              MOVE.B  #227,D7
              TRAP  #14
              BSR   FDIG1           ; GET TIME DIGITS FROM CLOCK
              BRA   SEL
NO2:          CMPI.B  #1,D0
              BNE   NO1
              BSR   STIM
NO1:          BRA   SEL
; ------------------------------------------------------------------------------------
;
;
GETCHAR:  MOVE.B  #247,D7           ;\
          TRAP  #14                 ; \GETS KEYPRESSED
          AND.L  #$0000000F,D0 ; /
          RTS
; ------------------------------------------------------------------------------------
;
;
OUTPT1:   MOVE.B  #0,2+PIAA     ; SELECT PIAA DATA DIRECTION REGISTER
          MOVE.B  #$FF,PIAA     ; MAKE PIAA DIRECTION = OUTPUT
          MOVE.B  #4,2+PIAA     ; SELECT PIAA DATA REGISTER
          RTS
; ------------------------------------------------------------------------------------
;
;
OUTPT2:   MOVE.B  #0,2+PIAB     ; SELECT PIAB DATA DIRECTION REGISTER
          MOVE.B  #$FF,PIAB     ; MAKE PIAB DIRECTION = OUTPUT
          MOVE.B  #$04,2+PIAB ; SELECT PIAB DATA REGISTER
          RTS
; ------------------------------------------------------------------------------------
;
;
INPT1:    MOVE.B  #0,2+PIAA
          MOVE.B  #$00,PIAA     ; MAKE PIAA DIRECTION = INPUT
          MOVE.B  #4,2+PIAA     ; SELECT PIAA DATA REGISTER
          RTS
; ------------------------------------------------------------------------------------
```

```
;
;   GET TIME DIGITS FROM CLOCK AND DISPLAY
; ----------------------------------------------------------------------------------
;
FDIG1:      BSR   INPT1         ; SET PIAA FOR INPUT
            BSR   GETTIMH       ; GET HOURS
            MOVE.B  D0,D3       ; MOVE HOURS INTO D3
            BSR   GETTIMM       ; GET MINUTES
            MOVE.B  D0,D2       ; MOVE MINUTES INTO D2
            BSR   GETTIMS       ; GET SECONDS
            MOVE.B  D0,D5       ; MOVE SECONDS INTO D5
            MOVE.B  D3,D0       ; \
            BSR   DISP          ; \
            MOVE.B  #$3A,D0     ; \DISPLAYS HOURS FOLLOWED BY COLON
            MOVE.B  #248,D7     ; /
            TRAP  #14           ; /
            MOVE.B  D2,D0       ; \
            BSR   DISP          ; \
            MOVE.B  #$3A,D0     ; \DISPLAYS MINUTES
                                  ; FOLLOWED BY COLON
            MOVE.B  #248,D7     ; /
            TRAP  #14           ; /
            MOVE.B  D5,D0       ; \DISPLAYS SECONDS
            BSR   DISP          ; /
            RTS
; ----------------------------------------------------------------------------------
;
;   SUBROUTINE THAT READS THE HOURS, MINUTES, AND SECONDS
;                       FROM THE CLOCK CIRCUIT
; ----------------------------------------------------------------------------------
;
GETTIMS:    NOP                 ; READ SECONDS INTO D0
            MOVE.B  PIAA,D0     ; CLEAR BUFFERS
            MOVE.B  #$85,PIAB   ; SELECT ADDRESS OF SECONDS COUNTER
            MOVE.B  #$84,PIAB   ; MAKE READ LOW
            MOVE.B  PIAA,D0     ; READ SECONDS
            NOP                 ; MAKE UP TIME
            MOVE.B  #$85,PIAB   ; MAKE READ HIGH
            RTS
; ----------------------------------------------------------------------------------
;
GETTIMM:    NOP                 ; READ MINUTES INTO D0
            MOVE.B  PIAA,D0     ; CLEAR BUFFERS
            MOVE.B  #$87,PIAB   ; SELECT ADDRESS OF MINUTES COUNTER
            MOVE.B  #$86,PIAB   ; MAKE READ LOW
            MOVE.B  PIAA,D0     ; READ MINUTES
            NOP                 ; MAKE UP TIME
            MOVE.B  #$87,PIAB   ; MAKE READ HIGH
            RTS
; ----------------------------------------------------------------------------------
;
GETTIMH:    NOP                 ; READ HOURS INTO D0
            MOVE.B  PIAA,D0     ; CLEAR BUFFERS
            MOVE.B  #$89,PIAB   ; SELECT ADDRESS OF HOURS COUNTER
            MOVE.B  #$88,PIAB   ; MAKE READ LOW
            MOVE.B  PIAA,D0     ; READ HOURS
            NOP                 ; MAKE TIME
            MOVE.B  #$89,PIAB   ; MAKE READ HIGH
            RTS
; ----------------------------------------------------------------------------------
```

```
;
;   THIS SUBROUTINE READS DAYS, MONTH, AND DAY OF WEEK FROM
;                               THE CLOCK CIRCUIT
; --------------------------------------------------------------------------------------
GETDAY:   NOP                     ; READ DAYS FROM CLOCK
          MOVE.B   PIAA,D0        ; CLEAR BUFFERS
          MOVE.B   #$8D,PIAB      ; SELECT ADDRESS OF DAYS COUNTER
          MOVE.B   #$8C, PIAB     ; MAKE READ LOW
          MOVE.B   PIAA,D0        ; READ DAYS
          NOP                     ; MAKE TIME
          MOVE.B   #$8D,PIAB      ; MAKE READ HIGH
          RTS

; --------------------------------------------------------------------------------------
;
GETMON:   NOP                     ; READ MONTH FROM CLOCK
          MOVE.B   PIAA,D0        ; CLEAR BUFFERS
          MOVE.B   #$8F,PIAB      ; SELECT ADDRESS OF MONTH COUNTER
          MOVE.B   #$8E,PIAB      ; MAKE READ LOW
          MOVE.B   PIAA,D0        ; READ MONTH
          NOP                     ; MAKE TIME
          MOVE.B   #$8F,PIAB      ; MAKE READ HIGH
          RTS

; --------------------------------------------------------------------------------------
;
GETDOW:   NOP                     ; READ DAY OF WEEK FROM CLOCK
          MOVE.B   PIAA,D0        ; CLEAR BUFFERS
          MOVE.B   #$8B,PIAB      ; SELECT ADDRESS OF DAY OF WEEK
          MOVE.B   #$8A,PIAB      ; MAKE READ LOW
          MOVE.B   PIAA,D0        ; READ DAY OF WEEK
          NOP                     ; MAKE TIME
          MOVE.B   #$8B,PIAB      ; MAKE READ HIGH
          RTS
; --------------------------------------------------------------------------------------
;   THIS SUBROUTINE DISPLAYS THE DATE ON THE SCREEN
;
; --------------------------------------------------------------------------------------
FDIG2:    BSR   INPT1             ; SET PIAA FOR INPUT
          BSR   GETDAY            ; GET DAYS
          MOVE.B   D0,D3          ; MOVE DAYS INTO D3
          BSR   GETMON            ; GET MONTHS
          MOVE.B   D0,D2          ; MOVE MONTHS INTO D2
          BSR   GETDOW            ; GET DAY OF WEEK
          MOVE.B   D0,D5          ; MOVE DAY OF WEEK INTO D5
          MOVE.B   D3,D0          ; \
          BSR   DISP              ;   \
          MOVE.B   #$3A,D0        ;     \DISPLAYS DAYS FOLLOWED BY COLON
          MOVE.B   #248,D7        ;     /
          TRAP  #14               ;   /
          MOVE.B   D2,D0          ; \
          BSR   DISP              ;   \
          MOVE.B   #$3A,D0        ;     \DISPLAYS MONTHS
                                  ;         FOLLOWED BY COLON
          MOVE.B   #248,D7        ;     /
          TRAP  #14               ;   /
          MOVE.B   D5,D0          ; \DISPLAYS DAY OF WEEK
          BSR   DISP              ; /
          RTS
; --------------------------------------------------------------------------------------
```

```
;
DISP:       MOVE.B   D0,D4      ; \
            AND.L    #$0F,D4    ;  \
            AND.L    #$F0,D0    ;   \
            DIVS     #16,D0     ;           \DISPLAYS HIGH DIGIT ON SCREEN
            AND.L    #$0F,D0    ;       /
            ADD.B    #$30,D0    ;      /
            MOVE.B   #248,D7    ;  /
            TRAP     #14        ; /
            ADD.B    #$30,D4    ;  \
            MOVE.B   D4,D0      ;           \DISPLAYS LOW DIGIT ON SCREEN
            MOVE.B   #248,D7    ;  /
            TRAP     #14        ; /
            RTS
; ----------------------------------------------------------------------------------------------------
;
STIM:       MOVE.W   #5,D2      ; \
            MOVE.L   #TIME,A5   ;  \
            MOVE.L   #TIMEE,A6  ;   \DISPLAYS MESSAGE ON SCREEN
            MOVE.B   #227,D7    ;  /
            TRAP     #14        ; /
; ----------------------------------------------------------------------------------------------------
;   THIS SUBROUTINE IS TO SET THE CLOCK
; ----------------------------------------------------------------------------------------------------
STI:        MOVE.B   #$06,D6
STO:        BSR   GETCHAR       ; GET KEYPRESS
            MULU.W   #16,D0     ; \D3 = HIGH DIGIT
            MOVE.B   D0,D3      ; /
            BSR   GETCHAR       ; GET KEYPRESS
            ADD.W    D3,D0      ; D0 = # TO BE SENT TO CLOCK
            SUBI.B   #$01,D6
            BEQ   SENTOW        ; SET DAY OF WEEK LAST
            CMPI.B   #$05,D6
            BEQ   SENTH         ; SET HOURS FIRST
            CMPI.B   #$04,D6
            BEQ   SENTM         ; SET MINUTES SECOND
            CMPI.B   #$03,D6
            BEQ   SENTS         ; SET SECONDS THIRD
            CMPI.B   #$02,D6
            BEQ   SENTD         ; SET DAYS FOURTH
            CMPI.B   #$01,D6
            BEQ   SENTMO        ; SET MONTH FIFTH
            BRA   STO
SENTM:      BSR   SENDM         ; \
            BRA   STO
SENTH:      BSR   SENDH         ;  \
            BRA   STO
SENTS:      BSR   SENDS         ;   \
            BRA   STO           ;           TO GET TO THE PROPER
                                ;               SUBROUTINE
SENTD:      BSR   SENDD         ;   /
            BRA   STO
SENTMO:     BSR   SENDMO        ;  /
            BRA   STO
SENTOW:     BSR   SENDOW        ; /
            RTS
; ----------------------------------------------------------------------------------------------------
```

```
;
;   THESE SUBROUTINES SEND THE # IN D0 TO THE PROPER COUNTER
;                                      OF THE CLOCK
; -----------------------------------------------------------------------------------------------------
SENDH:    NOP                         ; SET HOURS
          BSR   OUTPT1                ; ST PIAA AS OUTPUT
          MOVE.B  #$00,PIAA           ; CLEAR BUFFERS
          MOVE.B  #$89,PIAB           ; SELECT ADDRESS OF HOURS COUNTER
          MOVE.B  D0,PIAA             ; MOVE DATA ON BUS
          MOVE.B  #$09,PIAB           ; WRITE HOURS
          NOP                         ; MAKE TIME
          MOVE.B  #$89,PIAB           ; MAKE WRITE HIGH
          RTS
; -----------------------------------------------------------------------------------------------------
;
;SENDM:    NOP                        ; SET MINUTES
          BSR   OUTPT1                ; MAKE PIAA OUTPUT
          MOVE.B  #$00,PIAA           ; CLEAR BUFFERS
          MOVE.B  #$87,PIAB           ; SELECT ADDRESS OF MINUTES COUNTER
          MOVE.B  D0,PIAA             ; MOVE DATA ON BUS
          MOVE.B  #$07,PIAB           ; WRITE MINUTES
          NOP                         ; MAKE TIME
          MOVE.B  #$87,PIAB           ; MAKE WRITE HIGH
          RTS
; -----------------------------------------------------------------------------------------------------
;
SENDS:    NOP                         ; SET SECONDS
          BSR   OUTPT1                ; MAKE PIAA OUTPUT
          MOVE.B  #$00,PIAA           ; CLEAR BUFFERS
          MOVE.B  #$85,PIAB           ; SELECT ADDRESS OF SECONDS COUNTER
          MOVE.B  D0,PIAA             ; MOVE DATA ON BUS
          MOVE.B  #$05,PIAB           ; WRITE SECONDS
          NOP                         ; MAKE TIME
          MOVE.B  #$85,PIAB           ; MAKE WRITE HIGH
          NOP
          RTS
; -----------------------------------------------------------------------------------------------------
;
SENDD:    NOP                         ; SET DAY USING THE SAME PATTERN
          BSR   OUTPT1
          MOVE.B  #$00,PIAA
          MOVE.B  #$8D,PIAB
          MOVE.B  D0,PIAA
          MOVE.B  #$0D,PIAB
          NOP
          MOVE.B  #$8D,PIAB
          NOP
          RTS
; -----------------------------------------------------------------------------------------------------
;
SENDMO:   NOP                         ; SET MONTH USING THE SAME PATTERN
          BSR   OUTPT1
          MOVE.B  #$00,PIAA
          MOVE.B  #$8F,PIAB
          MOVE.B  D0,PIAA
          MOVE.B  #$0F,PIAB
          NOP
          MOVE.B  #$8F,PIAB
          NOP
          RTS
; -----------------------------------------------------------------------------------------------------
```

```
          ;
SENDOW:   NOP                    ; SET DAY OF WEEK
                                 ; USING THE SAME PATTERN
          BSR   OUTPT1
          MOVE.B  #$00,PIAA
          MOVE.B  #$8B,PIAB
          MOVE.B  D0,PIAA
          MOVE.B  #$0B,PIAB
          NOP
          MOVE.B  #$8B,PIAB
          NOP
          RTS
; ----------------------------------------------------------------------------
          ;
RTIME:    MOVE.L  #STIME,A5   ; \
          MOVE.L  #STEE,A6    ;  \DISPLAYS MESSAGE ON SCREEN
          MOVE.B  #227,D7     ;  /
          TRAP  #14           ; /
          RTS
; ----------------------------------------------------------------------------
          ;
RDATE:    MOVE.L  #DATE,A5    ; \
          MOVE.L  #DATEE,A6   ;  \DISPLAYS MESSAGE ON SCREEN
          MOVE.B  #227,D7     ;  /
          TRAP  #14           ; /
          RTS
; ----------------------------------------------------------------------------
          ;
          ;                        MESSAGES
; ----------------------------------------------------------------------------
MENU:     .DB LF,LF,"SELECT OPTION: ",CR,LF
          .DB "   1) Set Time & Date ",CR,LF
          .DB "   2) Display Time ",CR,LF
          .DB "   3) Display Date ",CR,LF
          .DB "   4) Return to Tutor Monitor ",CR,LF,LF,LF
MENUE:    NOP
TIME:     .DB " Time is input in 24 hour format as HHMMSS",CR,LF,LF,LF
          .DB " Date                            as DDMMWW",CR,LF
TIMEE:    NOP
STIME:    .DB " The present time is: (HH:MM:SS) "
STEE:     NOP
DATE:     .DB " The present date is: (DD:MM:WW) "
DATEE:    NOP
          .END
```

Another challenging project, one that makes an excellent final exam for a programming/interfacing course, is an interfacing project that simulates a traffic intersection.

Example 8.1

This is a traffic intersection controller project that demonstrates mastery of 24 bits of output and 8 bits of input flowcharting, report preparation, and more.

To Do:

1. Provide at least four different traffic controlling situations, two of which sense the lane switches.

2. Make the situation as lifelike as possible, but shorten the timing cycles by a factor of 10 for demonstration purposes. Software timing loops are sufficient, but the PI/T timer IC can be used if desired.
3. Provide keyboard input to select the desired traffic simulation, along with a video display of the appropriate message.
4. Provide the capability to return to the operating system.
5. Incorporate the ability to debug your program software without the traffic intersection hardware being present.

Required: A complete report consisting of a verbal description of the task to be performed, a flowchart of the traffic intersection operations, a flowchart of the controlling software, a complete source code listing with comments, a memory map, and a schematic of the setup.

Provide a short description of your procedures for program debugging using the 68000 tutor/monitor program without the actual traffic intersection.

The report is to be done utilizing a computer word processor.

The assignment is to be demonstrated to the instructor, in a formal professional presentation.

Grading will be based on completeness, originality, neatness, clarity of the report, and on the presentation.

Figures 8-9 and 8-10 show the traffic intersection layout and a schematic of the circuit used. Construction of the intersection is relatively simple. The LEDs can be mounted in a hollow rectangular block, and mounted on or above the roadway. The roadway switches are small reed relays mounted in cutout grooves in the road (plywood) surface, covered with tape. The cars are small ceramic magnets. Note that a single current-limiting resistor can be used with satisfactory results.

Originally built for use with Z-80 systems, it was adapted for use on the Motorola ECB by Mohammed El Abdellaoui, to whom the author is grateful. The Motorola ECB has two easily assessible input/output ports, utilizing the 68230 PI/T. Its initialization was described briefly in an earlier example. To get two additional ports, a 6821 PIA was added to the ECB. A schematic of the added PIA is shown in Figure 8-11.

Initialization of the 68230 and 6821 are relatively simple. To set both ports of the 6821 as outputs, to set one port of the 68230 as an input and the second port as an output, we use the following sequence:

```
; ----------------------------------------------------------------
; Input/Output Port Initialization for 68230 PI/T I/O IC
;
        .EQU   PB2,$10013     ; PORTA - ROAD SWITCHES
        .EQU   PA2,$10011     ; PORTB - RED LIGHTS
        MOVE.B  #$00,$10005    ; MAKE PORTA AN INPUT PORT
        MOVE.B  #$FF,$10007    ; MAKE PORTB AN OUTPUT PORT
        MOVE.B  #$80,$1000F    ; SET SUBMODE FOR PORTA
        MOVE.B  #$80,$1000D    ; SET SUBMODE FOR PORTB
; ----------------------------------------------------------------
; Input/Output Port Initialization for 6821 PIA I/O IC
;
        .EQU   PA1,$30001     ; DATA REGISTER (A) - YELLOW LIGHTS
        .EQU   PB1,$30005     ; DATA REGISTER (B) - GREEN LIGHTS
        MOVE.B  #0,2+PA1       ; SELECT PIAA DATA DIRECTION REGISTER
        MOVE.B  #$FF,PA1       ; MAKE PIAA DIRECTION = OUTPUT
        MOVE.B  #4,2+PA1       ; SELECT PIAA DATE REGISTER
; ----------------------------------------------------------------
        MOVE.B  #0,2+PB1       ; SELECT PIAB DATA DIRECTION REGISTER
        MOVE.B  #$FF,PB1       ; MAKE PIAB DIRECTION = OUTPUT
        MOVE.B  #$04,2+PB1     ; SELECT PIAB DATA REGISTER
; ----------------------------------------------------------------
```

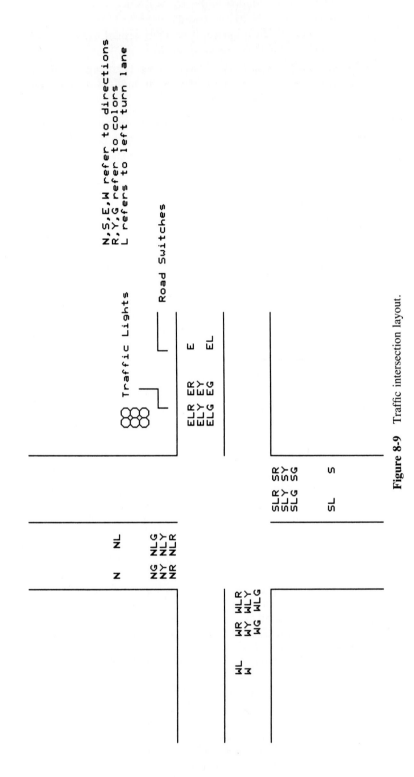

Figure 8-9 Traffic intersection layout.

208

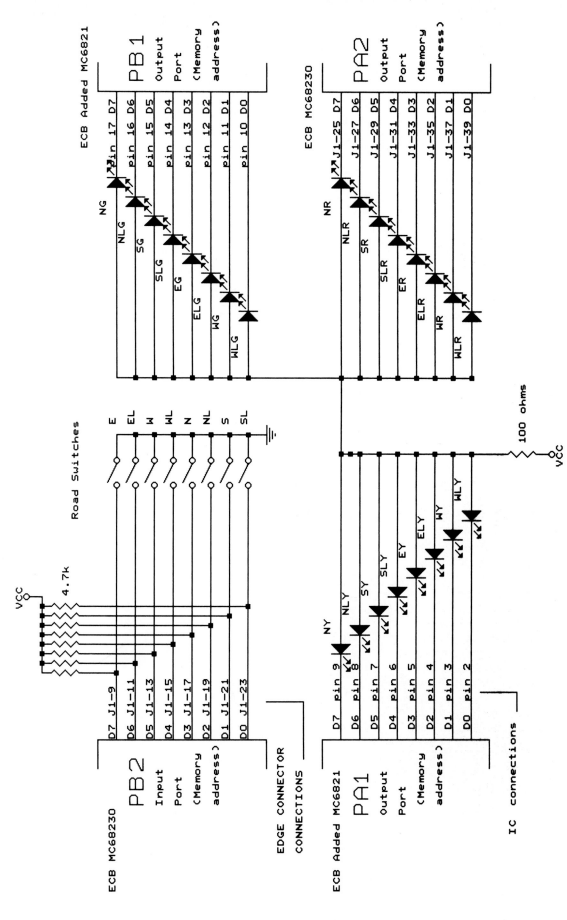

Figure 8-10 Traffic intersection schematic diagram.

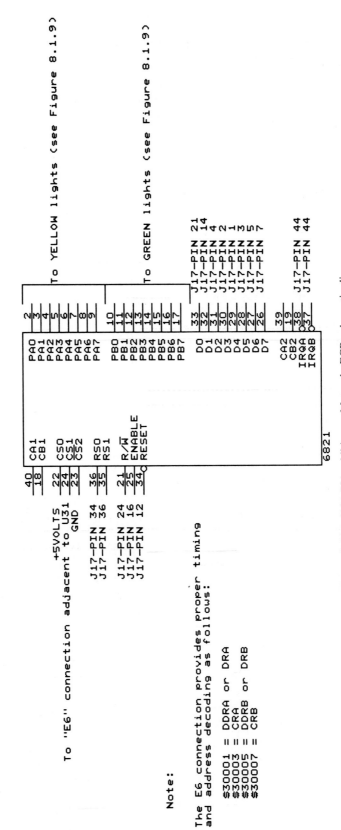

Figure 8-11 6821 PIA addition to Motorola ECB schematic diagram.

```
;
;
;A TABLE SHOWING HOW THE LIGHTS AND SWITCHES ARE CONNECTED
; BITS   7    6    5    4    3    2    1    0
;-------------------------------------------------------------------
; PA1   NY   NLY  SY   SLY  EY   ELY  WY   WLY
; PB1   NG   NLG  SG   SLG  EG   ELG  WG   WLG
; PA2   NR   NLR  SR   SLR  ER   ELR  WR   WLR
;
; PB2   E    EL   W    WL   N    NL   S    SL      (SWITCHES)
;
;Programming Example:
;
;     MOVE.B  #$0F,PB1; turn N & S GREENs ON
;     MOVE.B  #$0F,PA2; turn E & W REDs   ON
;     MOVE.B  #$FF,PA1; turn   ALL YELLOWs OFF
```

Since the input/output assignments do not change during the program, these steps are performed once at the beginning of the program. We then READ from address PB2, and WRITE to addresses PB1, PA1, and PB1.

To preserve the challenge of this assignment, we omit Mohammed's excellent program and leave that task to the student.

8.2 INTERFACING WITH THE 6850 ASYNCHRONOUS COMMUNICATION INTERFACE ADAPTER (ACIA)

Like the PIA, the 6850 ACIA, has been around since the 6800 µP and remains popular for many applications. It, too, is an 8-bit device, but in this case it takes a byte of data from a memory location and sends it out to an external device in a **serial** fashion a bit at a time.

The ACIA is an asynchronous device, as opposed to a synchronous one. Comparing asynchronous and synchronous in simple terms, synchronous is faster, harder to program, and more critical in timing considerations. This means that asynchronous is simpler, making it ideal from a beginner's standpoint. It is a perfectly natural choice for the majority of µC applications, by the way.

While both forms of serial data transmission involve transmission and reception of a single bit at a time, synchronous data either uses a separate clock line to synchronize the bits on both ends or it transmits clock synchronizing bits prior to sending a long series of data bits. Asynchronous data, on the other hand, does not send clocking information. Instead, it *frames* each byte of data with other bits of information that allow synchronization on the receiving end. This allows for a little bit of instability or error in the clock on the receiving end. To get a better understanding, the format of an asynchronous character needs to be discussed.

8.2.1 Asynchronous Data Character Format

Most but not all asynchronous serial data is transmitted utilizing a voltage format known as RS-232. Complete discussion of this standard is not covered here. Figure 8-12 shows the voltage levels associated with this transmission. A logic 1 and logic 0 are shown, along with levels know as *mark* and *space*. The idle time, when no characters are being sent is called *marking time*.

To transmit the ASCII character E, for example, we refer to an ASCII chart to see that its (7-bit) binary representation is 010 0101. Figure 8-13 depicts a typical way of

211

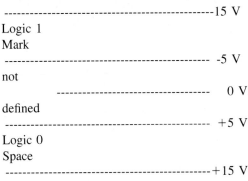

Figure 8-12 RS-232 voltage levels.

transmitting it, a bit at a time. The vertical axis represents the binary value (or RS-232 voltage level), and the horizontal axis represents time, starting with transmission of the character. At time = 0 a start bit, which is *always* a 0, is transmitted. Then the 7 bits of the ASCII E character are sent, starting with the *least* significant bit first. We have chosen to use no parity and one stop bit. Without going into the details, the use of the parity bit, the choice of the number of data bits, and the number of stop bits usually depends on the application. We have chosen the most efficient form, one in which 2 out of the 9 total bits are overhead required to allow accurate detection of the character on the reception end. We are wasting 2/9 or 22.2% of the transmission time for each character.

Associated with this character format is the **bit time,** or time duration of each bit. The reciprocal of this bit time is called the **baud** rate. Typical baud rates used are 110, 300, 600, 1200, 2400, 4800, 9600, and 19,200. The baud rate is **not** defined as the number of words per minute, or vice versa! It is left as an exercise for the student to show that this relation **can be accurate** for some formats. Using the example above, at a baud rate of 300 (3.3333-ms bit duration), assuming that the average length of each word is six characters, show that 300 baud is almost equal to 300 words per minute.

To get a general idea of how the ACIA decodes the received character, realize that it is programmed for the same baud rate, the same number of data bits, the same parity, and same number of stop bits as the transmission end. When it detects the beginning of the start bit, it waits a period equal to one-half of a bit. It then determines whether a 0 or 1 is present at this instant. It continues checking the condition of the line at periods equal to the bit duration for the remainder of the character. Hopefully, any possible error in accuracy of the receiving clock will not be significant enough by the end of a single character. As our upcoming ACIA example will show, the complexities of this data transmission/reception are taken care of by the ACIA with a few simple commands. Look at the ACIA as an "automatic serial-to-parallel converter."

The 6850 is a 24-pin IC containing essentially an 8-bit serial port that can be used for simultaneous input and output to an external device. Figures 8-14 and 8-15 show the pin assignments for the ACIA and a software programmer's model.

A brief description of the pins connected to the 68000 system:

1: V_{ss} is the **ground** pin.
3, 4: clock input pins, used to set the **serial baud rate.**
12: V_{cc} is the 5-V dc **power** pin.
13: **read/write** pin. The 68000 will take it high during read operations and low during writes.

(pin description continued)

8, 9, 10: used to **select** or activate the ACIA when a selected portion of memory is addressed. They, in effect, determine the specific memory address to which the ACIA responds.

14: **clock** or timing signal pin to synchronize to the 68000.

15 to 22: eight **data** lines used to transfer data between the ACIA and the 68000.

11: **register select** lines used to select the ACIA's four internal registers.

7: **interrupt** line allowing the ACIA to interrupt normal 68000 instruction processing.

Next, the external pins to which we will be connecting.

5, 23, 24: **handshaking** lines used for sophisticated operations.

2, 6: lines used to **receive** and **send** serial **data.** Note that these lines are TTL compatible, and appropriate TTL to RS-232 and RS-232 to TTL interface ICs must be used.

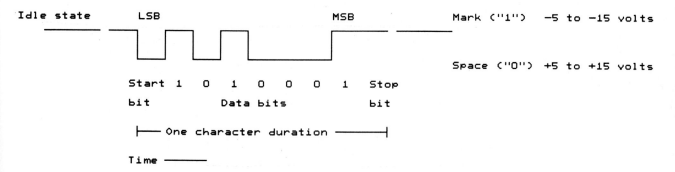

Figure 8-13   RS-232 representation of an ASCII "E".

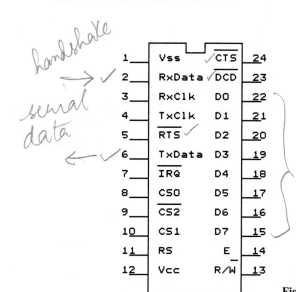

Figure 8-14   6850 ACIA pin assignment.

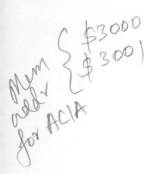

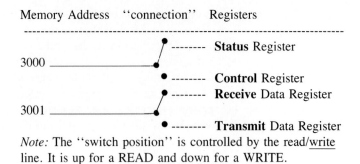

Memory Address   "connection"   Registers

3000 _____ -------- **Status** Register

                             -------- **Control** Register
                             -------- **Receive** Data Register

3001 _____

                             -------- **Transmit** Data Register

Note: The "switch position" is controlled by the read/write line. It is up for a READ and down for a WRITE.

Figure 8-15   ACIA programming model.

As with the PIA, the ACIA uses an innovative way to address the internal registers with a minimum number of hardware connections. **To the 68000, the ACIA looks like just two memory locations.**

Looking at the programming model, dwelling on the names of the internal registers of the ACIA, we could deduce that we would only want to WRITE to the Control Register and to the Transmit Register, and that we would only want to READ from the Status Register and from the Receive Register. So, with the combination of the RS register select line and the R/W line we are able to address the four registers.

8.2.2   Initializing the Control Register

At first glance, setting up the Control Register is a bit intimidating, but it becomes relatively easy to understand by taking it a bit at a time. To program the ACIA, initially we write the proper control byte to the Control Register and do not have to change it during our application. Figure 8-16 shows the layout of the Control Register bits.

Since the ACIA does not have a reset pin, it does a partial internal reset during power-up. To complete the process, we need to write the following byte initially to the Control Register: xxxxxx11, where the $\times$'s are don't cares. After this, the proper setup byte is sent. Figure 8-17 explains the Control Register bits.

The example to follow will use the following serial data format:

Baud rate = 2400 (The bit duration for each bit is 1/2400 = 416.66 μs.)
8 data bits
No parity
1 stop bit

To determine the proper configuration of the Control Register we must check the clock frequency being used on the specific 68000 system's ACIA being used. For example, the ECB uses a MC14411 baud rate generator IC, which produces frequencies 16 times the desired baud rate. The control register is set for divide by 16, and the appropriate jumpers are selected on the computer board.

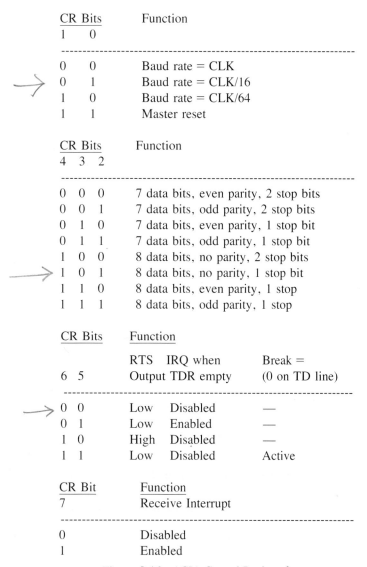

CR Bits		Function
1	0	
0	0	Baud rate = CLK
0	1	Baud rate = CLK/16
1	0	Baud rate = CLK/64
1	1	Master reset

CR Bits			Function
4	3	2	
0	0	0	7 data bits, even parity, 2 stop bits
0	0	1	7 data bits, odd parity, 2 stop bits
0	1	0	7 data bits, even parity, 1 stop bit
0	1	1	7 data bits, odd parity, 1 stop bit
1	0	0	8 data bits, no parity, 2 stop bits
1	0	1	8 data bits, no parity, 1 stop bit
1	1	0	8 data bits, even parity, 1 stop
1	1	1	8 data bits, odd parity, 1 stop

CR Bits		Function		
		RTS	IRQ when	Break =
6	5	Output	TDR empty	(0 on TD line)
0	0	Low	Disabled	—
0	1	Low	Enabled	—
1	0	High	Disabled	—
1	1	Low	Disabled	Active

CR Bit	Function
7	Receive Interrupt
0	Disabled
1	Enabled

Figure 8-16   ACIA Control Register format.

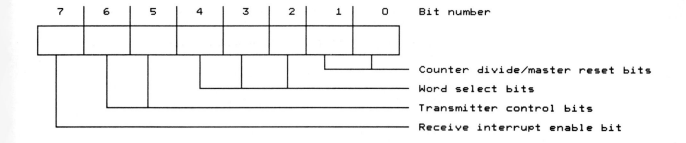

Figure 8-17   ACIA Control Register bit description.

This means that we need to use 01 (binary) for the two least significant bits of the Control Register. For bits 4, 3, and 2 we need to use 101 (binary). For bits 5 and 6 we choose 00. We want to operate in the normal (noninterrupt) mode and want to set the RTS line low (we will not be using it), and to set up for no break signal transmission. Bit 7 needs to be 0 for noninterrupt operation.

The first part of our serial data transmission project:

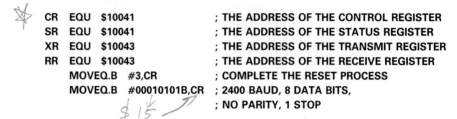

```
        CR  EQU  $10041        ; THE ADDRESS OF THE CONTROL REGISTER
        SR  EQU  $10041        ; THE ADDRESS OF THE STATUS REGISTER
        XR  EQU  $10043        ; THE ADDRESS OF THE TRANSMIT REGISTER
        RR  EQU  $10043        ; THE ADDRESS OF THE RECEIVE REGISTER
            MOVEQ.B  #3,CR           ; COMPLETE THE RESET PROCESS
            MOVEQ.B  #00010101B,CR   ; 2400 BAUD, 8 DATA BITS,
                                     ; NO PARITY, 1 STOP
```

8.2.3 Interpreting the Status Register

This register contains 8 bits of information telling us the status of the data transmission. Figure 8-18 shows the layout of the Status Register bits. Fortunately, for our elementary application, we are concerned only with a couple of the Control Register bits. Concerned only with bits 0 and 1, we need to know when the ACIA is ready to accept another byte of data for transmission and when it has an incoming byte for us to read. We will not be using interrupts and will try to operate by ignoring the possible error conditions.

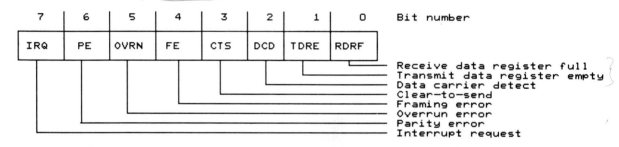

Status Register (SR)

Figure 8-18   ACIA Status Register format.

EXAMPLE 8.2: ACIA EXAMPLE

Connecting two 68000 systems together utilizing the serial interface is relatively simple. The following was done as a student project by Moses Lavien using the Motorola ECB. It could be adapted with a little effort for use on the Micro Designs MAX 68000.

The program is very rudimentary, allowing one station to send a line of text while the opposite end receives the text. The menu selection must be changed to reverse the direction of communication. With a little ingenuity in programming the program could be modified to allow *simultaneous* transmission and reception between both ends. In other words, one end could be sending text for display at the other end at the same time as it is receiving text from the other end (being displayed in another portion of the video screen). Situations requiring characters to be sent continuously as fast as possible would necessitate addition of interrupt-driven routines to avoid loss of characters being transmitted or received.

Figures 8-19 and 8-20 show the schematic diagram and the flowchart. The program follows.

ECB Connections

Figure 8-1, Schematic diagram for Example 8.2.

217

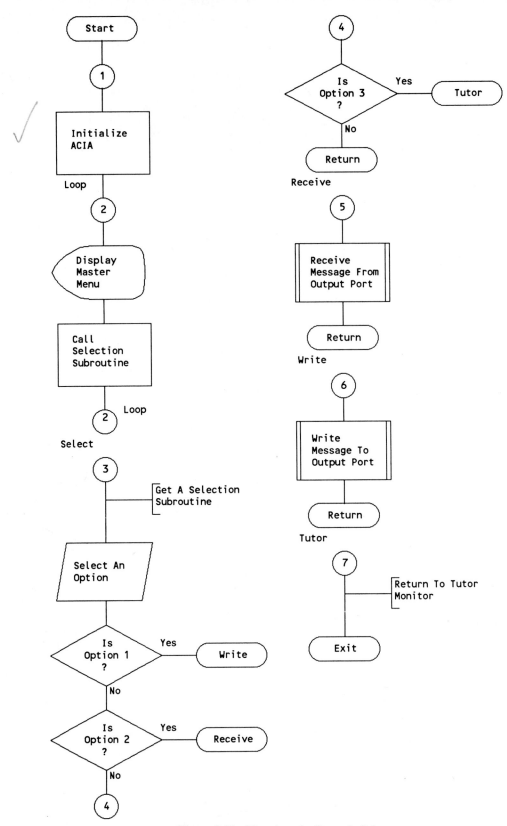

Figure 8-20   Flowchart for Example 8.2.

```
*-*-*-*-*-*-*-*-*-*-*-*-*-*-*-*-*-*-*-*-*-*-*-*-*-*-*-*-*-*-*-*-*-*-*-*-*-*-*-*-*-
; Example 8.2
; Created for use on the Motorola ECB Computer
; ********************************************************************
; This is a MOTOROLA SERIAL DATA COMMUNICATION PROGRAM intended to
; establish a communication link between two
; Motorola (68000) computer
; systems. Generally, this is an interactive program. Upon
; execution of this program, a menu will be displayed asking the
; user to select one of the following options:
;
;                                    1) Send data to Motorola computer 1
;                                    2) Receive data from Motorola computer 2
;                                    3) Return to tutor monitor
; Upon making a selection, an appropriate message indicating the
; task being performed will be displayed.
;
;
; -------------------------------------------------------------------
;
;                         ACKNOWLEDGMENT
;
;                         =============
;
;
; Professor Barry was very helpful in the successful completion
; of this project. Besides teaching me the technique of writing
; assembly language programs to do serial data communication, he
; also provided the guidance that was necessary to complete this
; project.
;
;                         Moses Dayan Lavien
;                    (ELET 3281) COMPUTER DESIGN
;
;
; -------------------------------------------------------------------
           .COMMAND   +AM2      ; REQUIRED FOR PROPER HEX OUTPUT FORMAT
           .ORG   $1000             ; the program starts at this memory location
; ********************************************************************
; This portion of the program sets up the PIA and other control
; registers.
; ********************************************************************
.EQU   ACIACR,$10041                 ; control register
.EQU   ACIASR,$10041                 ; status register
.EQU   ACIATDR,$10043                ; transmit data register
.EQU   ACIARDR,$10043                ; receive data register
.EQU   CR,$0D                        ; carriage return
.EQU   LF,$0A                        ; ASCII line feed
.EQU   REDO,03                       ; ASCII reset code
.EQU   CTRLREG,00010101B             ; control register baud rate
.EQU   RXFULL,0                      ; receive buffer full flag
.EQU   TXEMPTY,1                     ; set bit when ready
.EQU NULL,0                          ;
.EQU TUTOR,$8146                     ; monitor return address
;
              LEA.L $7FFE,A7         ; initialize user
                                       ; stack pointer

              BSR INIT               ; reset and set baud rate
LOOP:         BSR MENU               ; go to the main menu
                                       ; and select an
              BSR SELECT             ; option
              BRA LOOP               ; keep looping
;
```

```
; -------------------------------------------------------------------------------
; This portion of the program allows the user to make a selection
; from list of options. The operation performed depends on the
; selection made.
; *****************************************************************************
SELECT:     MOVE.B #247,D7                    ; input single character
            TRAP #14                          ;
            CMPI.B #'1',D0                     ; is selection 1? If not 1,
            BNE SKIP1                          ; was option 2 selected?
            BSR SENDDAT                        ; If option 1, send data.
            BRA EXITIT
SKIP1:      CMPI.B #'2',D0                     ; is it option 2?
                                                 ; If option 2,
            BNE SKIP2                          ; prepare to receive data.
            BSR RECEVE                         ;
            BRA EXITIT
SKIP2:      CMPI.B #'3',D0                     ; check for the third
                                                 ; option. If the
            BNE EXITIT                         ; the third option was
                                                 ; selected, go
            JMP TUTOR                          ; tutor.
EXITIT:     RTS
; *****************************************************************************
; -------------------------------------------------------------------------------
INIT:       MOVE.B #REDO,ACIACR               ; reset the ACIA
            MOVE.B #CTRLREG,ACIACR            ; set up the ACIA baud rate
            RTS                               ; return to the main program
; *****************************************************************************
;
; This portion of the program transmits data to the Motorola
; (68000) COMPUTERS.
; *****************************************************************************
;
SENDDAT:    LEA MSG2,A5                       ; point to the third and
            LEA MSG3,A6                       ; fourth messages
            JSR PUTSTR                        ; print them and go to the
NXTCHAR:    JSR GETCHAR                       ; character getting and
                                                 ; character
            JSR PUTCHAR                       ; putting routines
                                                 ; and wait a
            JSR WAIT                          ; while
            MOVE.B D0,ACIATDR                 ; for a carriage return.
            CMP #$0D,D0                       ; check for carriage return
            BEQ SENDUN                        ; if carriage return,
                                                 ; go to the
            BRA NXTCHAR                       ; main menu. Else go to
                                                 ; the nextchar
SENDUN:     RTS                               ; routine and repeat
                                                 ; the process.
;
; -------------------------------------------------------------------------------
; *****************************************************************************
;
; This portion of the program controls the reception of data
; from the Motorola (68000) computers.
; *****************************************************************************
```

```
                ;
        RECEVE: LEA MSG4,A5                       ; point to the fifth and
                LEA MSG5,A6                       ; sixth message
                JSR PUTSTR                        ; save them on the
                                                  ; stack pointer and
        CATCH:  MOVE.W #$FFFF,D7                  ; initialize D7 as an
                                                  ; output port
        TESTIT: BTST.B #RXFULL,ACIASR             ; test the status
                                                  ; register bit and
                BNE TIMOUT                        ; reload D0 if not equal
                SUBQ.W #1,D7                      ; decrement D7 and
                                                  ; retest the
                BRA TESTIT                        ; status register bit
        TIMOUT: MOVE.B ACIARDR,D0                 ; get data register address
                CMP.B #$0D,D0                     ; check for carriage return
                BEQ ENDREC                        ; if carriage, return to the
                JSR PUTCHAR                       ; main menu,
                                                  ; select an option
                JSR WAIT                          ; and wait
                BRA TESTIT                        ; while waiting,
                                                  ; retest the status
        ENDREC: RTS                               ; register and display
                                                  ; the main menu
                ;
                ; -----------------------------------------------------------------
                ;
        WAIT:   MOVE.W #$3FFF,D2                  ; D2 is a count
                                                  ; down register
        AGAIN:  SUBQXW #1,D2                      ; decrement it until it
                BNE AGAIN                         ; reaches zero
                RTS                               ; return to the main program
                ;
                ; -----------------------------------------------------------------
        PUTCHAR: MOVEM.L D0/D1/D7/A0,-(A7)        ; save registers
                                                  ; on the stack
                MOVE.B #248,D7                    ; output single character
                TRAP #14                          ;
                MOVEM.L (A7)+,D0/D1/D7/A0          ; restore registers
                                                  ; after use
                RTS                               ; return to the main program
                ;
                ; -----------------------------------------------------------------
                ;
        PUTSTR: MOVEM.L D0/D1/D7/A0/A5/A6,-(A7)   ; save registers on stack
                MOVE.B #227,D7                    ; and output CR, and
                                                  ; LF to output port
                TRAP #14                          ;
                MOVEM.L (A7)+,D0/D1/D7/A0/A5/A6    ; restore registers
                                                  ; after use
                RTS                               ; return to the main program
                ;
                ; -----------------------------------------------------------------
                ;
        GETCHAR: MOVEM.L D1/D7/A0,-(A7)           ; save registers
                                                  ; on the stack
```

```
                    MOVE.B #247,D7                  ; input single character
                    TRAP #14                        ;
                    AND.L #$0000007F,D0             ; mask the upper bits
                    MOVEM.L (A7)+,D1/D7/A0          ; restore registers
                                                       ; after use

                    RTS
;
; -----------------------------------------------------------------------------
;
; This portion of the program displays the main menu from which
; the user can make a selection.
;
; -----------------------------------------------------------------------------
;
MENU:               LEA MSG0,A5                     ; point to the first message
                    LEA MSG1,A6                     ; point to the second message
                    JSR PUTSTR                      ; print them and return
                    RTS                             ; return to the main program
;
; -----------------------------------------------------------------------------
;
; *****************************************************************************
;
MSG0:               .DB CR,LF,LF,LF
                    .DB " MOTOROLA (68000) TO MOTOROLA (68000) SERIAL DATA"
                    .DB " COMMUNICATION"
                    .DB "S.",CR,LF,LF
                    .DB " Your selection please:",CR,LF
                    .DB " 1) Send data to MOTOROLA COMPUTER #1." ,CR,LF
                    .DB " 2) Receive data from MOTOROLA COMPUTER #2." ,CR,LF
                    .DB " 3) Return to tutor monitor",CR,LF,LF,NULL
MSG1:               .DB CR,LF
;
MSG2:               .DB " sending data to MOTOROLA COMPUTER #1........."
                    .DB CR,LF,LF,NULL
MSG3:               .DB NULL
;
MSG4:               .DB " receiving data from MOTOROLA COMPUTER #2......"
                    .DB CR,LF,LF,NULL
MSG5:               .DB NULL
                    .END
```

8.3 CONSTRUCTION OF A SIMPLE 68008 SYSTEM

EXAMPLE 8.3

While construction of a 68000 is beyond the scope of this book, the 8-bit version, the 68008 is relatively simple to build without complex design considerations. It is identical to the 68000 both inside and instruction-wise, but interfaces easily to 8-bit RAM, EPROM, and I/O devices.

The schematic shown in Figure 8-21 is a very simple but usable computer circuit that can be used for experimentation. To use it, the program must be programmed into an EPROM, however.

Following is the beginning of the program that would be needed to initialize the computer properly:

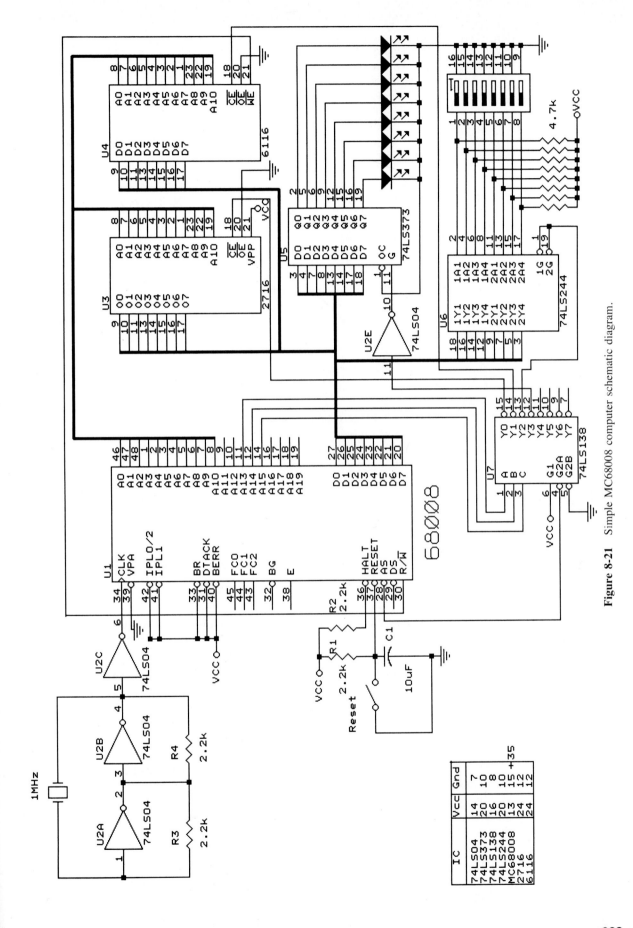

Figure 8-21 Simple MC68008 computer schematic diagram.

```
; Example 8.3
; Simple 68008 Computer
.ORG   $0000
.DL   $07FF          ; set the stack pointer
.DL   $0004          ; the location of the first instruction
        NOP          ; this is the first instruction
                     ; a typical program that transfers the DIP switch
                     ; conditions to the LEDs
                     ; replace with a more meaningful program!
                     ;
LOOP:MOVE.B   $4000,$6000 ; move the input to the output
BRA   LOOP             ; keep looping
                     ;
.END
```

The 68000, upon startup or reset, automatically takes the longword stored at location $00000000 and uses the value stored there for the Stack Pointer. The longword stored at location $00000004 is put into the Program Counter, and then computer execution starts at the address pointed to by the Program Counter. These address *vectors* are described in Chapter 9.

A few words about the circuit. The condition of the input DIP switches can be determined by doing a read from memory location $4000, and the LEDs will be activated by writing to memory location $6000. The program, stored in EPROM, resides between $0000 and $07FF, and the 2K RAM IC starts at $2000. The following table gives a brief introduction as to how the address decoding is done by the connection of the 68008's address lines A13, A14, and A15 to a 74LS138 3-of-8 decoder. A single output of the decoder is activated (goes low) depending on the inputs to the decoder.

Decoder output	Address bits																
	15	14	13	12	11	10	09	08	07	06	05	04	03	02	01	00	
Y0	0	0	0	0	0	0	0	0	0	0	0	0	0	0	0	0	= $0000
	0	0	0	1	1	1	1	1	1	1	1	1	1	1	1	1	= $1FFF
Y1	0	0	1	0	0	0	0	0	0	0	0	0	0	0	0	0	= $2000
	0	0	1	1	1	1	1	1	1	1	1	1	1	1	1	1	= $3FFF
Y2	0	1	0	0	0	0	0	0	0	0	0	0	0	0	0	0	= $4000
	0	1	0	1	1	1	1	1	1	1	1	1	1	1	1	1	= $5FFF
Y3	0	1	1	0	0	0	0	0	0	0	0	0	0	0	0	0	= $6000
	0	1	1	1	1	1	1	1	1	1	1	1	1	1	1	1	= $7FFF

Y0 is connected to the EPROM, which has only 2048 locations (0000 to 7FFF) and Y1 to the RAM, which also has 2048 locations. Y2 is connected to the input DIP switches, and reading from any address between $4000 and $5FFF will give the conditions of the switches; similarly for the output displays at $6000 to $7FFF. Only when system expansion is necessary do we need to be more concerned with the decoding details. Again, this is not a design text, and this circuit is for simple laboratory experimentation.

An idea for experimentation done previously by a student was the interfacing of the input port (the DIP switches) to the printer output port of a PC. Communication with the 68008 can then be done by printing to the printer port using BASIC, and so on. Special recognition should be given to three students (Khaleeq Ahmed, Lee Atkins, and Mike Gharagozloo), who worked much longer on this project than was expected. Although this circuit had been used on several previous occasions, this group got more than they bargained for during their construction and debugging. Operation was erratic, to say the least. ICs were changed, an oscilloscope was put to use along with a logic probe, but reliable operation was impossible. One troubleshooting technique used was (1) to remove the 68008, (2) connect pins 12 and 13 of the 74LS138 to ground to enable the 74LS244 input

IC and the 74LS373 output IC, and (3) use different configurations of the DIP switches and observe the corresponding results on the LEDs and on the corresponding data lines (D0 to D7) on the EPROM, RAM, and 68008 location. This will check out correct wiring of the data bus. Another technique used that did not help any, was modification of the EPROM to set the "error vector" addresses to the starting address of the program. The program counter value in location $00000004 was changed to $00000400, and the address $00000004 was stored in all of the vector locations described in Chapter 9. In other words, if an error occurred, the vector would send the program to the starting address of $00000400. This did not help the intermittent problems that were encountered.

The problem was finally traced to a defective protoboard, one that looked fairly new but had several locations were IC pins and wires did not make good contact. Figure 8-22 shows the circuit built with one of the ICs relocated in order to get reliable operation.

This example does not involve any input or output, but is provided as another challenging project.

EXAMPLE 8.4: DATA ACQUISITION AND SMOOTHING PROJECT

Your computer is connected to a missile tracking system. Through a direct memory access (DMA) technique, it provides elevation data of a missile that rises to some maximum altitude and then falls back to earth. It provides data every second, putting the elevation in hex feet (0 to 255) in successive memory locations. Graphically:

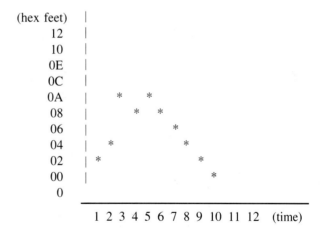

It appears that our system provided a faulty data point at time = 3. The rocket designer has told us that the rocket cannot more than double (or halve) its distance traveled upward (or downward) for each second.

Your task is to write a program that smooths out the bad data points, replacing them with the average of the preceding and the next data point (three-point smoothing). To make smoothing easier, we will assume that the FIRST and last DATA points are accurate. The smoothing technique has to have the ability to smooth out bad data points ANYWHERE in the data set, not just where you happen to spot them, and the program must work with any set of data (up to 100 points). To make things easy, we will assume that it starts at 0,0 (0 ft at 0 s), and terminates when two successive 0's are received. Also find the time of maximum elevation and its value (time = 5 s, elevation = 0 A in the example above) and the average velocity during the ascent and during the descent (anywhere on the smoothed curve).

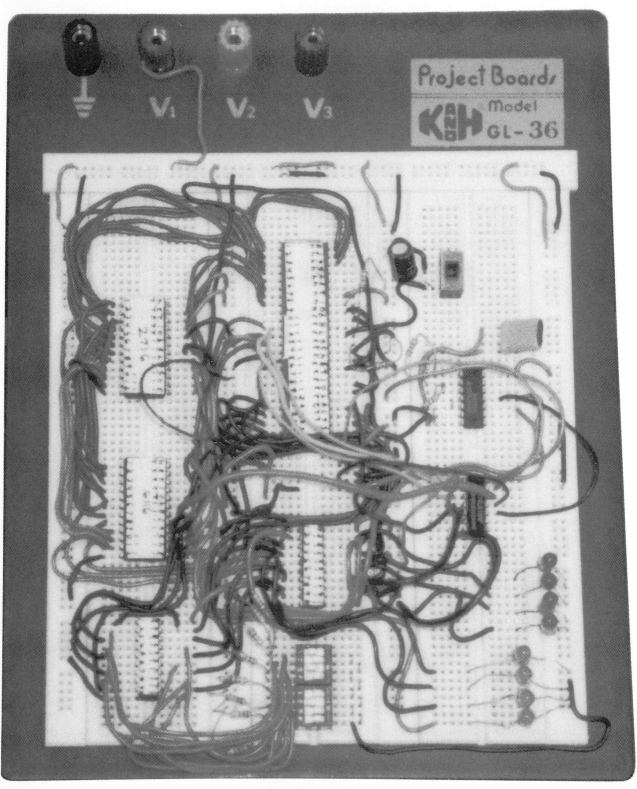

Figure 8-22   Photograph of 68008 computer.

Using the example above, taking the values at time $= 4$ and time $= 2$, we have $08 - 04 = 4$ ft, $4 - 2 = 2$ s, so the average velocity is $4/2 = 2$ ft per second. The data will be stored in memory, starting at $4000, and we let the contents of $4000 represent the value at time $= 0$, $4001 has the value for time $= 1$, and so on.

Following are two data sets that can be used for experimentation:

```
;
;    Data for last project
;  .ORG at an address such as xx00
DATA1:  .DB 000, 001, 002, 004, 007, 020, 020, 030, 040, 065, 070, 075, 080, 100
;time        0    1    2    3    4    5    6    7    8    9    10   11   12   13
;
        .DB 105, 110, 120, 124, 250, 170, 184, 190, 210, 000, 220, 230, 232, 240
;time       14   15   16   17   18   19   20   21   22   23   24   25   26   27
;
        .DB 255, 240, 220, 100, 200, 180, 170, 140, 135, 130, 000, 120, 110, 100
;time       28   29   30   31   32   33   34   35   36   37   38   39   40   41
;
        .DB 080, 070, 065, 055, 050, 040, 030, 025, 020, 010, 005, 000, 000, 000
;time       42   43   44   45   46   47   48   49   50   51   52   53   54   55
;
        .DB 000, 000, 000, 000, 000, 000, 000, 000, 000, 000, 000, 000, 000, 000
;time       56   57   58   59   60   61   62   63   64   65   66   67   68   69
;
        .DB 000, 000, 000, 000, 000, 000, 000, 000, 000, 000, 000, 000, 000, 000
;time       70   71   72   73   74   75   76   77   78   79   80   81   82   83
;
        .DB 000, 000, 000, 000, 000, 000, 000, 000, 000, 000, 000, 000, 000, 000
;time       84   85   86   87   88   89   90   91   92   93   94   95   96   97
;
        .DB 000, 000
;time       98   99
;
; Second set of data
;   .ORG xx00
;
DATA2:  .DB 000, 001, 002, 004, 005, 006, 009, 013, 016, 020, 024, 036, 048, 052
;time        0    1    2    3    4    5    6    7    8    9    10   11   12   13
;
        .DB 054, 158, 020, 064, 068, 070, 076, 190, 085, 088, 094, 096, 100, 104
;time       14   15   16   17   18   19   20   21   22   23   24   25   26   27
;
        .DB 105, 110, 018, 120, 124, 125, 126, 050, 135, 136, 138, 144, 148, 156
;time       28   29   30   31   32   33   34   35   36   37   38   39   40   41
;
        .DB 158, 162, 080, 166, 168, 170, 255, 180, 183, 185, 190, 194, 198, 208
;time       42   43   44   45   46   47   48   49   50   51   52   53   54   55
;
        .DB 209, 212, 214, 216, 254, 252, 250, 245, 243, 243, 242, 000, 240, 238
;time       56   57   58   59   60   61   62   63   64   65   66   67   68   69
;
        .DB 230, 229, 226, 255, 220, 214, 210, 200, 190, 180, 175, 170, 175, 160
;time       70   71   72   73   74   75   76   77   78   79   80   81   82   83
;
        .DB 155, 150, 000, 146, 140, 138, 132, 129, 125, 120, 110, 100, 080, 040
;time       84   85   86   87   88   89   90   91   92   93   94   95   96   97
;
        .DB 020, 000, 000
;time       98   99   100
;
```

The data can easily be sketched using a *spreadsheet* program such as Quattro or the shareware program known "As Easy As." Figure 8-23 shows the *raw* and *smooth* data sets.

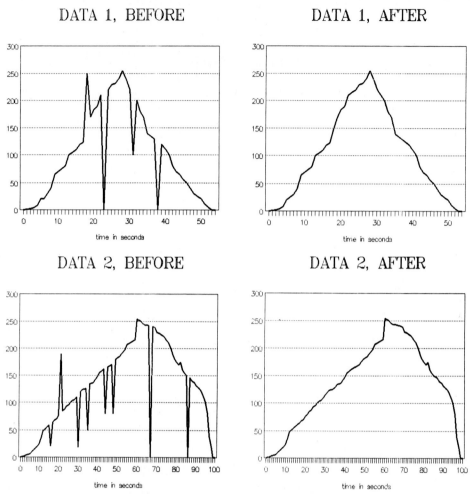

Figure 8-23 Data smoothing project data.

chapter 9

exception processing and interrupt handling

An *exception* is something that causes the postmaster to alter the normal path (avoiding a vicious dog, for instance, if delivering on foot). On the 68000, exceptions caused by external events are called *interrupts*. There are two types of exceptions on the 68000: input/output (I/O) devices cause interrupts, and internal operations, such as the TRAP instructions and program errors (division by zero, etc.) cause interrupts.

The postmaster needs to know where to go to process the various interrupts. In other words, the Program Counter needs to be reset to the proper address for the portion of memory containing the program to be executed during the interrupt condition. The 68000 uses 255 longword addresses to store these interrupt addresses called *vectors*. The vectors point to the start of the interrupt routines and must be stored in memory starting at address 000000. Table 9-1 shows how these interrupts are stored. The beginner need not be overwhelmed by these. Only one or two will be used in our applications.

TABLE 9-1

Exception Vector Assignments.

Vector number	Address	Function
0	0	Initial value for Supervisor Stack Pointer
1	4	Initial value for Program Counter
2	8	Bus error (nonexistent memory)
3	C	Address (boundary) error
4	10	Illegal instruction
5	14	Divide by zero error
6	18	CHK instruction
7	1C	TRAPV instruction
8	20	Privilege violation
9	24	Trace
10	28	Line 1010 emulator
11	32	Line 1011 emulator
12–14	30–38	Unassigned, reserved
15	3C	Uninitialized interrupt vector
16–23	40–5F	Unassigned, reserved
24	60	Spurious interrupt
25	64	Level 1 interrupt autovector
26	68	Level 2 interrupt autovector
27	6C	Level 3 interrupt autovector

TABLE 9-1 (continued)

Exception Vector Assignments.

28	70	Level 4 interrupt autovector
29	74	Level 5 interrupt autovector
30	78	Level 6 interrupt autovector
31	7C	Level 7 interrupt autovector ← *highest*
32–47	80–BF	TRAP #n instruction vectors
48–63	C0–FF	Unassigned, reserved
64–255	100–3FF	User interrupt vectors

To maintain order during what could be a large number of interrupts, occurring almost simultaneously, an *interrupt priority* feature is used. The 68000 has three interrupt input pins (IPL0, IPL1, and IPL2). They represent the priority level from 1 to 7. A 0 value indicates that no interrupt is in progress. To establish a priority, bits 8 to 10 in the Status Register contains what is called the *interrupt mask.* Containing binary values 000 to 111, they represent seven levels of interrupts. If the mask is 010 (= 2 decimal), interrupt requests of 2 and below are masked out (ignored), and levels 3 to 7 would be processed immediately, suspending a level 2 interrupt if one was in progress. Level 7 is the highest level, and it is always processed, even if the interrupt mask level is set to 7. It is *non-maskable.*

Care must be taken in the setting of the interrupt mask in the Status Register. Remember, interrupts, or interrupt service routines as the program segments are called, can occur at any time during program execution. The mask is normally set with a pair of instructions. The first one disables any possible interrupts (except the nonmaskable), so the next instruction can set the mask to the desired level. The instructions would be.

```
OR.W   #0000011100000000B,SR   (or OR.W   #$0700,SR)   to disable interrupts)
EOR.W  #0000010000000000B,SR   (or OR.W   #$0400,SR)   to set mask to 3)
```

Note that we use the exclusive OR instruction to clear some of the bits to get the desired mask. For the instructions above, the mask goes from 111 to 011, setting a level 3 mask.

The interrupt mask initialization as above is placed in the appropriate portion of the program. It does not alter program execution but simply enables possible interrupts having a level below the mask level. Program execution continues normally until an interrupt is encounterd. Following is the interrupt processing sequence:

1. Upon receipt of an interrupt, the 68000 first completes the current instruction.
2. The current contents of the PC and Status Register (system and user bytes) are pushed onto the SUPERVISOR'S stack.
3. The T bit in the Status Register is turned off and the S bit is turned on. This turns off the trace feature and puts the processor in the supervisory mode. For external interrupts, the interrupt mask is also updated, set to the level of the impending interrupt.
4. For a bus error or addressing error exception, some more information is pushed onto the stack.
5. The PC is loaded from the appropriate vector and execution begins at that address.

For example, if VECTOR 64, which is stored in memory addresses 00000100, 00000101, 00000102, and 00000103, contained the address 00008000, and its interrupt was received, after the proper pushing of values onto the stack, the PC would be loaded with the address 00008000 and program execution would resume there. Location

00000800 contains the start of a program that handles the interrupt. At its end is a RTE instruction to inform the processor to return from the interrupt and to restore its initial PC value for normal program resumption.

RESET is an important *exception* or interrupt. It occurs during power-up or during pressing of the system reset button. When reset occurs, the contents of interrupt VECTOR 0 (memory locations 00000000 to 00000003) are loaded into the Supervisor Stack Pointer Register **and** the contents of interrupt VECTOR 1 (locations 00000004 to 00000007) are loaded into the Program Counter. **Program execution starts at the location that is stored in memory locations 00000004 to 00000007 upon reset.**

Incorporation of interrupt routines into programs depends highly on the hardware configuration of the computer system. Following is a brief discussion on how experimentation could be done on the Motorola ECB.

The *ABORT* button on the ECB is a level 7 (nonmaskable) interrupt and is VECTOR 31. Referring to the previous table, we see that its vector location is the longword memory address of $0000007C. Let's create a short program that loops, awaiting an interrupt, and when the ABORT button is pressed, the contents of an input port location is transferred to an output port location.

The reset button on the ECB can still be used to regain control, resetting the computer to an initial boot-up condition.

EXAMPLE 9.0

```
; Example 9.0
; *****************************************************************
; For the Motorola ECB computer
; Loop continuously, awaiting an interrupt (the abort button)
; Upon interrupt, the contents of an input port location is
; transferred to an output port location, and normal program
; execution is resumed
;
        .ORG   $1000
        .COMMAND  +AM1              ; Sets proper hex format for .OBJ file
        .EQU   OUTPUT,$10011        ; The ECB output port address
        .EQU   INPUT,$10013         ; The ECB input port address
        LEA.L  $7FFE,A7             ; Set stack pointer
        MOVE.B  #$80,$1000D         ; Enables I/O on the 68230 PIT IC
        MOVE.L  #INTRUP,$07C        ; Set the interrupt vector
; Note:   The interrupt MASK does not have to be set since
;         we are using the highest level (7).
; ----------------------------------------------------------------
        LOOP:  NOP                  ; Beginning of looping program
               BRA   LOOP           ; continue looping, awaiting an
                                    ; interrupt
;
; Interrupt Service Routine
;
INTRUP:  MOVE.B  INPUT, OUTPUT      ; Transfer the input contents to the
                                    ; output
         RTE                        ; Return to looping program
         .END
```

Execution of the program is straightforward. The input switch conditions are transferred to the output LEDs each time the ABORT button is pressed. For more sophisticated applications, the first two instructions of the interrupt service routine would be as shown previously, setting the interrupt mask to the level needed to prevent additional interrupts. It should be noted that interrupts can occur during an interrupt service routine if each of the interrupt routines take care in the preservation of the data and address registers that are in use.

A typical *shell* for interrupt routines with a level of less than 7:

```
OR.W    #$0700,SR           ; Disable all interrupts temporarily
MOVEM.L  D0-D7/A0-A7,(A7)+  ; Save register contents
                            ;     on the stack
EOR.W   #$xxxx,SR           ; Set the mask to the highest level of
                            ; interrupts that are expected during
                            ; this interrupt routine
NOP                         ; Continue interrupt routine
                            ; etc.
OR.W    #$0700,SR           ; Disable all interrupts temporarily
MOVEM   -(A7),D0-D7/A0-A7   ; Restore registers
EOR.W   #$xxxx,SR           ; Set the mask to the highest level of
                            ; interrupts that are expected after
                            ; this interrupt routine
RTE                         ; Exit interrupt routine
```

Note that there is no need to save all of the registers as was done above. Modify the two instructions, saving only the registers used during the interrupt routine itself.

A typical shell for a nonmaskable interrupt service routine that may occur in the middle of another interrupt routine:

```
MOVEM.L  D0-D7/A0-A7,(A7)+ ; Save register contents
                           ;     on the stack
NOP                        ; Continue interrupt routine
                           ; etc.
MOVEM   -(A7),D0-D7/A0-A7  ; Restore registers
RTE                        ; Exit interrupt routine
```

Note that we save all of the registers in this case since we have no idea of what registers were already in use prior to the nonmaskable interrupt. The interrupt routine can use any/all of the registers, and then they are restored prior to resumption of the normal portion of the program.

Users of the Motorola ECB will find a series of TRAP 14 interrupt service routines that can be used to perform a variety of functions. They provide input characters from the keyboard, provide display of text on the screen, provide hex to ASCII conversion, and more. Relatively simple to use, we look at them as a series of built-in subroutines available for our use.

In conclusion, creation of interrupt routines and their debugging is the most challenging aspect of computer programming. Additional applications include use of the serial or parallel ports, where it is desired to capture incoming data that may occur at any time.

chapter 10

implementation of text examples

To emphasize the needs for hands-on learning, provisions have been made to allow execution of the text examples via several means. The main thrust of this book is toward use of the Motorola Educational Computer Board (ECB), the PseudoSam student assembler and simulator, and the Micro Board Designs MAX 68000. Several other hardware and software products are available for use with MS-DOS personal computers.

Using a MS-DOS Computer

Two basic assumptions are made in Sections 10.1 and 10.3 describing the application of the text examples on the ECB and MAX 68000. First, a basic understanding of the MS-DOS operating system normally used on personal computers is assumed. Second, assumed is familiarity with some type of ASCII text editor. The examples will normally be created with an editor and then uploaded to the ECB or MAX 68000 for execution, debugging, and so on.

These two assumptions are big ones for those with no or little previous exposure to computers! Neophytes (and their instructors) should exploit the vast resource of software known as shareware. Commercial software is of course available but should be evaluated by the more seasoned computerists before introduction to beginners. Available through shareware channels are numerous DOS tutorial disks, program aids that smooth operation of a computer, and useful utility programs such as the cross-assembler and simulator used with this book. The software is very economical and allows temporary evaluation prior to its actual purchase. Contributions and/or registration of shareware is expected for those software packages found to be useful.

Disks can be set up with automatic and simple menu selection programs, allowing even the introductory student fairly smooth operation. The most popular shareware disk menuing program is called AutoMenu. A disk presently used by the author is simply inserted, a text editor is selected from the menu allowing creation of a source file, the assembler is selected to assemble the program, and then selection is made to either run the program on the software simulator or upload it to the ECB or to the MAX 68000. To conserve disk space on a typical 5¼-inch 360K disk, an older version of DOS such as version 2.11 is used.

A copy of this disk and the accompanying documentation and registration information for Automenu is available from the author if provided with (1) a formatted system disk and a formatted data disk, (2) a disk mailer suitable for return of the disks, and (3) return postage (typically, 3 ounces). Acceptable media: 5¼-inch 360K- or 1.2M-bytes, 3½-inch 720K- or 1.44M-byte disks.

Figure 10-1 shows the typical way of using a personal computer connected to a 68000 development system.

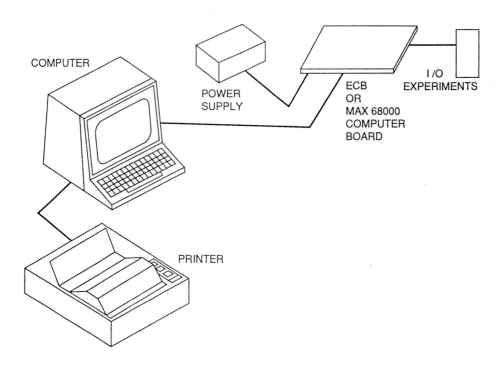

COMPUTER

POWER
SUPPLY

ECB
OR
MAX 68000
COMPUTER
BOARD

I/O
EXPERIMENTS

PRINTER

Alternative Systems

The text can also be used with other non-MS-DOS computers. The **Apple Macintosh, Atari ST, Sinclair QL, Commodore Amiga,** and **NeXT** computers are all 680x0-based computers. Commercial and shareware assemblers are available, allowing the text examples to be run. Depending on the hardware I/O configurations, the industrial control examples may or may not be adaptable. The Motorola ECB and the Micro Board Designs MAX 68000 both are operated via a serial RS-232 link and can be utilized with ANY computer having a RS-232 interface. They both have very basic assemblers, capable of running the text example, with minor modifications. Full enjoyment and learning will not be realized unless the examples are actually run on some hardware or software system.

10.1 RUNNING THE TEXT EXAMPLES ON THE MOTOROLA 68000 EDUCATIONAL COMPUTER BOARD

This educational or *evaluation* computer was the first introduced after introduction of the MC68000 in 1979. It is a simplified development system that is intended for use by students and designers to develop hardware applications and assembly language programs for the 68000. In this section we provide a very brief summary of the ECB's characteristics to assist in purchase selection. For information on the ECB, contact Fritz Wilson, Motorola Technical Training, P.O. Box 21007, Phoenix, AZ 85036, (602) 994-6900.

Subsequent sections will have the steps needed for implementation of the examples. Minor modifications will be needed in some cases, and minor enhancements will be outlined in some. The ECB manual must be referred to for help with its commands.

The ECB is a stand-alone 68000 system that requires (1) a power supply providing +5, +12, and −12 V, and (2) a terminal or computer having a RS-232 interface. It was introduced when computers were far more expensive than a terminal and has provision for loading/storing programs using a cassette tape interface and for printing programs with a Centronics-compatible parallel printer interface. **In this book we discuss use of the ECB with a MS-DOS computer only.** Neither the tape nor the printer interface will be used for their intended functions.

Programs can/will be stored on the MS-DOS computer and up/downloaded to the ECB. Since the printer interface will not be connected to a printer, it will prove very useful for numerous I/O experiments. Files will be printed with the printer connected to the MS-DOS computer.

The ECB's features include:

1. A MC68000 MPU running at 4 MHz
2. 32K bytes of dynamic RAM, arranged as 16K by 16 bits
3. 16K bytes of firmware ROM/EPROM monitor
4. Two serial RS-232 interfaces (one is connected to the MS-DOS computer)
5. An 16-bit parallel port, useful as a printer interface or for I/O
6. A cassette tape serial I/O interface
7. Self-contained firmware providing a monitor, debugger, disassembler, and assembler functions
8. A 24-bit programmable timer interface circuit
9. A wire-wrap area for custom circuitry
10. Reset and abort function switches
11. Connection points for external system expansion

The TUTOR firmware, which is called a monitor program, allows use of the ECB simply with a terminal if desired. TUTOR acts as an operating system and provides the basic functions. Normal operation is to:

1. Upload the HEX code representing the program into the ECB's memory.
2. Set the Program Counter (PC) to the first instruction of the program.
3. If problems are anticipated with the program, the instructions are traced or executed one at a time. The 68000's registers can be examined as the program is single-stepped.
4. If sections of the program appear to be completely functional, **breakpoints** can be set to allow rapid execution through the functional parts of the program up to the breakpoint, then single-stepped.
5. The program can be executed at full speed with no breakpoints, if desired. In the event of problems, return to the TUTOR monitor can be accomplished by hitting the abort button.

10.1.1 General Procedures for Getting Started with the Examples

The ECB has 16K RAM between $0900 and $7FFF. For convenience we will start our programs at $1000, and occasionally use the space between $0900 and $09FF for temporary storage. The 68230 PIT I/O IC is located between $10011 (output) and $10013 (input). Examples will show its use. Refer to the ECB manual for details.

Due to differences in the memory map and the random selection of addresses used in the text, minor modifications will have to be made to the programs before use with the ECB. The examples described below will assume that the following steps were completed initially.

1. Create an *ASseMbly source file* named EX-x-x.**ASM** with an ASCII file editor (i.e., EX-6-1.**ASM** for the first example).
    (a) Include the assembler directive **.COMMAND +AM2** on the first line of each program. This provides the proper hex format for the ECB.
    (b) Include, on the next line, the assembler directive **.ORG $1000** which will be the address of the first instruction.
    (c) Include the assembler directive **.END** for the last line of your source file.

2. Assemble the source file to create both **listing (.LST)** and hex **object (.OBJ)** files. This is done by entering **A68K EX-x-x.ASM.** If no errors are reported, proceed. Ohterwise, view the EX-x-x.**LST** file (from DOS, TYPE EX-x-x.LST) or print the EX-x-x.**LST** file (from DOS, COPY EX-x-x.LST PRN) to determine the errors. Refer to Section 10.2 or to the assembler documentation (on the disk) for assistance.

    Note: **Two important rules for the creation of a source program:**

    (a) The **first** column of each line is reserved for either a ; to signify a comment line or the first letter of a **label.** Skip over the first column, otherwise.

    (b) Spaces can be used for reading clarity, except between the two operands, which must be separated only by a **comma.**

3. Obtain a hard copy of the LISTING file by entering COPY EX-x-x.**LST** PRN, **with a printer attached to the MS-DOS computer.**

4. Connect and turn on the ECB.

5. Establish communication with the ECB, utilizing a terminal program. Hitting the red abort button on the ECB should display (on your MS-DOS computer) a formatted display of the registers; while hitting the black reset button displays the TUTOR prompt. If readable displays are not obtained, check the connections and baud rate (speed) of the ECB and the terminal program being used. Refer to the ECB manual for details.

6. Upload, using the ASCII file transfer mode, the **OBJ** *object file* from the MS-DOS computer to the ECB (i.e., EX-x-x.**OBJ**). It is imperative that the hex file, with the **.OBJ extension,** is the uploaded file. File transfer is initiated by first entering **LO** while communicating with the ECB, then enter the proper sequence for the terminal program for an ASCII upload. Upon completion of the upload, hit the enter key to get the ECB monitor prompt.

7. To check for a successful upload, you can enter **MD 1000 xx;DI,** where xx is the number of bytes (hex) that you wish to view. The monitor will disassemble and display your program.

8. To get a list of commands available, enter **HE.**

9. Set the program counter by entering **.PC 1000.**

10.1.2 Running Example 6.1

Our program:

```
          .command   +am2
          .org   $1000
   ;
   ; Example 6.1
   ; Move a byte from one memory location to another (0900 to 0910)
   ;
          MOVE.B   $0900,$0910
          .end
```

To see the effects of this program, enter the following:

```
   MM   0900
   AB          Deposits AB in location 0900
   12          Deposits 12 in location 0901
   34          Deposits 34 in location 0902
   56          etc. (not required, but used to
   78          show how other locations are not affected)
```

.		To exit from memory modify mode
MD	0900	Display new contents (note contents of 0910)
DF		Shows registers, check PC for correct value
T		Trace (single-step) through the first instruction
.		To exit from trace mode
MD	0900	Memory display, shows data source unaffected and
		shows new data in destination (0910)
DF		Note that PC is at $100A, past the end of our program

Be sure to reset the PC to 1000 before running the program again.

To experiment with the ECB's TUTOR built-in assembler, let's modify the program to move the longword starting at 0900 (12345678) to location 0910. Enter:

MM 1000;DI	Memory modify in assembler mode
space bar then	Must skip column 1 since no label!
MOVE.L $0900,$0910	
.	Exit from memory modify
MM 0910	Clean out destination, to see effects of move
FF	
FF	
.	Exit from memory modify
MD 0900	Memory display, check to see if 12345678 is there
.	Exit memory display
.PC 1000	Set PC
T	Execute the instruction
MD 0910	The destination should contain 12345678 longword

10.1.3   Running Example 6.6

Our program:

```
              .command   +am2
              .org   $1000
          ;
          ; Example 6.6
          ; Add two bytes of memory (locations 0300 and 0902),
          ; leaving the sum in a Data Register
          ;
              MOVE.B   $0900,D3
              ADD.B    $0902,D3
              .end
```

To see the effects of this program, enter the following:

MM	0900	
AB		Deposits AB in location 0900
12		Deposits 12 in location 0901 (not used or affected)
34		Deposits 34 in location 0902
.		To exit from memory modify
MD	0900	Memory display to see new contents
DF		Check PC for correct value (note D3 contents)
T		Trace (single-step) through the first instruction
ENTER key		Trace the next instruction
.		To exit from trace mode
MD	0900	Memory display, shows data sources unaffected
DF		Note the sum AB + 34 = DF in D3

10.1.4 Running Example 6.10

Our program:

```
        .command   +am2
        .org  $1000
;
; Example 6.10
; Subtract the contents of memory location (0902) from
; another (0900), leave result in another location (0904)
;
        MOVE.W  $0900,D0   ; get the salary data
        SUB.W   $0902,D0   ; subtract the tax from it
        MOVE.W  D0,$0904   ; store the result
        .end
```

To see the effects of this program, enter the following:

MM	**0900**	
56		Deposits 56 in location 0900
78		Deposits 78 in location 0901
12		Deposits 12 in location 0902
34		Deposits 34 in location 0903
.		To exit from memory modify
MD	**0900**	Memory display to see new contents
DF		Check PC for correct value
T		Trace (single-step) through the first instruction
ENTER	**key**	Trace next instruction
ENTER	**key**	Trace next instruction
.		To exit from trace mode
MD	**0900**	Memory display, shows data sources unaffected; note results in 0904 (5678 − 1234 = 4444)
DF		Note that PC is at $1012, past the end of the program; note that the result is also in D0

10.1.5 Running Example 6.14

Our program:

```
        .command   +am2
        .org  $1000
;
; Example 6.14
; Sum the byte contents of five memory locations (starting at 0900),
; store sum in another memory address (0910)
;
        MOVEQ.L  #0,D0     ; clear out the data register
        ADD.B   $0900,D0   ; add the first number to D0
        ADD.B   $0901,D0   ; add the next number
        ADD.B   $0902,D0   ; add the next number
        ADD.B   $0903,D0   ; add the next number
        ADD.B   $0904,D0   ; add the last number
        MOVE.B  D0,$0910   ; store the sum
        .end
```

To see the effects of this program, enter the following:

```
MM    0900
12                 Deposits 12 in location 0900
34                 Deposits 34 in location 0901
56                 Deposits 56 in location 0902
78                 Deposits 78 in location 0903
90                 Deposits 90 in location 0904
.                  To exit from memory modify
MD    0900         Memory display to see new contents
DF                 Check PC for correct value
T                  Trace (single-step) through the first instruction
ENTER  key         Six more times to complete program
.                  To exit from trace mode
MD    0900         Memory display, shows data sources unaffected;
                   note sum (= A4) in location 0910. The actual sum is 1A4,
                   the excess beyond FF is lost, since a byte-sized
                   operation was done.
DF                 Note that PC is at $1026, past the end of the program;
                   note that the sum is also in D0
```

10.1.6   Running Example 6.16

Our program:

```
                .command   +am2
                .org   $1000
        ;
        ; Example 6.16 (final version of Example 6.15)
        ; Sum the byte contents of 256 memory locations (starting at 0900),
        ; store the result in another memory location (1100)
        ; Exit (or loop) correctly when done
        ;
                MOVEQ.L  #$0,D0      ; clear out the accumulator
                MOVE.W   #$FF,D1     ; initialize the countdown register
                MOVEA.L  #$0900,A0   ; point to the source of data
        LOOP:   ADD.B    (A0)+,D0    ; add to D0 the data pointed to,
                                     ; bump pointer
                DBF   D1,LOOP        ; get the next data to add
                MOVE.B   D0,$1100    ; move the sum to the output address
        HERE:   JMP   HERE           ; a way of stopping our program
                .end
```

To see the effects of this program, enter the following:

```
MM    0900
12                 Deposits 12 in location 0900
34                 Deposits 34 in location 0901
56                 Deposits 56 in location 0902;
                   continue modifying up to 256 locations, if desired
.                  To exit from memory modify
MD    0900         Memory display to see new contents
MD    1100         Memory display of sum storage location before addition
DF                 Check PC for correct value
```

GO		Runs the program nonstop;
		press the red abort button to stop
MD	**0900**	Memory display, shows data sources unaffected
MD	**1100**	Memory display of sum
DF		Note that PC is at $1018, at the end of our program;
		note that the sum (which will depend on the contents of all
		256 locations) is in D0

10.1.7   Running Example 6.19

Our program:

```
                .command  +am2
                .org   $1000
        ;
        ; Example 6.19
        ; Sum the BCD contents (two digits/byte) of five memory locations
        ; (starting at 0900),
        ;    store a result that can be between 0 and 9999₁₀ in two
        ; memory locations (0910 and 0911)
          ; in BCD format
        ;
                MOVE.W   #4,CCR         ; clear the X bit, set the Z bit
                MOVEQ.L  #0,D0          ; clear out the sum storage
                MOVEQ.L  #4,D4          ; initialize the countdown
                                        ; counter to one LESS
                MOVEQ.L  #0,D3          ; clean out the carry sum
                MOVEQ.L  #0,D2          ; to be used as the simple
                                        ; carry incrementer
                MOVEA.L  #$0900,A0      ; point to the beginning of the input data
LOOP:           MOVE.B   (A0)+,D1       ; get the first number to be added,
                                        ; bump pointer
                ABCD.B   D1,D0          ; add to the sum
                BCC   NOPE              ; take care of a sum > 99
                                        ; otherwise proceed
                ABCD.B   D2,D3          ; add one to the carry sum
NOPE:           DBF  D4,LOOP            ; keep looping until done
                MOVE.B   D3,$0910       ; store the upper two BCD digits
                MOVE.B   D0,$0911       ; store the lower two BCD digits
HERE:           JMP  HERE               ; loop when done
                .end
```

To see the effects of this program, enter the following:

MM	**0900**	
12		Deposits 12 in location 0900
34		Deposits 34 in location 0901
56		Deposits 56 in location 0902
78		Deposits 78 in location 0903
90		Deposits 90 in location 0904
.		To exit from memory modify
MD	**0900**	Memory display to see new contents
DF		Check PC for correct value
GO		Runs the program nonstop;
		press the RED abort button to stop
MD	**0910**	Memory display, shows data sources unaffected
		and BCD sum (= 0270) in locations 0910 and 0911
DF		Note that PC is at the end of the program;
		note that the sum is in D3 and D0

10.1.8 Running Example 6.20

Our program:

```
                    ; Example 6.20
                    ; Add some six-digit BCD data stored in memory
                    ; (starting at 0900), store sum in other
                    ; memory locations (09F0–09F3) as an eight-digit BCD number
                    ;
                    ; $0900–0903 contains the first six-digit BCD number (leading 0's)
                    ; $0904–0907              second
                    ; etc.
                    ; $09F0–09F3 contains the possible eight-digit SUM
                    ; $09F4–09F7 contains the NUMBER of scores to be added
                    ;
                    .command  +am2
                    .ORG  $0900
START:              .RS  256                ; reserve space for the numbers
                    .ORG  $09F0
SUM:                .RS  4                  ; reserve space for the SUM
NUM:                .DL  0004               ; the number of numbers is stored here
;
                    .ORG  $1000             ; the beginning of the program
;
                    CLR.L  D7               ; clear out the number counter
                    MOVE.L  NUM,D7          ; get the number of numbers to add
                    MULU.W  #4,D7           ; multiply by 4 since each
                                            ; number is 4 bytes long
                    ADD.L  #START,D7        ; D7 now points to the address
                                            ; PAST the last number
                    MOVEA.L  D7,A0          ; put the pointer in an Address Register
                    MOVE.W  #4,CCR          ; clear the X bit, set the Z bit,
                                            ; for BCD arithmetic
                    LEA.L  SUM+4,A1         ; point to one past the sum
                                            ; storage location
                    CLR.L  SUM              ; clear out the longword sum
                                            ; storage location
                    MOVE.L  NUM,D2          ; initialize the countdown counter
                    SUB.L  #1,D2            ; correct for the DBF quirk, so
                                            ; countdown will be correct
LOOP:               ABCD.B  -(A0),-(A1)     ; decrement pointers, add the
                                            ; least significant digits
                    ABCD.B  -(A0),-(A1)     ; repeat to add the middle two digits
                    ABCD.B  -(A0),-(A1)     ; repeat to add the most significant digits
                    ABCD.B  -(A0),-(A1)     ; all done, take care of potential carry
                    LEA.L  SUM+4,A1         ; reset the sum pointer
                    DBF  D2,LOOP            ; keep looping until all numbers are added
DONE:               BRA  DONE
                    .END
```

To see the effects of this program, enter the following:

MM	0900	
00		
01		
23		
45		Deposits BCD number 00012345 in locations 0900–0903
00		
67		
89		
90		Deposits BCD number 00678990
00		
12		
34		
56		Deposits BCD number 00123456
00		
78		
90		
10		Deposits BCD number 00789010
.		To exit from memory modify
MD	0900	Memory display to see new contents
DF		Check PC for correct value
GO		Runs the program nonstop; press the RED abort button to stop
MD	0900	Memory display, shows data sources unaffected
MD	09F0	Displays BCD sum (= 1,603,801); note the number of numbers (= 4) that are stored in 21F04–21F07
RD		Note that PC is at the end of the program

If you wish to run the program with more or fewer numbers, change the value stored in 09F4–09F7 to reflect the number of numbers. Reset the PC to 1000, then enter GO. The six-digit BCD numbers are actually set up as eight digits, with two leading 0's.

10.1.9 Running Example 6.22

Our program:

```
; Example 6.22
;
; Sort an array of numbers (bytes) into ascending order
;
              .ORG   $0900
START:        .RS  16                  ; reserve space for numbers
NUM:          .DW  16                  ; the number of numbers in the array
;
              .command   +am2
              .ORG   $1000
;
; Pre-load an array with descending numbers
              LEA.L   START,A0         ; point to beginning of array
              MOVE.B  #$FF,D0          ; start filling array with FF
              MOVE.W  NUM,D1           ; load the countdown counter
              SUBQ.W  #1,D1            ; take care of DBF quirk
LOOP:         MOVE.B  D0,(A0)+         ; store the number
              SUBQ.B  #1,D0            ; decrease it by one
              DBF.W   D1,LOOP          ; loop until done
```

```
        ; Beginning of bubble sort
AGAIN:      CLR.L   D7            ; clear the swap flag
            LEA.L   START,A0      ; point to BEGINNING of array
            LEA.L   NUM-1,A2      ; point to END of array
READ1:      MOVE.B  (A2),D0       ; get pointed to
                                  ; (first number) to compare
            SUBQ.L  #1,A2         ; decrement pointer
            CMPA.L  A2,A0         ; pointing past beginning of array?
            BHI  CHECK            ; if so, see if swap flag is set
            MOVE.B  (A2),D2       ; if not, get pointed to
                                  ; (second number) to compare
            CMP.B   D2,D0         ; compare the numbers,
                                  ; is the first > second?
            BHI  READ1            ; if so, don't swap, go
                                  ; read another number to compare
            MOVE.B  D2,1(A2)      ; if not, swap the numbers
            MOVE.B  D0,(A2)       ; swap the numbers
            MOVEQ.L #1,D7         ; set swap flag
            BRA  READ1            ; read another number to compare
CHECK:      CMPI.L  #1,D7         ; is swap flag set?
            BEQ  AGAIN            ; if so, do sort again
HERE:       BRA  HERE            ; if not, sorting is complete
            .END
```

To see the effects of this program, enter the following:

```
BR   101A      Sets a breakpoint (temporary stop)
DF             Check PC for correct value
GO             Run the program down to the breakpoint
MD   0900      Memory display to see descending array
NOBR           Remove the breakpoint
GO             Runs the program nonstop to the end;
               press the RED abort button to stop
MD   0900      Memory display to show sorted ascending array
```

If problems are encountered debugging a program, including times when a break-point is used, it may be necessary to hit the BLACK reset button to restore the program properly. In some cases, the program will have to be reloaded into the 68000.

10.1.10   Running Example 7.1

Our program:

```
; Example 7.1 - Assume that bit 0 of memory location $10011 is attached to
; a LED [connect a LED between pins 39 and 40 (GND) of connector J1]
; Create a program that blinks the LED approximately once per second
; Assume a computer clock speed of 4 MHz
;
; Blink LED connected to bit 0 of location $10011 at a rate of one per
; second
; Also beeps once a second
;
            .command  +am2
            .ORG  $1000
            .EQU  OUTPUT,$10011
            ; assign an address to a label
```

```
                    ; added instruction needed to initialize the 68230 PIT for actual output
                              MOVE.B   #$80,$1000D
                    ;
        TOP:          MOVE.B   #0,OUTPUT         ; turn LED OFF
                      MOVE.W   #$FFFF,D0        ; initialize countdown counter to
                                                ; 65536
        OFFLOOP:      NOP                        ; do nothing, burns up some time
                      NOP
                      NOP
                      NOP
                      NOP                        ;
                      NOP                        ;
                      DBF   D0,OFFLOOP          ; decrement counter, burn up time
                      MOVE.B   1,OUTPUT          ; turn LED on
                      MOVE.W   #$FFFF,D0         ; reinitialize countdown counter
        ONLOOP:       NOP
                      NOP
                      NOP
                      NOP                        ;
                      NOP                        ;
                      NOP                        ;
                      DBF   D0,ONLOOP           ; decrement counter, burn up time
                      JSR   MAKBEEP
                      BRA   TOP                  ; keep looping indefinitely
        MAKBEEP:      MOVE.B   #$07,D0           ; ASCII char for a beep
                      MOVE.B   #248,D7           ; output 1 character
                      TRAP   #14                 ; to computer (terminal)
                      RTS
                      .END
```

To see the effects of this program, enter the following:

```
        DF              Check PC for correct value
        GO              Run the program continuously;
                        the LED should blink about once per second and
                        the PC speaker should beep
```

10.1.11   Running Example 7.3

Our Program:

```
        ; Example 7.3
        ; Turn ON LED-3 when switch-4 is CLOSED
        ; Turn ON LED-7 when switch-1 AND switch-2 are CLOSED
        ; Toggle LED-2 for each monentary closing of switch-0
        ;
                      .command   +am2
                      .ORG   $1000
                      .EQU   INPUT,$10013      ;assign an address to a label
                      .EQU   OUTPUT,$10011     ;assign an address to a label
        ; If desired, connect a SPST switch between pins 23 and 24,
        ;          a SPST                between pins 21 and 22,
        ;          a SPST                between pins 19 and 20,
        ;          a LED                 between pins 39 and 40,
        ;          a LED                 between pins 37 and 38,
        ;          a LED                 between pins 35 and 36 on J1
        ;
```

```
          ; instruction added to initialize the 68230 PIT for actual input and output
                        MOVE.B   #$80,$1000D
          ;
                        MOVE.B   #0,OUTPUT        ; turn off all LEDs
          TOP:          BTST.B   #4,INPUT         ; test bit 4 of the input location
                        BEQ  CLOSED               ; branch if Z bit is set
                                                  ; (switch closed)
          OPEN:         BCLR  #3,OUTPUT           ; turn LED-3 OFF (switch is open)
                        BRA  TOP2                 ; check next situation
          CLOSED:       BSET  #3,OUTPUT           ; turn LED-3 ON (switch is closed)
          ; Test second situation
          TOP2:         MOVE.B   INPUT,D0         ; get the switch conditions
                        AND.B   #00000110B,D0     ; mask off unwanted bits
                        BEQ  CLOSED2              ; branch if Z bit set
                                                  ; (if BOTH are closed)
          OPEN2:        BCLR  #7,OUTPUT           ; turn LED off
                                                  ; since BOTH are not closed
                        BRA  TOP3                 ; check next situation
          CLOSED2:      BSET  #7,OUTPUT           ; turn LED ON
                                                  ; (both switches are closed)
          ; Test third situation
          ;                                       Switch 0 must make its changes between
          TOP3:         MOVE.B   INPUT,D0; get the switch conditions _____here___
                        AND.B   #01,D0            ; mask out unwanted bits
                                                  ; (= 00000001)              
                        MOVE.B   D0,D1            ; save the switch condition
                        MOVE.B   INPUT,D0         ; read the switches      ___here___
                        AND.B   #01,D0            ; mask out unwanted bits
                        CMP.B   D0,D1             ; compare now with previous
                        BEQ  NOCHG                ; skip over since
                                                  ; there was no change
          CHANGE:       BCHG  #2,OUTPUT           ; change LED-2
          NOCHG:        BRA  TOP                  ; loop back to top of program
                        .END
```

To see the effects of this program, enter the following:

> **DF**   Check PC for correct value
> **GO**   Run the program continuously;
>   activate the switches and check for proper
>   LED activity

If problems are encountered, set the PC to 1000, then T (trace) through the program an instruction at a time. Check for proper instruction branching, remembering that a closed switch (connected to ground) is a logic 0, and a BEQ instruction will occur when the switch is closed.

10.1.12   Running Example 7.5

Our program:

```
          ; Example 7.5
          ; Turn a buzzer ON if the stall indicator is ON, the gear is DOWN,
          ; and the low-rpm indicator is ON
          ; or if
          ; The gear is UP AND the high-airspeed indicator is ON
          ;
```

```
                .command  +am2
                .ORG   $1000
                .EQU   INPUT,$100013        ; assign an address to a label
                .EQU   OUTPUT,$10011        ; assign an address to a label
; If desired, connect switches and LED as indicated in the text
; Connect the switches to port A and the LED port B
;
; added instructions to make 68230 functional
;
                MOVE.B  #$80,$1000D
;
                BCLR.B  #7,OUTPUT            ; turn the buzzer off initially
                MOVE.B  #00000110B,D1        ; situation 2 mask
TOP             MOVE.B  #00001011B,D0        ; situation 1 mask
                MOVE.B  INPUT,D2             ; get the input condition
                AND.B   D0,D2                ; mask out undesired bits
                BCHG    #1,D0                ; invert the gear-down bit
                                             ; to reflect desired test
                CMP.B   D0,D2                ; look for a match
                BEQ   BUZZ                   ; turn on buzzer for situation 1
                MOVE.B  NPUT,D2              ; read switches again
                AND.B   D1,D2                ; mask out undesired bits
                CMP.B   D1,D2                ; look for a match
                BEQ   BUZZ                   ; turn on buzzer for situation 2
                BCLR.B  #7,OUTPUT            ; turn off buzzer
                                             ; since no situation
                BRA   TOP                    ; keep looping
BUZZ            BSET    #7,OUTPUT            ; turn on buzzer
                BRA   TOP                    ; keep looping
                .END
```

To see the effects of this program, enter the following:

DF   Check PC for correct value
GO   Run the program continuously;
         activate the switches and check for proper
         LED activity

If problems are encountered, set the PC to 1000, then T (trace) through the program an instruction at a time. Check for proper instruction branching, remembering that a closed switch (connected to ground) is a logic 0, and a BEQ instruction will occur when the switch is closed.

10.1.13   Running Example 7.7

Our program:

```
: Example 7.7
; Continually monitor incoming data, which can be changing,
; and provide the following information every 10 seconds:
;1) The number of locations reporting a pressure of 0
;2) The number of locations reporting a pressure of 255
;3) The addresses of the locations reporting a pressure of 255
;4) The average pressure for the 50 sensors
```

```
                ;5) The maximum pressure reported
                ;6) The minimum pressure reported
                ;
                ; A beep will occur after each data-gathering period
                ;
                        .command   +am2
        .ORG  $1000
                LEA.L   $7FFE,A7          ; Load the Stack Pointer
        START:  CLR.L   D0                ; Init the MAXIMUM to ZERO
                CLR.L   D3                ; Clear upper portion of D3
                MOVE.B  #255,D3           ; Init the MINIMUM to 255
        ;
                CLR.L   D1                ; Clear the ZERO counter
                CLR.L   D2                ; Clear the 255 counter
                CLR.L   D4                ; Clear the SUM totalizer
        ;
                CLR.L   D7                ; Init the ZERO CHECKER
                CLR.L   D6                ; Clear upper portion of D6
                MOVE.B  #255,D6           ; Init the 255 CHECKER
        ;
                LEA.L   DATA,A0           ; Point to beginning of data
                LEA.L   HIADRS,A1         ; Point to beginning
                                          ; of '255' addresses
        ; End of initialization
        ; Check for a ZERO
        LOOP:   CMP.B   (A0),D7           ; Is data = 0?
                BNE.S   OTZERO            ; Nope
                ADDQ.B  #1,D1             ; Yes, increment ZERO counter
        ; Check for a new maximum
        NOTZERO: CMP.B  (A0),D0           ; Is it greater than old maximum?
                BCC.B   OTMAX             ; Nope
                MOVE.B  (A0),D0           ; Yes, store new MAXIMUM
        ; Check for a new minimum
        NOTMAX: CMP.B   (A0),D3           ; Is it less than old minimum?
                BCS.B   NOTMIN            ; Nope
                MOVE.B  (A0),D3           ; Yes, store new MINIMUM
        ; Check for a 255
        NOTMIN: CMP.B   (A0),D6           ; Is it 255?
                BNE.S   NOT255            ; Nope
                ADDQ.B  #1,D2             ; Yes, increment 255 storage
                MOVE.L  A0,(A1)+          ; Save address of 255 pressure
        ; Add number to sum
        NOT255: ADD.B   (A0),D4           ; Not = 255, add to sum
                BCC     CONT              ;
                ADD.L   #256,D4           ; Byte sum was > 255,
                                          ; add carry to D4
        CONT:   ADDA.L  #1,A0             ; Point to next address
                CMPA.L  #DATAEND,A0       ; Done with 50 addresses?
                BNE.S   LOOP              ; Nope, loop again
        ; Find the average
                DIVU.W  #50,D4            ; Divide by 50, result in D4
        ; Store the results in memory locations (for use by display routines)
                MOVE.W  D4,AVERAGE        ; Store average
                MOVE.B  D0,MAXIMUM
                MOVE.B  D1,ZEROS
                MOVE.B  D2,TWO55
                MOVE.B  D3,MINIMUM
```

```
; Call subroutine to display results
                BSR   DISPLAY              ;
                                           ;
; Call subroutine to kill 10 seconds
                BSR   BURNTIME
EXIT:           BRA   START                ; Start cycle all over again
;
; Display routine for the ECB only
;
DISPLAY:        LEA.L  MSG1,A5             ; point to string beginning
                LEA.L  MSGE,A6             ; point to string end+1
                MOVE.B #227,D7             ; display string routine
                TRAP  #14
        ;
                MOVE.B  MINIMUM,D0         ; get minimum (hex) value
                BSR  CONVDISP              ; convert to ASCII and display
                MOVE.B  MAXIMUM,D0         ; get maximum (hex) value
                BSR  CONVDISP
                MOVE.B  ZEROS,D0           ; get number of 0's
                BSR  CONVDISP
                MOVE.B  TWO55,D0           ; get number of 255's
                BSR  CONVDISP
                MOVE.B  AVERAGE,D0         ; get average (most significant
                BSR  CONVDISP              ; byte)
                MOVE.B  AVERAGE+1,D0       ; get average (least significant
                BSR  CONV                  ; byte)
                BSR  BEEP                  ; produce a beep
                RTS
; Average is two bytes, and to display the digits without dividing blanks,
; note how the BSR CONV is used to skip down into the CONVDISP
; subroutine, skipping the blank-producing routine
;
CONVDISP:       BSR  BLANKIT               ; display a few blanks
;
CONV:           LEA.L  MI,A6               ; point to storage
                MOVE.B #233,D7             ; converts two hex digits to ASCII
                TRAP  #14                  ; results are now in MI, MI+1 loc.
        ;
                LEA.L  MI,A5
                LEA.L  MIE,A6
                MOVE.B #227,D7             ; displays value (in ASCII)
                TRAP  #14
                RTS
;
MI:             .RS  4                     ; temporary storage for the ASCII
MIE:            NOP                        ; characters
;
BEEP:           LEA.L  BEEPER,A5           ; produces a beep and a
                LEA.L  BEEPE,A6            ; carriage return/linefeed
                MOVE.B #227,D7             ; on screen
                TRAP  #14
;
BURNTIME:       MOVE.L #$0FFFF,D0          ; kills about 10 seconds
LOOP1:          NOP
                NOP
                SUB.L  #1,D0
                BNE  LOOP1
                RTS
;
```

```
MSG1:       .DB   "        Pressure Monitoring System",$0A,$0D
            .DB   "MIN   MAX  #0's  #255's AVERAGE",$0A$0D
            .DB   "------------------------",$0A,$0D
            .ALIGN     ; used to ensure next instruction is on an even address
MSG1E:      NOP                              ; dummy instruction, used as a pointer
;
BLANK:      .DB  "                "  ; some blanks to separate the digits
            .ALIGN
BLANKE:     NOP
BEEPER:     .DB  07
            .ALIGN
BEEPE:      NOP
;
            .ORG   $0900                    ; storage locations for the results
MAXIMUM:    .RS   1
ZEROS:      .RS   1
TWO55:      .RS   1
MINIMUM:    .RS   1
AVERAGE:    .RS   2
            .ORG   $0910                    ; the starting address
                                            ; of the input data (decimal numbers)
; Note: For some reason the assembler assumes digits starting with 0 to be
;          octal numbers, the numbers 00 to 07 are the same
; (octal, decimal, hex)
;          a $ was added to 08 and 09 to make them hex (or decimal)
; Numbers greater than 10 (decimal) are properly converted to hex numbers
; for storage
DATA:       .DB   01,02,03,04,05,06,07,$08,$09,10,11,12,13,14
            .DB   15,16,17,18,19,20,21,22,23,24,25,26,27,28
            .DB   29,30,31,32,33,34,35,36,37,38,39,40,41,42
            .DB   43,44,45,255,46,255,47,255
DATAEND:    .RS   2
; End of data
; Storage area for addresses having 255's starts here
HIADRS:     .RS   200
            .END
```

To see the effects of this program, enter the following:

    DF   Check PC for correct value

    GO   Run the program continuously

    The display and sound subroutines use routines built into the ECB's monitor (EPROM). The program will scroll with the same numbers since the *input data* is not changing. To check it for different data values, modify the memory contents starting at 0910, reset the PC, and run the program again.

10.2 RUNNING THE TEXT EXAMPLES USING THE PSEUDOSAM/PSEUDOMAX ASSEMBLER/SIMULATOR

    Several commercial and public domain 68000 cross-assemblers and simulators for MS-DOS PCs are available. The ones used with this book are special student versions of a commercially available software from **PseudoCorp.** For information on their commercial software, contact them at **PseudoCorp,** 716 Thimble Shoals Boulevard, Suite E, Newport News, VA 23606, (804) 873-1947.

    The input to the **PSeudoSam** assembler is a file containing instructions, assembler directives, and comments. It is recommended that the source lines be no longer than 80 characters, as this will guarantee that the lines of the listing file do not exceed 132 characters in length. The assembler treats uppercase and lowercase identically. When the listing file is printed, it will have up to 132 characters per line. If normal (8½-inch) paper is used, the printer should be commanded into the compressed mode to allow printing of the 132 characters.

Each line of the source code consists of the following fields:

> LABEL: OPERATION   OPERAND,OPERAND,. . . ;COMMENT

The fields may be separated by any combination of spaces and tabs. Except for the comment field and quoted strings, there must be no spaces or tabs within a field.

Labels should begin in column 1, and when a label is encountered in the source code, it is defined to have a value equal to the current location (program) counter. The operation field must not begin in the column 1, because the operation would be confused with a label.

The operand field may or may not be required, depending on the instruction or directive being used. If present, the field consists of one or more comma-separated items *with no intervening spaces or tabs.* (There may be spaces or tabs within an item, but only within quoted strings.)

The comment field usually consists of everything on a source line after the operand field. A ; is normally used to signify a comment, and it may contain any characters desired. A comment may also be inserted in the source file in another way: A ; at the beginning of the line or after the label field will cause the rest of the line to be ignored (i.e., treated as a comment).

Symbols appear in the source code as labels, constants, and operands. The first character of a symbol must be either a letter (A–Z) or a period (".''). The remaining characters may be letters, dollar signs ("$''), periods (".''), or underscores ("_''). A symbol may be of any length, but only the first 16 characters are significant. Remember that capitalization is ignored, so symbols that are capitalized differently are really the same.

Only a few of the **assembler directives** will be used in this book. Refer to the documentation file on the disk for full details.

	.ORG	$xxxx	- Set origin of program to location $xxxx
	.EQU	label,$xxxx	- Equate 'label' to address $xxxx
label	.DB	$xx	- Define byte (hex) to 'label'
label	.DB	"ASCII text"	- Store ASCII text (in hex)
label	.DW	$xxxx	- Define word (hex) to 'label'
label	.DL	$xxxx	- Define longword (Hex) to "label"
	.RS	xx	- Reserve xx bytes for storage
	.END		- End of source file,
			the last line of the source program

10.2.1 General Procedures for Getting Started with the Examples

The **PSeudoMax simulator** has RAM between $0000 and $7FFF. Due to differences in the memory map, and the random selection of addresses used in the text, minor modifications will have to be made to the programs before use with the simulator. In the examples described below we assume that the following steps were completed initially.

1. Create an *assembly source file* named EX-x-x.**ASM** with an ASCII file editor (i.e., Ex-6-1.**ASM** for the first example).

    (a) Include the assembler directive **.COMMAND +AM2** on the first line of each program. This provides the proper hex format for the simulator.

(b) Include, on the next line, the assembler directive **.ORG $1000,** which will be the address of the first instruction.

(c) Include the assembler directive **.END** for the last line of the source file.

2. Assemble the source file to create both **listing (.LST)** and hex **object (.OBJ)** files. This is done by entering **A68K EX-x-x.ASM.** If no errors are reported, proceed. Otherwise, view the EX-x-x.**LST** file (from DOS, TYPE EX-x-x.LST) or print the EX-x-x.**LST** file (from DOS, COPY EX-x-x.LST PRN) to determine the errors. Refer to the introduction to this section or to the assembler documentation (on the disk) for assistance.

   Note: **Two important rules for the creation of a source program:**

   (a) The **first** column of each line is reserved for either a ; to signify a comment line, or the first letter of a **label.** Skip over the first column otherwise.

   (b) Spaces can be used for reading clarity, *except* between the two operands, which must be separated only by a **comma.**

3. Obtain a hard copy of the LISTING file by entering COPY EX-x-x.**LST** PRN, **with a printer attached to the MS-DOS computer.**

4. Run **S68K** from the DOS prompt.

There are several ways to set up a smooth-running disk with the assembler and simulator. One is to use a shareware program called AutoMenu.

Another useful shareware utility called KEY-FAKE allows you to "jam" keystrokes into the next program that is executed. To use KEY-FAKE you create a simple start-up batch file (with the editor). Call it **GO.BAT.** It would look as follows:

KEY-FAKE '"L" "O" "%1.OBJ" 13 "R"
S68K

To start it up, enter **GO filename,** where "filename" is the file of the program you wish to simulate. Leave off the .OBJ extension; the batch file adds it for you. This will automatically start up the simulator, load the program, and prepare the program for running.

If you do not use KEY-FAKE, simply enter **S68K** to load the simulator. Then enter:

1. For the first time you run the simulator, it is a good idea to enter a ! to display the memory attributes. The following screen should be displayed, after a brief delay:

```
Memory Attributes
00000000   00007FFF  +r +e +w −io −b −n −x
00008000   00FFFFFF  −r −e −w −io −b +n −x
```

2. Then enter **L** (to load the object file).
3. Then enter **O** (to denote that it is an object file).
4. Then enter the **filename.OBJ** (include the .OBJ extension).
5. Assuming that you have originated your program at $1000, you may want to enter **.MW1=1000** to make sure that the program has loaded. Your program's hex codes should be visible.
6. Press the **CTRL** and **HOME** keys to reset the simulator.
7. Enter **PC=1000** (assuming that you originated the program at $1000).
8. If you need to preload any memory locations with data, enter @ **xxxx,** where xxxx is the starting address you would like to edit. Enter the desired data, using the arrow keys to move through the window, and then hit the enter key to quit.
9. If you would like to view the destination of the data transfers, enter **.MW1=xxxx** using the desired address.
10. Press the **left arrow** to **single step** through one instruction.
11. Repeat for each step of the program.

12. Do not continue stepping past the end of the program.

If you wish to run the program at **full speed,** press the **right arrow** key and the end key to stop. Hit the **CTRL** and **end** keys to return to the main menu, then enter **Q** to quit.

10.2.2 Running Example 6.1

Our program:

```
        .command   +am2
        .org   $1000
    ;
    ; Example 6.1
    ; Move a byte from one memory location to another (7000 to 7010)
    ;
        MOVE.B   $7000,$7010
        .end
```

To see the effects of this program, enter the following:

@ **7000**	
AB	Deposits AB in location 7000
12	Deposits 12 in location 7001
34	Deposits 34 in location 7002
56	etc. (not required, but used to
78	show how other locations are not affected)
Enter Key	To exit from memory modify mode
.MW1 = 7000	Display new contents (note contents of 7010); check PC for correct value
LEFT arrow	Trace (single-step) through the first instruction; memory display shows data source unaffected and shows new data in destination (7010); note that PC is at $1000A, past the end of the program

Be sure to reset the PC to 1000 before running the program again.

10.2.3 Running Example 6.6

Our program:

```
        .command   +am2
        .org   $1000
    ;
    ; Example 6.6
    ; Add 2 bytes of memory (locations 7000 and 7002),
    ; leaving the sum in a data register
    ;
        MOVE.B   $7000,D3
        ADD.B    $7002,D3
        .end
```

To see the effects of this program, enter the following:

@ **7000**	
AB	Deposits AB in location 7000
12	Deposits 12 in location 7001 (not used or affected)
34	Deposits 34 in location 7002
Enter key	To exit from memory modify

.MW1 = 7000	Memory display to see new contents;
	check PC for correct value (note D3 contents)
Left arrow	Trace (single-step) through the first instruction
Left arrow	Trace next instructions;
	memory display shows data sources unaffected;
	note the sum AB + 34 = DF in D3

10.2.4 Running Example 6.10

Our program:

```
        .command   +am2
        .org   $1000
;
; Example
; Subtract the contents of memory location (7002) from
; another (7000), leave result in another location (7004)
;
        MOVE.W   $7000,D0    ; get the salary data
        SUB.W   $7002,D0     ; subtract the tax from it
        MOVE.W   D0,$7004    ; store the result
        .end
```

To see the effects of this program, enter the following:

@ 7000	
56	Deposits 56 in location 7000
78	Deposits 78 in location 7001
12	Deposits 12 in location 7002
34	Deposits 34 in location 7003
ENTER key	To exit from memory modify
.MW1 = 7000	Memory display to see new contents;
	check PC for correct value
Left arrow	Trace (single-step) through the first instruction
Left arrow	Trace next instruction
Left arrow	Trace next instruction;
	memory display shows data sources unaffected;
	note results in 7004 (5678 − 1234 = 4444);
	note that the result is also in D0

10.2.5 Running Example 6.14

Our program:

```
        .command   +am2
        .org   $1000
;
; Example 6.14
; Sum the byte contents of five memory locations (starting at 7000),
; store sum in another memory address (7010)
;
        MOVEQ.L   #0,D0      ; clear out the Data Register
        ADD.B   $7000,D0     ; add the first number to D0
        ADD.B   $7001,D0     ; add the next number
        ADD.B   $7002,D0     ; add the next number
        ADD.B   $7003,D0     ; add the next number
        ADD.B   $7004,D0     ; add the last number
        MOVE.B   D0,$7010    ; store the sum
        .end
```

To see the effects of this program, enter the following:

```
@ 7000
12              Deposits 12 in location 7000
34              Deposits 34 in location 7001
56              Deposits 56 in locations 7002
78              Deposits 78 in location 7003
90              Deposits 90 in location 7004
Enter key       To exit from memory modify
.MW1 = 7000     Memory display to see new contents;
                check PC for correct value
Left arrow      Trace (single-step) through the first instruction
Left arrow      SIX more times to complete program;
                memory display shows data sources unaffected;
                note sum (= A4) in location 7010. The actual sum is 1A4,
                the excess beyond FF is lost, since a byte-sized
                operation was done. Note that the sum is also in D0.
```

10.2.6   Running Example 6.16

Our program:

```
        .command   +am2
        .org   $1000
;
; Example 6.16 (final version of Example 6.15)
; Sum the byte contents of 256 memory locations (starting at 7000),
; store the result in another memory location (7100)
; Exit (or loop) correctly when done
;
        MOVEQ.L  #$0,D0       .; clear out the accumulator
        MOVE.W   #$FF,D1      ; initialize the countdown register
        MOVEA.L  #$7000,A0    ; point to the source of data
LOOP:   ADD.B  (A0)+,D0       ; to D0 add the data pointed to,
                              ; bump pointer
        DBF  D1,LOOP          ; get the next data to add
        MOVE.B  D0,$7100      ; move the sum to the output address
HERE:   JMP  HERE             ; a way of stopping the program
        .end
```

To see the effects of this program, enter the following:

```
@ 7000
12              Deposits 12 in location 7000
34              Deposits 34 in location 7001
56              Deposits 56 in location 7002;
                continue modifying up to 256 locations, if desired
Enter key       To exit from memory modify
.MW1 = 7000     Memory display to see new contents
.MW1 = 7100     Memory display of sum storage location before
                addition;
                check PC for correct value
RIGHT arrow     Runs the program nonstop
                It will take a while to complete, and can be speeded
                up by pressing the F8 key to disable screen updates.
                Press F9 to resume screen updates, and press the
                END key to stop program execution. Note that the sum
                (which will depend on the contents of all
                256 locations) is in D0.
```

10.2.7 Running Example 6.19

Our program:

```
            .command   +am2
            .org   $1000
;
; Example 6.19
; Sum the BCD contents (two digits/byte) of five memory locations
; (starting at 7000),
;   store a result that can be between 0 and 999 10 in two
; memory locations (7010 and 7011)
;   in BCD format.
;
            MOVE.W   #4,CCR       ; clear the X bit, set the Z bit
            MOVEQ.L  #0,D0        ; clear out the sum storage
            MOVEQ.L  #4,D4        ; initialize the countdown
                                  ; counter to 1 LESS
            MOVEQ.L  #0,D3        ; clean out the carry sum
            MOVEQ.L  #0,D2        ; to be used as the simple
                                  ; carry incrementer
            MOVEQ.L  #$7000,A0    ; point to the beginning of the input data
LOOP:       MOVE.B   (A0)+,D1     ; get the first number to be added,
                                  ; bump pointer
            ABCD.B   D1,D0        ; add to the sum
            BCC   NOPE            ; take care of a sum > 99
                                  ; otherwise proceed
            ABCD.B   D2,D3        ; add one to the carry sum
NOPE:       DBF   D4,LOOP         ; keep looping until done
            MOVE.B   D3,$7010     ; store the upper two BCD digits
            MOVE.B   D0,$7011     ; store the lower two BCD digits
HERE:       JMP   HERE            ; loop when done
            .end
```

To see the effects of this program, enter the following:

```
@ 7000
12                 Deposits 12 in location 7000
34                 Deposits 34 in location 7001
56                 Deposits 56 in location 7002
78                 Deposits 78 in location 7003
90                 Deposits 90 in location 7004
Enter key          To exit from memory modify
.MW1 = 7000        Memory display to see new contents;
                   check PC for correct value
Right arrow        Runs the program nonstop;
                   press the END key to stop;
                   memory display shows data sources unaffected
                   and BCD sum (= 0270) in locations 7010 and 7011
RD                 Note that the sum is in D3 and D0
```

10.2.8   Running Example 6.20

Our program:

```
                  ; Example 6.20
                  ; Add some six-digit BCD data stored in memory
                  ; (starting at 7000), store sum in other
                  ; memory locations (71F0–71F3) as an eight-digit BCD number
                  ;
                  ; $7000–7003 contains the first six-digit BCD number (leading 0's)
                  ; $7004–7007              second
                  ; etc.
                  ; $71F0–71F3 contains the possible eight-digit sum
                  ; $71F4–71F7 contains the number of scores to be added
                  ;
                  .command   +am2
                  .ORG  $7000
START:  .RS  256                    ; reserve space for the numbers
                  .ORG  $71F0
SUM:    .RS  4                      ; reserve space for the sum
NUM:    .DL  0004                   ; the number of numbers is stored here
;
.ORG $1000 ; the beginning of the program
;
                  CLR.L   D7          ; clear out the number counter
                  MOVE.L  NUM,D7      ; get the number of numbers to add
                  MULU.W  #4,D7       ; multiply by 4 since each
                                      ; number is 4 bytes long
                  ADD.L   #START,D7   ; D7 now points to the address
                                      ; past the last number
                  MOVEA.L  D7,A0      ; put the pointer in an address register
                  MOVE.W  #4,CCR      ; clear the X bit, set the Z bit,
                                      ; for BCD arithmetic
                  LEA.L   SUM+4,A1    ; point to one past the sum
                                      ; storage location
                  CLR.L   SUM         ; clear out the longword sum
                                      ; storage location
                  MOVE.L  NUM,D2      ; initialize the countdown counter
                  SUB.L   #1,D2       ; correct for the DBF quirk so
                                      ; countdown will be correct
LOOP:   ABCD.B  –(A0),–(A1)          ; decrement pointers, add the
                                      ; least significant digits
                  ABCD.B  –(A0),–(A1) ; repeat to add the middle two digits
                  ABCD.B  –(A0),–(A1) ; repeat to add the most significant digits
                  ABCD.B  –(A0),–(A1) ; all done, take care of potential carry
                  LEA.L   SUM+4,A1    ; reset the sum pointer
                  DBF  D2,LOOP        ; keep looping until all numbers are added
DONE:   BRA  DONE
                  .END
```

To see the effects of this program, enter the following:

```
@ 7000
00
01
23
45          Deposits BCD number 00012345 in locations 7000–7003
```

00	
67	
89	
90	Deposits BCD number 00678990
00	
12	
34	
56	Deposits BCD number 00123456
00	
78	
90	
10	Deposits BCD number 00789010
Enter key	To exit from memory modify
.MW1 = 7000	Memory display to see new contents; check PC for correct value
Right arrow	Runs the program nonstop; press the END key to stop memory display, shows data sources unaffected
.MW1 = 71F0	Displays BCD sum (= 1,603,801); note that number of numbers (= 4) is stored in 71F4–71F7
RD	Note that PC is at the end of the program

If you wish to run the program with more or fewer numbers, change the value stored in 21F04 to 21F07 to reflect the number of numbers. Reset the PC to 1000, then press the right arrow. The six-digit BCD numbers are actually set up as eight digits, with two leading 0's.

10.2.9 Running Example 6.22

Our program:

```
; Example 6.22
;
; Sort an array of numbers (bytes) into ascending order
;
        .ORG   $7000
START:  .RS   16              ; reserve space for numbers
NUM:    .DW   16              ; the number of numbers in the array
;
        .command  +am2
        .ORG   $1000
;
; Preload an array with descending numbers
        LEA.L   START,A0      ; point to beginning of array
        MOVE.B  #$FF,D0       ; start filling array with FF
        MOVE.W  NUM,D1        ; load the countdown counter
        SUBQ.W  #1,D1         ; take care of DBF quirk
LOOP:   MOVE.B  D0,(A0)+      ; store the number
        SUBQ.B  #1,D0         ; decrease it by one
        DBF.W   D1,LOOP       ; loop until done
; Beginning of bubble sort
AGAIN:  CLR.L   D7            ; clear the swap flag
        LEA.L   START,A0      ; point to beginning of array
        LEA.L   NUM-1,A2      ; point to end of array
READ1:  MOVE.B  (A2),D0       ; get pointed to
                              ; (first number) to compare
```

```
                SUBQ.L  #1,A2        ; decrement pointer
                CMPA.L  A2,A0        ; pointing past beginning of array?
                BHI  CHECK           ; if so, see if swap flag is set
                MOVE.B  (A2),D2      ; if not, get pointed to
                                     ; (second number) to compare
                CMP.B  D2,D0         ; compare the numbers,
                                     ; is the first > second?
                BHI  READ1           ; if so, don't swap,
                                     ; read another number to compare
                MOVE.B  D2,1(A2)     ; if not, swap the numbers
                MOVE.B  D0,(A2)      ;         swap the numbers
                MOVEQ.L  #1,D7       ; set swap flag
                BRA  READ1           ; read another number to compare
        CHECK:  CMPI.L  #1,D7        ; is swap flag set?
                BEQ  AGAIN           ; if so, do sort again
        HERE:   BRA  HERE            ; if not, sorting is complete
                .END
```

To see the effects of this program, enter the following:

!101A=+b   Sets a breakpoint (temporary stop);
                    check PC for correct value

.MW1 = 7000   Memory display to see descending array

Right arrow   Run the program down to the breakpoint
                    Press the F8 key to speed the program up. When the
                    breakpoint is reached, execution will stop and the
                    screen is updated.

Right arrow   Runs the program nonstop to the end;
                    press the END key to stop

10.2.10   Running Example 7.1

Our program:

```
                ; Example 7.1 - Assume that bit 0 of memory location $7000 is attached
                ; to a LED
                ; Create a program that blinks the LED approximately once per second
                ;
                ;
                ; Blink LED connected to bit 0 of location $7000 at a rate of one per
                ; second
                ;
                ;
                .command   +am2
                .ORG  $1000
                .EQU  OUTPUT,$7000   ; assign an address to a label
        ;
        TOP:    MOVE.B  #0,OUTPUT    ; turn LED off
                MOVE.W  #$FFFF,D0    ; initialize countdown counter to
                                     ; 65536
        OFFLOOP: NOP                 ; do nothing, burns up some time
                NOP
                NOP
                NOP
                NOP                         ;
                NOP                         ;
                DBF  D0,OFFLOOP      ; decrement counter, burn up time
                MOVE.W  #$FFFF,D0    ; do it again
```

```
         OFLOOP:   NOP
                   NOP
                   NOP
                   NOP
                   NOP
                   NOP
                   DBF   D0,OFLOOP
                   MOVE.B  #1,OUTPUT    ; turn LED on
                   MOVE.W  #$FFFF,D0    ; reinitialize countdown counter
         ONLOOP:   NOP
                   NOP
                   NOP
                   NOP                  ;
                   NOP                  ;
                   NOP                  ;
                   DBF   D0,ONLOOP      ; decrement counter, burn up time
                   MOVE.W  #$FFFF,D0
         ONLOP:    NOP
                   NOP
                   NOP
                   NOP
                   NOP
                   NOP
                   DBF   D0,ONLOP
                   JSR   MAKBEEP
                   BRA   TOP            ; keep looping indefinitely
                   .END
```

To see the effects of this program, enter the following:

```
         ! 7000=+io    Sets address 7000 for
                       input/output;
                       check PC for correct value
         Right arrow   Run the program continuously
```

The program will stop and display the value being output to location $7000 each time that instruction is executed. Press the ENTER key to continue. It will take a very long time for the simulator to count down from $FFFF. You may want to stop execution (press the END key), then enter **D0=000F** to reduce D0's contents some. Then press the right arrow to resume.

10.2.11   Running Example 7.3

Our program:

```
         ; Example 7.3
         ; Turn ON LED-3 when switch-4 is CLOSED
         ; Turn ON LED-7 when switch-1 AND switch-2 are CLOSED
         ; Toggle LED-2 for each momentary closing of switch-0
         ;
                   .command  +am2
                   .ORG   $1000
                   .EQU   INPUT,$7000    ; assign an address to a label
                   .EQU   OUTPUT,$7001   ; assign an address to a label
         ;
                   MOVE.B  #0,OUTPUT     ; turn off all LEDs
         TOP:      BTST.B  #4,INPUT      ; test bit 4 of the input location
                   BEQ   CLOSED         ; branch if Z bit is set
                                        ; (switch closed)
```

```
OPEN:      BCLR  #3,OUTPUT            ; turn LED-3 OFF (switch is open)
           BRA  TOP2                  ; check next situation
CLOSED:    BSET  #3,OUTPUT            ; turn LED-3 ON (switch is closed)
; Test second situation
TOP2:      MOVE.B  INPUT,D0           ; get the switch conditions
           AND.B  #00000110B,D0       ; mask off unwanted bits
           BEQ  CLOSED2               ; branch if Z bit set
                                      ; (if both are closed)
OPEN2:     BCLR  #7,OUTPUT            ; turn LED OFF
                                      ; since BOTH are not closed
           BRA  TOP3                  ; check next situation
CLOSED2:   BSET  #7,OUTPUT            ; turn LED ON
                                      ; (both switches are closed)
; Test third situation
;                      Switch 0 must make its changes between
TOP3:
           MOVE.B  INPUT,D0 ; get the switch conditions      __here__
           AND.B  #01,D0               ; mask out unwanted bits
                                       ; (= 00000001)
           MOVE.B  D0,D1     ←         ; save the switch condition
           MOVE.B  INPUT,D0  ←         ; read the switches again ____here___
           AND.B  #01,D0               ; mask out unwanted bits
           CMP.B  D0,D1                ; compare now with previous
           BEQ  NOCHG                  ; skip over since
                                       ; there was no change
CHANGE:    BCHG  #2,OUTPUT            ; change LED-2
NOCHG:     BRA  TOP                    ; loop back to top of program
           .END
```

To see the effects of this program, enter the following:

.MW1 = 7000	Sets a memory window for display
!7000,7001=+io	Sets two locations for input/output
	The program will pause for each input or output operation. Refer to the text example and input the various switch situations (in hex), checking for proper outputs.
	Check PC for correct value
Right arrow	Run the program continuously

10.2.12 Running Example 7.5

Our program:

```
; Example 7.5
; Turn a buzzer ON if the stall indicator is ON, the gear is DOWN,
; and the LOW-rpm indicator is ON
; or if
; The gear is UP AND the high-airspeed indicator is ON
;
           .command   +am2
           .ORG  $1000
           .EQU   INPUT,$7000        ; assign an address to a label
           .EQU   OUTPUT,$7001       ; assign an address to a label
;
           BCLR.B  #7,OUTPUT         ; turn the buzzer off initially
           MOVE.B   #00000110B,D1    ; situation 2 mask
```

```
        TOP:    MOVE.B  #00001011B,D0   ; situation 1 mask
                MOVE.B  INPUT,D2        ; get the input condition
                AND.B   D0,D2           ; mask out undesired bits
                BCHG    #1,D0           ; invert the gear-down bit
                                        ; to reflect desired test
                CMP.B   D0,D2           ; look for a match
                BEQ  BUZZ               ; turn on buzzer for situation 1
                MOVE.B  INPUT,D2        ; read switches again
                AND.B   D1,D2           ; mask out undesired bits
                CMP.B   D1,D2           ; look for a match
                BEQ  BUZZ               ; turn on buzzer for situation 2
                BCLR.B  #7, OUTPUT      ; turn off buzzer
                                        ; since no situation
                BRA  TOP                ; keep looping
        BUZZ:   BSET  #7,OUTPUT         ; turn on buzzer
                BRA  TOP                ; keep looping
                .END
```

To see the effects of this program, enter the following:

> **.MW1 = 7000**   Check PC for correct value
> **Right arrow**   Run the program continuously

10.2.13 Running Example 7.7

Our program:

```
        ; Example 7.7
        ; Continually monitoring incoming data, which can be changing,
        ; and provide the following information every 10 seconds:
        ;1) The number of locations reporting a pressure of 0
        ;2) The number of locations reporting a pressure of 255
        ;3) The addresses of the locations reporting a pressure of 255
        ;4) The average pressure for the 50 sensors
        ;5) The maximum pressure reported
        ;6) The minimum pressure reported
        ;
        ; A beep will occur after each data-gathering period
        ;
                .command   +am2
                .ORG   $1000
                LEA.L   $7FFE,A7        ; Load the Stack Pointer
        START:  CLR.L   D0              ; Init the MAXIMUM to ZERO
                CLR.L   D3              ; Clear upper portion of D3
                MOVE.B  #255,D3         ; Init the MINIMUM to 255
        ;
                CLR.L   D1              ; Clear the ZERO counter
                CLR.L   D2              ; Clear the 255 counter
                CLR.L   D4              ; Clear the SUM totalizer
        ;
                CLR.L   D7              ; Init the ZERO CHECKER
                CLR.L   D6              ; Clear upper portion of D6
                MOVE.B  #255,D6         ; Init the 255 checker
        ;
                LEA.L   DATA,A0         ; Point to beginning of data
                LEA.L   HIADRS,A1       ; Point to beginning
                                        ; of '255' addresses
        ; End of initialization
        ; Check for a ZERO
```

```
LOOP:       CMP.B   (A0), D7          ; Is data = 0?
            BNE.S   NOTZERO           ; Nope
            ADDQ.B  #1,D1             ; Yes, increment ZERO counter
; Check for a new MAXIMUM
NOTZERO:    CMP.B   (A0),D0           ; Is it greater than old maximum?
            BCC.B   NOTMAX            ; Nope
            MOVE.B  (A0),D0           ;Yes, store new MAXIMUM
; Check for a new MINIMUM
NOTMAX:     CMP.B   (A0),D3           ; Is it less than old minimum?
            BCS.B   NOTMIN            ; Nope
            MOVE.B  (A0),D3           ; Yes, store new MINIMUM
; Check for a 255
NOTMIN:     CMP.B   (A0),D6           ; Is it 255?
            BNE.S   NOT255            ; Nope
            ADDQ.B  #1,D2             ; Yes, increment 255 storage
            MOVE.L  A0, (A1)+         ; Save address of 255 pressure
; Add number to SUM
NOT255:     ADD.B   (A0),D4           ; Not = 255, add to sum
            BCC     CONT              ;
            ADD.L   #256,D4           ; Byte sum was > 255,
                                      ; add carry to D4
CONT:       ADDA.L  #1,A0             ; Point to next address
            CMPA.L  #DATAEND,A0       ; Done with 50 addresses?
            BNE.S   LOOP              ; Nope, loop again
; Find the average
            DIVU.W  #50,D4            ; Divide by 50, result in D4
; Store the results in memory locations (for use by display routines)
            MOVE.W  D4,AVERAGE        ; Store average
            MOVE.B  D0,MAXIMUM
            MOVE.B  D1,ZEROS
            MOVE.B  D2,TWO55
            MOVE.B  D3,MINIMUM
; Call subroutine to display results
            BSR   DISPLAY             ;
                                      ;
; Call subroutine to kill 10 seconds
            BSR   BURNTIME
EXIT:       BRA   START               ; Start cycle all over again
;
;
DISPLAY:    NOP
            RTS                       ; Return to main program
BURNTIME:   MOVE.L  #$0FFFFF,D0       ; kills about 10 seconds
LOOP1:      NOP
            NOP
            SUB.L   #1,D0
            BNE   LOOP1
            RTS
;
            .ORG  $7000               ; Storage locations for the results
MAXIMUM: .RS  1
ZEROS:      .RS  1
TWO55:      .RS  1
MINIMUM: .RS  1
AVERAGE: .RS  2
            .ORG  $7100               ; The starting address
                                      ; of the input data (decimal numbers)
```

```
; Note: For some reason the assembler assumes digits starting with 0 to be
;        octal numbers. The numbers 00 to 07 are the same
; (octal, decimal, hex)
; a $ was added to 08 and 09 to make them hex (or decimal)
; Numbers greater than 10 (decimal) are properly converted to
; hex numbers for storage
DATA:       .DB   01,02,03,04,05,06,07,$08,$09,10,11,12,13,14
            .DB   15,16,17,18,19,20,21,22,23,24,25,26,27,28
            .DB   29,30,31,32,33,34,35,36,37,38,39,40,41,42
            .DB   43,44,45,255,46,255,47,255
DATAEND:    .RS   2
; End of data
; Storage area for addresses having 255's starts here
HIADRS:     .RS   200
            .END
```

To see the effects of this program, enter the following:

```
                      Check PC for correct value
.MW1=7000             Set display window
!1084=+b              Set a breakpoint
Right arrow           Run the program continuously, then press the F8 key
                      When the program stops, enter
@7134                 to see the addresses of the 255's
```

Display routines are limited to hex displays in memory windows.

10.3 RUNNING THE TEXT EXAMPLES ON THE MICRO BOARD DESIGNS MAX 68000

This recently introduced educational computer is a cost-effective means of teaching micro-computer programming and interfacing. It is also a good alternative for industrial control applications. In this section we provide a very brief summary of the **MAX 68000**'s characteristics to assist in purchase selection. For information on the MAX 68000 and the other products they offer, contact **Micro Board Designs,** 5103 Eastbrooke Place, Williamsville, NY 14221, (716) 633-8613.

In subsequent sections we provide the steps needed for implementation of the examples. Minor modifications will be needed in some cases, and minor enhancements will be outlined in some. The MAX 68000 manual must be referred to for help with its commands.

The MAX 68000 is a stand-alone 68000 system that requires (1) a 5-V power supply (provided) and (2) a terminal or computer having a RS-232 interface. It has provisions for printing programs with its Centronics-compatible parallel printer interface. **In this book we discuss use of the MAX 68000 with a MS-DOS computer only.** Programs can/will be stored on the MS-DOS computer and up/downloaded to the MAX 68000. Since the printer interface will not be connected to a printer, it will prove very useful for numerous I/O experiments. Files will be printed with the printer connected to the MS-DOS computer.

The MAX 68000's features include:

1. A MC68000 MPU running at 8 MHz
2. 16K bytes of static RAM (expandable to 64K)
3. 16K bytes of firmware ROM/EPROM monitor (expandable to 128K)
4. Two serial RS-232 interfaces (one is connected to the MS-DOS computer)
5. A 68230 PIT I/O chip containing three 8-bit parallel ports

6. A 6821 PIA chip containing two 8-bit parallel ports (one is used for printer interface)
7. Self-contained firmware providing monitor, debugger, and disassembler functions
8. A 24-bit programmable timer interface circuit, useful for generating time delays or to measure time or frequency
9. Expansion bus connector with address decoding circuitry
10. Reset and abort function switches
11. Operation from a single 5-V power supply
12. Built-in routines for keyboard input, video display, number conversions, and so on.

Normal operation is to:

1. *Upload* the hex code representing the program into the MAX 68000's memory.
2. Set the Program Counter (PC) to the first instruction of the program.
3. If problems are anticipated with the program, the instructions are traced or executed one at a time. The 68000's registers can be examined as the program is single-stepped.
4. If sections of the program appear to be completely functional, breakpoints can be set to allow rapid execution up to the breakpoint through the functional parts of the program.
5. The program can be executed *at full speed* with no breakpoints if desired. In the event of problems, return to the monitor can be accomplished by hitting the ABORT button.

10.3.1 General Procedures for Getting Started with the Examples

The MAX 68000 has 16K RAM between $20000 and $23FFF. It uses some of the space at the top, so our programs will be set to use between $20000 and $21FFF, which should be more than enough for even the most challenging project. The 6821 PIA I/O IC is located between $E8001 and $E8007. Examples will show its use. Refer to the MAX 68000 manual for details.

Due to differences in the memory map and the random selection of addresses used in the book, minor modifications will have to be made to the programs before use with the MAX 68000. The examples described below will assume that the following steps were completed initially.

1. Create an *assembly source file* named EX-x-x.**ASM** with an ASCII file editor (i.e., EX-6-1.**ASM** for the first example).
   (a) Include the assembler directive **.COMMAND +AM2** on the first line of each program. This provides the proper hex format for the MAX 68000.
   (b) Include, on the next line, the assembler directive **.ORG $20000,** which will be the address of the first instruction.
   (c) Include the assembler directive **.END** for the last line of the source file.
2. Assemble the source file to create both **Listing (.LST)** and hex **OBJect (.OBJ)** files. This is done by entering **A68K EX-x-x.ASM.** If no errors are reported, proceed. Otherwise, view the EX-x-x.**LST** file (from DOS, TYPE EX-x-x.LST) or print the EX-x-x.**LST** file (from DOS, COPY EX-x-x.LST PRN) to determine the errors. Refer to Section 10.2 or to the assembler documentation (on the disk) for assistance.
   NOTE: **Two important rules for the creation of a source program:**
   (a) The **first** column of each line is reserved for either a ; to signify a comment line or the first letter of a **label.** Skip over the first column otherwise.
   (b) Spaces can be used for reading clarity, except between the two operands, which must be separated only by a **comma.**

3. Obtain a hard copy of the LISTING file by entering COPY EX-x-x.**LST** PRN, **with a printer attached to the MS-DOS computer.**
4. Connect and turn on the MAX 68000.
5. Establish communication with the MAX 68000, utilizing a terminal program. Hitting the RED ABORT button on the MAX 68000 should display (on your MS-DOS computer) a formatted display of the registers; while hitting the BLACK RESET button displays the **> MAX 68000 monitor** prompt. If readable displays are not obtained, check the connections and baud rate (speed) of the MAX 68000 and the terminal program being used. Refer to the MAX 68000 manual for details.
6. Upload, using the ASCII file transfer mode, the **OBJ** *object file* from the MS-DOS computer to the MAX 68000 (i.e., EX-x-x.**OBJ**). It is imperative that the hex file, with the **.OBJ extension,** is the uploaded file. File transfer is initiated by entering **LOAD** while communicating with the MAX 68000, then entering the proper sequence for the terminal program for an ASCII upload. Upon completion of the upload, hit the enter key to get the MAX 68000 monitor prompt.
7. To check for a successful upload, you can enter **DASM 20000 xx,** where xx is the number of bytes (hex) that you wish to view. The monitor will disassemble and display your program.
8. To get a list of commands available, enter **HELP.**
9. Set the Program Counter by entering **PC 20000.**

10.3.2   Running Example 6.1

Our program:

```
        .command   +am2
        .org   $20000
;
; Example 6.1
; Move a byte from one memory location to another (21000 to 21010)
;

        MOVE.B   $21000,$21010
        .end
```

To see the effects of this program, enter the following:

MM 21000	
AB	Deposits AB in location 21000
12	Deposits 12 in location 21001
34	Deposits 34 in location 21002
56	etc. (not required, but used to
78	show how other locations are not affected)
.	to exit from memory modify mode
VIEW 21000 10	Display new contents (note contents of 21010)
RD	Shows registers, check PC for correct value
TR	Trace (single-step) through the first instruction
.	to exit from trace mode
VIEW 21000 10	Memory display, shows data source unaffected and shows new data in destination (21010)
RD	Note that PC is at $2000A, past the end of the program

Be sure to reset the PC to 20000 before running the program again.

10.3.3 Running Example 6.6

Our program:

```
                .command   +am2
                .org   $20000
        ;
        ; Example 6.6
        ; Add 2 bytes of memory (locations 21000 and 21002),
        ; leaving the sum in a data register
        ;

                MOVE.B   $21000,D3
                ADD.B   $21002,D3
                .end
```

To see the effects of this program, enter the following:

MM21000	
AB	Deposits AB in location 21000
12	Deposits 12 in location 21001 (not used or affected)
34	Deposits 34 in location 21002
.	To exit from memory modify
VIEW 21000 10	Memory display to see new contents
RD	Check PC for correct value (note D3 contents)
TR	Trace (single-step) through the first instruction
ENTER key	Trace next instruction
.	To exit from trace mode
VIEW 21000 10	Memory display, shows data sources unaffected
RD	Note the sum AB + 34 = DF in D3

10.3.4 Running Example 6.10

Our program:

```
                .command   +am2
                .org   $20000
        ;
        ; Example 6.10
        ; Subtract the contents of memory location (21002) from
        ; another (21000), leave result in another location (21004)
        ;
                MOVE.W   $21000,D0   ; get the salary data
                SUB.W   $21002,D0   ; subtract the tax from it
                MOVE.W   D0,$21004   ; store the result
                .end
```

To see the effects of this program, enter the following:

MM 21000	
56	Deposits 56 in location 21000
78	Deposits 78 in location 21001
12	Deposits 12 in location 21002
34	Deposits 34 in location 21003
.	To exit from memory modify

VIEW 21000 10	Memory display to see new contents
RD	Check PC for correct value
TR	Trace (single-step) through the first instruction
ENTER key	Trace next instruction
ENTER key	Trace next instruction
.	To exit from trace mode
VIEW 21000 10	Memory display, shows data sources unaffected; note results in 21004 (5678 − 1234 = 4444)
RD	Note that PC is at $20012, past the end of the program; note that the result is also in D0

10.3.5 Running Example 6.14

Our program:

```
            .command   +am2
            .org  $20000
    ;
    ; Example 6.14
    ; Sum the byte contents of five memory locations (starting at 21000),
    ; store sum in another memory address (21010)
    ;
            MOVEQ.L  #0,D0        ; clear out the data register
            ADD.B   $21000,D0     ; add the first number to D0
            ADD.B   $21001,D0     ; add the next number
            ADD.B   $21002,D0     ; add the next number
            ADD.B   $21003,D0     ; add the next number
            ADD.B   $21004,D0     ; add the last number
            MOVE.B   D0,$21010    ; store the sum
            .end
```

To see the effects of this program, enter the following:

MM 21000	
12	Deposits 12 in location 21000
34	Deposits 34 in location 21001
56	Deposits 56 in location 21002
78	Deposits 78 in location 21003
90	Deposits 90 in location 21004
.	to exit from memory modify
VIEW 21000 10	Memory display to see new contents
RD	Check PC for correct value
TR	Trace (single-step) through the first instruction
ENTER key	Six more times to complete program
.	to exit from trace mode
VIEW 21000 10	Memory display, shows data sources unaffected Note sum (= A4) in location 21010. The actual sum is 1A4, the excess beyond FF is lost, since a byte-sized operation was done.
RD	Note that PC is at $20026, past the end of our program; note that the sum is also in D0

10.3.6 Running Example 6.16

Our program:

```
                .command   +am2
                .org   $20000
        ;
        ; Example 6.16 (final version of Example 6.15)
        ; Sum the byte contents of 256 memory locations (starting at 21000),
        ; store the result in another memory location (21100)
        ; Exit (or loop) correctly when done
        ;
                MOVEQ.L  #$0,D0           ; clear out the accumulator
                MOVE.W   #$FF,D1          ; initialize the countdown register
                MOVEA.L  #$21000,A0       ; point to the source of data
        LOOP:   ADD.B    (AO)+,D0         ; to D0, add the data pointed to
                                          ; bump pointer
                DBF  D1,LOOP              ; get the next data to add
                MOVE.B   D0,$21100        ; move the sum to the output address
        HERE:   JMP  HERE                 ; a way of stopping the program
                .end
```

To see the effects of this program, enter the following:

MM 21000	
12	Deposits 12 in location 21000
34	Deposits 34 in location 21001
56	Deposits 56 in location 21002;
	continue modifying up to 256 locations, if desired
.	To exit from memory modify
VIEW 21000 10	Memory display to see new contents
VIEW 21100 10	Memory display of sum storage location before addition
RD	Check PC for correct value
GO	Runs the program nonstop;
	press the RED ABORT button to stop
VIEW 21000 10	Memory display, shows data sources unaffected
VIEW 21100 10	Memory display of sum
RD	Note that PC is at $20018, at the end of our program;
	note that the sum (which will depend on the contents of all
	256 locations) is in D0

10.3.7 Running Example 6.19

Our program:

```
                .command   +am2
                .org   $20000
        ;
        ; Example 6.19
        ; Sum the BCD contents (two digits/byte) of five memory locations
        ; (starting at 21000),
        ;     store a result that can be between 0 and 9999₁₀ in two
        ; memory locations (21010 and 21011)
        ;     in BCD format
        ;
```

```
              MOVE.W   #4,CCR        ; clear the X bit, set the Z bit
              MOVEQ.L  #0,D0         ; clear out the sum storage.
              MOVEQ.L  #4,D4         ; initialize the countdown
                                     ; counter to one less
              MOVEQ.L  #0,D3         ; clean out the carry sum
              MOVEQ.L  #0,D2         ; to be used as the simple
                                     ; carry incrementer
              MOVEA.L  #$21000,A0    ; point to the beginning of the input data
LOOP:  MOVE.B   (A0)+,D1      ; get the first number to be added,
                                     ; bump pointer
              ABCD.B   D1,D0         ; add to the sum
              BCC   NOPE            ; take care of a sum > 99
                                     ; otherwise proceed
              ABCD.B   D2,D3         ; add one to the carry sum
NOPE:  DBF  D4,LOOP          ; keep looping until done
              MOVE.B   D3,$21010     ; store the upper two BCD digits
              MOVE.B   D0,$21011     ; store the lower two BCD digits
HERE:  JMP  HERE             ; loop when done
              .end
```

To see the effects of this program, enter the following:

```
MM 21000
12              Deposits 12 in location 21000
34              Deposits 34 in location 21001
56              Deposits 56 in location 21002
78              Deposits 78 in location 21003
90              Deposits 90 in location 21004
.               To exit from memory modify
VIEW 21000 10   Memory display to see new contents
RD              Check PC for correct value
GO              Runs the program nonstop;
                press the RED ABORT button to stop
VIEW 21000 10   Memory display, shows data sources unaffected
                and BCD sum (= 0270) in locations 21010 and 21011
RD              Note that PC is at the end of our program;
                note that the sum is in D3 and D0
```

10.3.8   Running Example 6.20

Our program:

```
; Example 6.20
; Add some six-digit BCD data stored in memory
; (starting at 21000), store sum in other
; memory locations (21F00–21F03) as an eight-digit BCD number
;
; $21000–21003 contains the first six-digit BCD number (leading 0's)
; $21004–21007             second
; etc.
; $21F00–21F03 contains the possible eight-digit sum
; $21F04–21F07 contains the number of scores to be added
;
.command  +am2
.ORG  $21000
```

```
START: .RS   256              ; reserve space for the numbers
       .ORG  $21F00
SUM:   .RS   4                ; reserve space for the sum
NUM:   .DL   0004             ; the number of numbers is stored here
;
       .ORG  $20000           ; the beginning of the program
;
       CLR.L   D7             ; clear out the number counter
       MOVE.L  NUM,D7         ; get the number of numbers to add
       MULU.W  #4,D7          ; multiply by 4 since each
                              ; number is 4 bytes long
       ADD.L   #START,D7      ; D7 now points to the address
                              ; past the last number
       MOVEA.L D7,A0          ; put the pointer in an address register
       MOVE.W  #4,CCR         ; clear the X bit, set the Z bit,
                              ; for BCD arithmetic
       LEA.L   SUM+4,A1       ; point to one past the sum
                              ; storage location
       CLR.L   SUM            ; clear out the longword sum
                              ; storage location
       MOVE.L  NUM,D2         ; initialize our countdown counter
       SUB.L   #1,D2          ; correct for the DBF quirk, so
                              ; countdown will be correct
LOOP:  ABCD.B  -(A0),-(A1)    ; decrement pointers, add the
                              ; least significant digits
       ABCD.B  -(A0),-(A1)    ; repeat to add the middle two digits
       ABCD.B  -(A0),-(A1)    ; repeat to add the most significant digits
       ABCD.B  -(A0),-(A1)    ; all done, take care of potential carry
       LEA.L   SUM+4,A1       ; reset the sum pointer
       DBF     D2,LOOP        ; keep looping until all numbers are
                              added
DONE:  BRA   DONE
       .END
```

To see the effects of this program, enter the following:

```
MM 21000
00
01
23
45              Deposits BCD number 00012345 in locations 21000–21003
00
67
89
90              Deposits BCD number 00678990
00
12
34
56              Deposits BCD number 00123456
00
78
90
10              Deposits BCD number 00789010
.               To exit from memory modify
VIEW 21000 10   Memory display to see new contents
RD              Check PC for correct value
```

GO	Runs the program nonstop;
	press the RED ABORT button to stop
VIEW 21000 10	Memory display, shows data sources unaffected
VIEW 21F00 10	Displays BCD sum (= 1,603,801);
	note number of numbers (= 4) is stored in 21F04–21F07
RD	Note that PC is at the end of the program

If you wish to run the program with more or fewer numbers, change the value stored in 21F04 to 21F07 to reflect the number of numbers. Reset the PC to 20000, then enter GO. The six-digit BCD numbers are actually set up as eight digits, with two leading 0's.

10.3.9   Running Example 6.22

Our program:

```
; Example 6.22
;
; Sort an array of numbers (bytes) into ascending order
;
          .ORG   $21000
START:    .RS    16         ; reserve space for numbers
NUM:      .DW    16         ; the number of numbers in the array
;
          .command   +am2
          .ORG   $20000
;
; Pre-load an array with descending numbers
          LEA.L   START,A0    ; point to beginning of array
          MOVE.B  #$FF,D0     ; start filling array with FF
          MOVE.W  NUM,D1      ; load the countdown counter
          SUBQ.W  #1,D1       ; take care of DBF quirk
LOOP:     MOVE.B  D0, (A0)+   ; store the number
          SUBQ.B  #1,D0       ; decrease it by one
          DBF.W   D1,LOOP     ; loop until done
; Beginning of bubble sort
AGAIN:    CLR.L   D7          ; clear the swap flag
          LEA.L   START,A0    ; point to beginning of array
          LEA.L   NUM-1,A2    ; point to end of array
READ1:    MOVE.B  (A2),D0     ; get pointed to
                              ; (first number) to compare
          SUBQ.L  #1,A2       ; decrement pointer
          CMPA.L  A2, A0      ; pointing past beginning of array?
          BHI   CHECK         ; if so, see if swap flag is set
          MOVE.B  (A2),D2     ; if not, get pointed to
                              ; (second number) to compare
          CMP.B   D2,D0       ; compare the numbers,
                              ; is the first > second?
          BHI   READ1         ; if so, don't swap, go
                              ; read another number to compare
          MOVE.B  D2,1 (A2)   ; if not, swap the numbers
          MOVE.B  D0, (A2)    ;       swap the numbers
          MOVEQ.L #1,D7       ; set swap flat
          BRA   READ1         ; read another number to compare
CHECK:    CMPI.L  #1,D7       ; is swap flag set?
          BEQ   AGAIN         ; if so, go do sort again
HERE:     BRA   HERE          ; if not, sorting is complete
          .END
```

To see the effects of this program, enter the following:

BR 2001A	Sets a breakpoint (temporary stop)
RD	Check PC for correct value
GO	Run the program down to the breakpoint
VIEW 21000 10	Memory display to see descending array
NOBR	Remove the breakpoint
GO	Runs the program nonstop to the end; press the red abort button to stop
VIEW 21000 10	Memory display to show sorted ascending array

If problems are encountered debugging a program, including times when a breakpoint is used, it may be necessary to hit the black reset button to restore the program properly. In some cases the program will have to be reloaded into the 68000.

10.3.10 Running Example 7.1

Our program:

```
; Example 7.1 - Assume that bit 0 of memory location $E8005 is attached to
; a LED [connect a LED between pins 9 and 20 (GND) of connector J7]
; Create a program that blinks the LED approximately once per second
; Assume a computer clock speed of 8 MHz
;
; Blink LED connected to bit 0 of location $E8005 at a rate of one per
; second
; Also BEEPs once a second
;
            .command   +am2
            .ORG   $20000
            .EQU   OUTPUT,$E8005       ; assign an address to a label
            .EQU   MAKBEEP,$484        ; address of the beep routine
; added instructions needed to initialize the 6821 PIA for actual output
            MOVE.B   #0,OUTPUT+2
            MOVE.B   #$FF,OUTPUT        ; makes port B of 6821 = all outputs
            MOVE.B   #04,OUTPUT+2
;
TOP:        MOVE.B   #0,OUTPUT          ; turn LED off
            MOVE.W   #$FFFF,D0          ; initialize countdown counter to
                                        ; 65536
OFFLOOP:    NOP                         ; do nothing, burns up some time
            NOP
            NOP
            NOP
            NOP                         ;
            NOP                         ;
            DBF   D0,OFFLOOP            ; decrement counter, burn up time
            MOVE.W   #$FFFF,D0          ; do it again
OFLOOP:     NOP
            NOP
            NOP
            NOP
            NOP
            NOP
            DBF   D0,OFLOOP
            MOVE.B   #1,OUTPUT          ; turn LED on
            MOVE.W   #$FFFF,D0          ; reinitialize countdown counter
```

```
ONLOOP:   NOP
          NOP
          NOP
          NOP                    ;
          NOP                    ;
          NOP                    ;
          DBF   D0,ONLOOP        ; decrement counter, burn up time
          MOVE.W  #$FFFF,D0
ONLOOP:   NOP
          NOP
          NOP
          NOP
          NOP
          NOP
          DBF   D0,ONLOP
          JSR   MAKBEEP
          BRA   TOP              ; keep looping indefinitely
          .END
```

To see the effects of this program, enter the following:

RD   Check PC for correct value
GO   Run the program continuously;
         the LED should blink about once per second and
         the PC speaker should beep

Note that the program is slightly different from the original Example 7.1. Since the MAX 68000 runs twice as fast as the ECB, the time delays were simply duplicated twice to give the required delay. A more elegant timing-delay routine could have been written. It is left to the student to check this routine for time accuracy.

10.3.11   Running Example 7.3

Our program:

```
; Example 7.3
; Turn ON LED-3 when switch-4 is CLOSED
; Turn ON LED-7 when switch-1 AND switch-2 are CLOSED
; Toggle LED-2 for each momentary closing of switch-0.
;
          .command   +am2
          .ORG   $20000
          .EQU   INPUT,$E8001      ; assign an address to a label
          .EQU   OUTPUT,$E8005     ; assign an address to a label
; If desired, connect a SPST switch between pins 5 and 20,
;                 a SPST       between pins 3 and 20,
;                 a SPST       between pins 2 and 20,
;                 a LED        between pins 11 and 20,
;                 a LED        between pins 16 and 20,
;                 a LED        between pins 10 and 20 on J7
;
; instructions added to initialize the 6821 PIA for actual input and output
          MOVE.B   #0,OUTPUT+2
          MOVE.B   #$FF,OUTPUT    ; makes port B of 6821 = output
          MOVE.B   #4,OUTPUT+2
```

```
              MOVE.B  #0,INPUT+2
              MOVE.B  #0,INPUT         ; makes port A of 6821 = input
              MOVE.B  #4,INPUT+2
;
              MOVE.B  #0,OUTPUT        ; turn off all LEDs
TOP:          BTST.B  #4,INPUT         ; test bit 4 of the input location
              BEQ  CLOSED              ; branch if Z bit is set
                                       ; (switch closed)
OPEN:         BCLR  #3,OUTPUT          ; turn LED-3 OFF (switch is open)
              BRA  TOP2                ; check next situation
CLOSED:       BSET  #3,OUTPUT          : turn LED-3 ON (switch is closed)
; Test second situation
TOP2:         MOVE.B  INPUT,D0         ; get the switch conditions
              AND.B  #00000110B,D0     ; mask off unwanted bits
              BEQ  CLOSED2             ; branch if Z bit set
                                       ; (if both are closed)
OPEN2:        BCLR  #7,OUTPUT          ; turn LED OFF
                                       ; since BOTH are not closed
              BRA  TOP3                ; check next situation
CLOSED2:      BSET  #7,OUTPUT          ; turn LED ON
                                       ; (both switches are closed)
; Test third situation
;                   Switch 0 must make its changes between
TOP3:         MOVE.B  INPUT,D0         ; get the switch conditions _____ here

              AND.B  #01,D0            ; mask out unwanted bits
                                       ; (= 00000001)
              MOVE.B  D0,D1            ; save the switch condition
              MOVE.B  INPUT,D0         ; read the switches again _____ here
              AND.B  #01,D0            ; mask out unwanted bits
              CMP.B  D0,D1             ; compare now with previous
              BEQ  NOCHG               ; skip over since
                                       ; there was no change
CHANGE:       BCHG  #2,OUTPUT          ; change LED-2
NOCHG:        BRA  TOP                 ; loop back to top of program
              .END
```

To see the effects of this program, enter the following:

RD Check PC for correct value
GO Run the program continuously;
        activate the switches and check for proper
        LED activity

If problems are encountered, set the PC to 20000, then TR (trace) through the program an instruction at a time. Check for proper instruction branching, remembering that a closed switch (connected to ground) is a logic 0 and a BEQ instruction will occur when the switch is closed.

10.3.12 Running Example 7.5

Our program:

```
; Example 7.5
; Turn a buzzer ON if the stall indicator is ON, the gear is DOWN,
; and the low-rpm indicator is ON
; or if
```

```
; The gear is up and the high-airspeed indicator is on
;
            .command   +am2
            .ORG    $20000
            .EQU    INPUT,$E8001       ; assign an address to a label
            .EQU    OUTPUT,$E8005      ; assign an address to a label
; If desired, connect switches and LED as indicated in the text
; Connect the switches to port A and the LED port B
;
; added instructions to make 6821 functional
;
            MOVE.B   #0,OUTPUT+2
            MOVE.B   #$FF,OUTPUT     ; makes port B = output
            MOVE.B   #4,OUTPUT+2
            MOVE.B   #0,INPUT+2
            MOVE.B   #0,INPUT        ; makes port A = input
            MOVE.B   #4,INPUT+2
;
            BCLR.B   #7,OUTPUT       ; turn the buzzer off initially
            MOVE.B   #00000110B,D1   ; situation 2 mask
TOP:        MOVE.B   #00001011B,D0   ; situation 1 mask
            MOVE.B   INPUT,D2        ; get the input condition
            AND.B    D0,D2           ; mask out undesired bits
            BCHG     #1,D0           ; invert the gear-down bit
                                     ; to reflect desired test
            CMP.B    D0,D2           ; look for a match
            BEQ   BUZZ               ; turn on buzzer for situation 1
            MOVE.B   INPUT,D2        ; read switches again
            AND.B    D1,D2           ; mask out undesired bits
            CMP.B    D1,D2           ; look for a match
            BEQ   BUZZ               ; turn on buzzer for situation 2
            BCLR.B   #7,OUTPUT       ; turn off buzzer
                                     ; since no situation
            BRA   TOP                ; keep looping
BUZZ:       BSET   #7,OUTPUT         ; turn on buzzer
            BRA   TOP                ; keep looping
            .END
```

To see the effects of this program, enter the following:

> **RD**   Check PC for correct value
> **GO**   Run the program continuously;
>   activate the switches and check for proper
>   LED activity

If problems are encountered, set the PC to 20000, then TR (trace) through the program an instruction at a time. Check for proper instruction branching, remembering that a closed switch (connected to ground) is a logic 0 and a BEQ instruction will occur when the switch is closed.

10.3.13   Running Example 7.7

Our program:

```
; Example 7.7
; Continually monitor incoming data, which can be changing,
; and provide the following information every 10 seconds:
```

```
; 1) The number of locations reporting a pressure of 0
; 2) The number of locations reporting a pressure of 255
; 3) The addresses of the locations reporting a pressure of 255
; 4) The average pressure for the 50 sensors
; 5) The maximum pressure reported
; 6) The minimum pressure reported
;
; A beep will occur after each data-gathering period
;
                .command   +am2
                .ORG   $20000
                LEA.L   $21FFE,A7        ; Load the Stack Pointer
START:          CLR.L   D0               ; Init the MAXIMUM to ZERO
                CLR.L   D3               ; Clear upper portion of D3
                MOVE.B  #255,D3          ; Init the MINIMUM to 255
;
                CLR.L   D1               ; Clear the ZERO counter
                CLR.L   D2               ; Clear the 255 counter
                CLR.L   D4               ; Clear the SUM totalizer
;
                CLR.L   D7               ; Init the ZERO CHECKER
                CLR.L   D6               ; Clear upper portion of D6
                MOVE.B  #255,D6          ; Init the 255 CHECKER
;
                LEA.L   DATA,A0          ; Point to beginning of data
                LEA.L   HIADRS,A1        ; Point to beginning
                                         ; of '255' addresses
; End of initialization
; Check for a zero
LOOP:           CMP.B   (A0),D7          ; Is data = 0?
                BNE.S   NOTZERO          ; Nope
                ADDQ.B  #1,D1            ; Yes, increment ZERO counter
; Check for a new maximum
NOTZERO:        CMP.B   (A0),D0          ; Is it greater than old maximum?
                BCC.B   NOTMAX           ; Nope
                MOVE.B  (A0),D0          ; Yes, store new MAXIMUM
; Check for a new minimum
NOTMAX:         CMP.B   (A0),D3          ; Is it less than old minimum?
                BCS.B   NOTMIN           ; Nope
                MOVE.B  (A0),D3          ; Yes, store new MINIMUM
; Check for a 255
NOTMIN:         CMP.B   (A0),D6          ; Is it 255?
                BNE.S   NOT255           ; Nope
                ADDQ.B  #1,D2            ; Yes, increment 255 storage
                MOVE.L  A0, (A1)+        ; Save address of 255 pressure
; Add number to sum
NOT255:         ADD.B   (A0),D4          ; Not = 255, add to sum
                BCC   ONT                ;
                ADD.L   #256,D4          ; Byte sum was > 255,
                                         ; add carry to D4
CONT:           ADDA.L  #1,A0            ; Point to next address
                CMPA.L  #DATAEND,A0      ; Done with 50 addresses?
                BNE.S   LOOP             ; Nope, loop again
; Find the average
                DIVU.W  #50,D4           ; Divide by 50, result in D4
; Store the results in memory locations (for use by display routines)
                MOVE.W  D4,AVERAGE       ; Store average
                MOVE.B  D0,MAXIMUM
                MOVE.B  D1,ZEROS
                MOVE.B  D2,TWO55
                MOVE.B  D3,MINIMUM
```

```
                ; Call subroutine to display results
                          BSR   DISPLAY              ;
                                                     ;
                ; Call subroutine to kill 10 seconds
                          BSR   BURNTIME
                EXIT:     BRA   START                ; Start cycle all over again
                ;
                ; Display routine for the MAX 68000 only
                ;
                          .equ  PUTSTR,$400          ; puts ASCII string pointed to by A0
                                                     ; on screen
                          .equ  PUTCHAR,$408         ; puts ASCII character in D0 on screen
                          .equ  NEWLINE,$414         ; provides a carriage return
                                                     ; and linefeed on screen
                          .equ  HEXTOA,$41C          ; converts hex byte in D0 to two
                                                     ; ASCII characters
                          . equ MAKBEEP,$484         ; makes a beep on PC speaker
                DISPLAY:  MOVEA.L #MSG1,A0           ; point to beginning of message
                          JSR   PUTSTR               ; display message
                          JSR   MAKBEEP
                          MOVE.B  MINIMUM,D0         ; get minimum value
                          JSR   HEXTOA               ; convert to ASCII
                          JSR   PUTCHAR              ; put most significant character
                                                     ; on screen
                          MOVE.B  D1,D0              ; get least significant character
                          JSR   PUTCHAR              : display it
                          BSR   BLANKS               ; put some blanks on screen
                          MOVE.B  MAXIMUM,D0         ; get the maximum
                          JSR   HEXTOA
                          JSR   PUTCHAR
                          MOVE.B  D1,D0
                          JSR   PUTCHAR
                          BSR   BLANKS
                          MOVE.B  ZEROS,D0           ; get the number of 0's
                          JSR   HEXTOA
                          JSR   PUTCHAR
                          MOVE.B  D1,D0
                          JSR   PUTCHAR
                          BSR   BLANKS
                          MOVE.B  TWO55,D0           ; get the number of 255's
                          JSR   HEXTOA
                          JSR   PUTCHAR
                          MOVE.B  D1,D0
                          JSR   PUTCHAR
                          BSR   BLANKS
                          MOVE.B  AVERAGE,D0         ; get the average
                          JSR   HEXTOA
                          JSR   PUTCHAR
                          MOVE.B  D1,D0
                          JSR   PUTCHAR
                          MOVE.B  AVERAGE+1,D0
                          JSR   HEXTOA
                          JSR   PUTCHAR
                          MOVE.B  D1,D0
                          JSR   PUTCHAR
                          JSR   NEWLINE
                          JSR   NEWLINE
                          RTS                        ; return to main program
```

```
BLANKS:     MOVEA.L  #BLANK,A0        ; puts some blanks between characters
            JSR   PUTSTR
            RTS
BURNTIME:   MOVE.L  #$0FFFFF,D0       ; kills about 10 seconds
LOOP1:      NOP
            NOP
            SUB.L  #1,D0
            BNE LOOP1
            RTS
;
MSG1:       .DB  "      Pressure Monitoring System ",$0A,$0D
            .DB  "MIN    MAX     #0's     #255's  AVERAGE ",$0A,$0D
            .DB  " -------------------------------------------------------------- ",$0A,$0D,$00

; NOTE: STRING MUST END WITH A $00
            .ALIGN
;
BLANK:      .DB  "      ",$00
            .ORG  $21000              ; storage locations for the results
MAXIMUM:  .RS   1
ZEROS:    .RS   1
TWO55:    .RS   1
MINIMUM:  .RS   1
AVERAGE:  .RS   2
            .ORG  $21010 ; The starting address
                        ; of the input data (decimal numbers)

; Note: For some reason the assembler assumes digits starting with 0 to be
; octal numbers. The numbers 00 to 07 are the same (octal, decimal, hex)
; a $ was added to 08 and 09 to make them hex (or decimal). Numbers greater
; than 10 (decimal) are properly converted to hex numbers for storage.
DATA:       .DB  01,02,03,04,05,06,07,$08,$09,10,11,12,13,14
            .DB  15,16,17,18,19,20,21,22,23,24,25,26,27,28
            .DB  29,30,31,32,33,34,35,36,37,38,39,40,41,42
            .DB  43,44,45,255,46,255,47,255
DATAEND:  .RS   2
; End of data
; Storage area for addresses having 255's starts here
HIADRS:   .RS   200
            .END
```

To see the effects of this program, enter the following:

RD   Check PC for correct value
GO   Run the program continuously

The display and sound subroutines use routines built into the MAX 68000's monitor (EPROM). The program will scroll with the same numbers since the *input data* is not changing. To check it for different data values, modify the memory contents starting at 21010, reset the PC, and run the program again.

MC68000 instruction execution times

D-1 INTRODUCTION

This Appendix contains listings of the instruction execution times in terms of external clock (CLK) periods. In this data, it is assumed that both memory read and write cycle times are four clock periods. A longer memory cycle will cause the generation of wait states which must be added to the total instruction time.

The number of bus read and write cycles for each instruction is also included with the timing data. This data is enclosed in parenthesis following the number of clock periods and is shown as: (r/w) where r is the number of read cycles and w is the number of write cycles included in the clock period number. Recalling that either a read or write cycle requires four clock periods, a timing number given as 18(3/1) relates to 12 clock periods for the three read cycles, plus 4 clock periods for the one write cycle, plus 2 cycles required for some internal function of the processor.

Note: The number of periods includes instruction fetch and all applicable operand fetches and stores.

D-2 OPERAND EFFECTIVE ADDRESS CALCULATION TIMING

Table D-1 lists the number of clock periods required to compute an instruction's effective address. It includes fetching of any extension words, the address computation, and fetching of the memory operand. The number of bus read and write cycles is shown in parenthesis as (r/w). Note there are no write cycles involved in processing the effective address.

TABLE D-1.

Effective Address Calculation Times

	Addressing Mode	Byte, Word	Long
	Register		
Dn	Data Register Direct	0(0/0)	0(0/0)
An	Address Register Direct	0(0/0)	0(0/0)
	Memory		
(An)	Address Register Indirect	4(1/0)	8(2/0)
(An)+	Address Register Indirect with Postincrement	4(1/0)	8(2/0)
−(An)	Address Register Indirect with Predecrement	6(1/0)	10(2/0)
d(An)	Address Register Indirect with Displacement	8(2/0)	12(3/0)
d(An, ix)*	Address Register Indirect with Index	10(2/0)	14(3/0)
xxx.W	Absolute Short	8(2/0)	12(3/0)
xxx.L	Absolute Long	12(3/0)	16(4/0)
d(PC)	Program Counter with Displacement	8(2/0)	12(3/0)
d(PC, ix)*	Program Counter with Index	10(2/0)	14(3/0)
#xxx	Immediate	4(1/0)	8(2/0)

*The size of the index register (ix) does not affect execution time.

D-3 MOVE INSTRUCTION EXECUTION TIMES

Tables D-2 and D-3 indicate the number of clock periods for the move instruction. This data includes instruction fetch, operand reads, and operand writes. The number of bus read and write cycles is shown in parenthesis as (r/w).

TABLE D-2.

Move Byte and Word Instruction Execution Times

Source	Destination								
	Dn	An	(An)	(An)+	−(An)	d(An)	d(An, ix)*	xxx.W	xxx.L
Dn	4(1/0)	4(1/0)	8(1/1)	8(1/1)	8(1/1)	12(2/1)	14(2/1)	12(2/1)	16(3/1)
An	4(1/0)	4(1/0)	8(1/1)	8(1/1)	8(1/1)	12(2/1)	14(2/1)	12(2/1)	16(3/1)
(An)	8(2/0)	8(2/0)	12(2/1)	12(2/1)	12(2/1)	16(3/1)	18(3/1)	16(3/1)	20(4/1)
(An)+	8(2/0)	8(2/0)	12(2/1)	12(2/1)	12(2/1)	16(3/1)	18(3/1)	16(3/1)	20(4/1)
−(An)	10(2/0)	10(2/0)	14(2/1)	14(2/1)	14(2/1)	18(3/1)	20(3/1)	18(3/1)	22(4/1)
d(An)	12(3/0)	12(3/0)	16(3/1)	16(3/1)	16(3/1)	20(4/1)	22(4/1)	20(4/1)	24(5/1)
d(An, ix)*	14(3/0)	14(3/0)	18(3/1)	18(3/1)	18(3/1)	22(4/1)	24(4/1)	22(4/1)	26(5/1)
xxx.W	12(3/0)	12(3/0)	16(3/1)	16(3/1)	16(3/1)	20(4/1)	22(4/1)	20(4/1)	24(5/1)
xxx.L	16(4/0)	16(4/0)	20(4/1)	20(4/1)	20(4/1)	24(5/1)	26(5/1)	24(5/1)	28(6/1)
d(PC)	12(3/0)	12(3/0)	16(3/1)	16(3/1)	16(3/1)	20(4/1)	22(4/1)	20(4/1)	24(5/1)
d(PC, ix)*	14(3/0)	14(3/0)	18(3/1)	18(3/1)	18(3/1)	22(4/1)	24(4/1)	22(4/1)	26(5/1)
#xxx	8(2/0)	8(2/0)	12(2/1)	12(2/1)	12(2/1)	16(3/1)	18(3/1)	16(3/1)	20(4/1)

*The size of the index register (ix) does not affect execution time.

TABLE D-3.

Move Long Instruction Execution Times

Source	Destination								
	Dn	An	(An)	(An)+	−(An)	d(An)	d(An, ix)*	xxx.W	xxx.L
Dn	4(1/0)	4(1/0)	12(1/2)	12(1/2)	12(1/2)	16(2/2)	18(2/2)	16(2/2)	20(3/2)
An	4(1/0)	4(1/0)	12(1/2)	12(1/2)	12(1/2)	16(2/2)	18(2/2)	16(2/2)	20(3/2)
(An)	12(3/0)	12(3/0)	20(3/2)	20(3/2)	20(3/2)	24(4/2)	26(4/2)	24(4/2)	28(5/2)
(An)+	12(3/0)	12(3/0)	20(3/2)	20(3/2)	20(3/2)	24(4/2)	26(4/2)	24(4/2)	28(5/2)
−(An)	14(3/0)	14(3/0)	22(3/2)	22(3/2)	22(3/2)	26(4/2)	28(4/2)	26(4/2)	30(5/2)
d(An)	16(4/0)	16(4/0)	24(4/2)	24(4/2)	24(4/2)	28(5/2)	30(5/2)	28(5/2)	32(6/2)
d(An, ix)*	18(4/0)	18(4/0)	26(4/2)	26(4/2)	26(4/2)	30(5/2)	32(5/2)	30(5/2)	34(6/2)
xxx.W	16(4/0)	16(4/0)	24(4/2)	24(4/2)	24(4/2)	28(5/2)	30(5/2)	28(5/2)	32(6/2)
xxx.L	20(5/0)	20(5/0)	28(5/2)	28(5/2)	28(5/2)	32(6/2)	34(6/2)	32(6/2)	36(7/2)
d(PC)	16(4/0)	16(4/0)	24(4/2)	24(4/2)	24(4/2)	28(5/2)	30(5/2)	28(5/2)	32(6/2)
d(PC, ix)*	18(4/0)	18(4/0)	26(4/2)	26(4/2)	26(4/2)	30(5/2)	32(5/2)	30(5/2)	34(6/2)
#xxx	12(3/0)	12(3/0)	20(3/2)	20(3/2)	20(3/2)	24(4/2)	26(4/2)	24(4/2)	28(5/2)

*The size of the index register (ix) does not affect execution time.

D-4 STANDARD INSTRUCTION EXECUTION TIMES

The number of clock periods shown in Table D-4 indicates the time required to perform the operations, store the results, and read the next instruction. The number of bus read and write cycles is shown in parenthesis as (r/w). The number of clock periods and the number of read and write cycles must be added respectively to those of the effective address calculation where indicated.

In Table D-4 the headings have the following meanings: An = address register operand, Dn = data register operand, ea = an operand specified by an effective address, and M = memory effective address operand.

TABLE D-4.

Standard Instruction Execution Times

Instruction	Size	op<ea>, An†	op<ea>, Dn	op Dn, <M>
ADD	Byte, Word	8(1/0) +	4(1/0) +	8(1/1) +
	Long	6(1/0) + **	6(1/0) + **	12(1/2) +
AND	Byte, Word	—	4(1/0) +	8(1/1) +
	Long	—	6(1/0) + **	12(1/2) +
CMP	Byte, Word	6(1/0) +	4(1/0) +	—
	Long	6(1/0) +	6(1/0) +	—
DIVS	—	—	158(1/0) + *	—
DIVU	—	—	140(1/0) + *	—
EOR	Byte, Word	—	4(1/0) ***	8(1/1) +
	Long	—	8(1/0) ***	12(1/2) +
MULS	—	—	70(1/0) + *	—
MULU	—	—	70(1/0) + *	—
OR	Byte, Word	—	4(1/0) +	8(1/1) +
	Long	—	6(1/0) + **	12(1/2) +
SUB	Byte, Word	8(1/0) +	4(1/0) +	8(1/1) +
	Long	6(1/0) + **	6(1/0) + **	12(1/2) +

NOTES:
+ add effective address calculation time
† word or long only
* indicates maximum value
** The base time of six clock periods is increased to eight if the effective address mode is register direct or immediate (effective address time should also be added).
*** Only available effective address mode is data register direct.
DIVS, DIVU — The divide algorithm used by the MC68000 provides less than 10% difference between the best and worst case timings.
MULS, MULU — The multiply algorithm requires 38 + 2n clocks where n is defined as:
  MULU: n = the number of ones in the <ea>
  MULS: n = concatanate the <ea> with a zero as the LSB; n is the resultant number of 10 or 01 patterns in the 17-bit source; i.e., worst case happens when the source is $5555.

D-5 IMMEDIATE INSTRUCTION EXECUTION TIMES

The number of clock periods shown in Table D-5 includes the time to fetch immediate operands, perform the operations, store the results, and read the next operation. The number of bus read and write cycles is shown in parenthesis as (r/w). The number of clock periods and the number of read and write cycles must be added respectively to those of the effective address calculation where indicated.

In Table D-5, the headings have the following meanings: # = immediate operand, Dn = data register operand, An = address register operand, and M = memory operand. SR = status register.

TABLE D-5.

Immediate Instruction Execution Times

Instruction	Size	op #, Dn	op #, An	op #, M
ADDI	Byte, Word	8(2/0)	—	12(2/1) +
	Long	16(3/0)	—	20(3/2) +
ADDQ	Byte, Word	4(1/0)	8(1/0) *	8(1/1) +
	Long	8(1/0)	8(1/0)	12(1/2) +
ANDI	Byte, Word	8(2/0)	—	12(2/1) +
	Long	16(3/0)	—	20(3/1) +
CMPI	Byte, Word	8(2/0)	—	8(2/0) +
	Long	14(3/0)	—	12(3/0) +
EORI	Byte, Word	8(2/0)	—	12(2/1) +
	Long	16(3/0)	—	20(3/2) +
MOVEQ	Long	4(1/0)	—	—
ORI	Byte, Word	8(2/0)	—	12(2/1) +
	Long	16(3/0)	—	20(3/2) +
SUBI	Byte, Word	8(2/0)	—	12(2/1) +
	Long	16(3/0)	—	20(3/2) +
SUBQ	Byte, Word	4(1/0)	8(1/0) *	8(1/1) +
	Long	8(1/0)	8(1/0)	12(1/2) +

+ add effective address calculation time
* word only

D-6 SINGLE OPERAND INSTRUCTION EXECUTION TIMES

Table D-6 indicates the number of clock periods for the single operand instructions. The number of bus read and write cycles is shown in parenthesis as (r/w). The number of clock periods and the number of read and write cycles must be added respectively to those of the effective address calculation where indicated.

TABLE D-6.

Single Operand Instruction Execution Times

Instruction	Size	Register	Memory
CLR	Byte, Word	4(1/0)	8(1/1) +
	Long	6(1/0)	12(1/2) +
NBCD	Byte	6(1/0)	8(1/1) +
NEG	Byte, Word	4(1/0)	8(1/1) +
	Long	6(1/0)	12(1/2) +
NEGX	Byte, Word	4(1/0)	8(1/1) +
	Long	6(1/0)	12(1/2) +
NOT	Byte, Word	4(1/0)	8(1/1) +
	Long	6(1/0)	12(1/2) +
S_{CC}	Byte, False	4(1/0)	8(1/1) +
	Byte, True	6(1/0)	8(1/1) +
TAS	Byte	4(1/0)	10(1/1) +
TST	Byte, Word	4(1/0)	4(1/0) +
	Long	4(1/0)	4(1/0) +

+ add effective address calculation time

D-7 SHIFT/ROTATE INSTRUCTION EXECUTION TIMES

Table D-7 indicates the number of clock periods for the shift and rotate instructions. The number of bus read and write cycles is shown in parenthesis as (r/w). The number of clock periods and the number of read and write cycles must be added respectively to those of the effective address calculation where indicated.

TABLE D-7.

Shift/Rotate Instruction Execution Times

Instruction	Size	Register	Memory
ASR, ASL	Byte, Word	6 + 2n(1/0)	8(1/1) +
	Long	8 + 2n(1/0)	—
LSR, LSL	Byte, Word	6 + 2n(1/0)	8(1/1) +
	Long	8 + 2n(1/0)	—
ROR, ROL	Byte, Word	6 + 2n(1/0)	8(1/1) +
	Long	8 + 2n(1/0)	—
ROXR, ROXL	Byte, Word	6 + 2n(1/0)	8(1/1) +
	Long	8 + 2n(1/0)	—

+ add effective address calculation time
n is the shift count

D-8 BIT MANIPULATION INSTRUCTION EXECUTION TIMES

Table D-8 indicates the number of clock periods required for the bit manipulation instructions. The number of bus read and write cycles is shown in parenthesis as (r/w). The number of clock periods and the number of read and write cycles must be added respectively to those of the effective address calculation where indicated.

TABLE D-8.

Bit Manipulation Instruction Execution Times

Instruction	Size	Dynamic		Static	
		Register	Memory	Register	Memory
BCHG	Byte	—	8(1/1) +	—	12(2/1) +
	Long	8(1/0) *	—	12(2/0) *	—
BCLR	Byte	—	8(1/1) +	—	12(2/1) +
	Long	10(1/0) *	—	14(2/0) *	—
BSET	Byte	—	8(1/1) +	—	12(2/1) +
	Long	8(1/0) *	—	12(2/0) *	—
BTST	Byte	—	4(1/0) +	—	8(2/0) +
	Long	6(1/0)	—	10(2/0)	—

+ add effective address calculation time
* indicates maximum value

D-9 CONDITIONAL INSTRUCTION EXECUTION TIMES

Table D-9 indicates the number of clock periods required for the conditional instructions. The number of bus read and write cycles is indicated in parenthesis as (r/w). The number of clock periods and the number of read and write cycles must be added respectively to those of the effective address calculation where indicated.

TABLE D-9.

Conditional Instruction Execution Times

Instruction	Displacement	Branch Taken	Branch Not Taken
B$_{CC}$	Byte	10(2/0)	8(1/0)
	Word	10(2/0)	12(2/0)
BRA	Byte	10(2/0)	—
	Word	10(2/0)	—
BSR	Byte	18(2/2)	—
	Word	18(2/2)	—
DB$_{CC}$	CC true	—	12(2/0)
	CC false	10(2/0)	14(3/0)

+ add effective address calculation time
* indicates maximum value

D-10 JMP, JSR, LEA, PEA, AND MOVEM INSTRUCTION EXECUTION TIMES

Table D-10 indicates the number of clock periods required for the jump, jump-to-subroutine, load effective address, push effective address, and move multiple registers instructions. The number of bus read and write cycles is shown in parenthesis as (r/w).

TABLE D-10.

JMP, JSR, LEA, PEA, and MOVEM Instruction Execution Times

Instr	Size	(An)	(An) +	− (An)	d(An)	d(An, ix) +	xxx.W	xxx.L	d(PC)	d(PC, ix) *
JMP	—	8(2/0)	—	—	10(2/0)	14(3/0)	10(2/0)	12(3/0)	10(2/0)	14(3/0)
JSR	—	16(2/2)	—	—	18(2/2)	22(2/2)	18(2/2)	20(3/2)	18(2/2)	22(2/2)
LEA	—	4(1/0)	—	—	8(2/0)	12(2/0)	8(2/0)	12(3/0)	8(2/0)	12(2/0)
PEA	—	12(1/2)	—	—	16(2/2)	20(2/2)	16(2/2)	20(3/2)	16(2/2)	20(2/2)
MOVEM M → R	Word	12 + 4n (3 + n/0)	12 + 4n (3 + n/0)		16 + 4n (4 + n/0)	18 + 4n (4 + n/0)	16 + 4n (4 + n/0)	20 + 4n (5 + n/0)	16 + 4n (4 + n/0)	18 + 4n (4 + n/0)
	Long	12 + 8n (3 + 2n/0)	12 + 8n (3 + 2n/0)		16 + 8n (4 + 2n/0)	18 + 8n (4 + 2n/0)	16 + 8n (4 + 2n/0)	20 + 8n (5 + 2n/0)	16 + 8n (4 + 2n/0)	18 + 8n (4 + 2n/0)
MOVEM R → M	Word	8 + 4n (2/n)	—	8 + 4n (2/n)	12 + 4n (3/n)	14 + 4n (3/n)	12 + 4n (3/n)	16 + 4n (4/n)	—	—
	Long	8 + 8n (2/2n)	—	8 + 8n (2/2n)	12 + 8n (3/2n)	14 + 8n (3/2n)	12 + 8n (3/2n)	16 + 8n (4/2n)	—	—

n is the number of registers to move
* is the size of the index register (ix) does not affect the instruction's execution time

D-11 MULTI-PRECISION INSTRUCTION EXECUTION TIMES

Table D-11 indicates the number of clock periods for the multi-precision instructions. The number of clock periods includes the time to fetch both operands, perform the operations, store the results, and read the next instructions. The number of read and write cycles is shown in parenthesis as (r/w).

In Table D-11, the headings have the following meanings: Dn = data register operand and M = memory operand.

TABLE D-11.

Multi-Precision Instruction Execution Times

Instruction	Size	op Dn, Dn	op M, M
ADDX	Byte, Word	4(1/0)	18(3/1)
	Long	8(1/0)	30(5/2)
CMPM	Byte, Word	—	12(3/0)
	Long	—	20(5/0)
SUBX	Byte, Word	4(1/0)	18(3/1)
	Long	8(1/0)	30(5/2)
ABCD	Byte	6(1/0)	18(3/1)
SBCD	Byte	6(1/0)	18(3/1)

D-12 MISCELLANEOUS INSTRUCTION AND EXECUTION TIMES

Table D-12 and D-13 indicate the number of clock periods for the following miscellaneous instructions. The number of bus read and write cycles is shown in parenthesis as (r/w). The number of clock periods plus the number of read and write cycles must be added to those of the effective address calculation where indicated.

TABLE D-12.

Miscellaneous Instruction Execution Times

Instruction	Size	Register	Memory
ANDI to CCR	Byte	20(3/0)	—
ANDI to SR	Word	20(3/0)	—
CHK	—	10(1/0) +	—
EORI to CCR	Byte	20(3/0)	—
EORI to SR	Word	20(3/0)	—
ORI to CCR	Byte	20(3/0)	—
ORI to SR	Word	20(3/0)	—
MOVE from SR	—	6(1/0)	8(1/1) +
MOVE to CCR	—	12(2/0)	12(2/0) +
MOVE to SR	—	12(2/0)	12(2/0) +
EXG	—	6(1/0)	—
EXT	Word	4(1/0)	—
	Long	4(1/0)	—
LINK	—	16(2/2)	—
MOVE from USP	—	4(1/0)	—
MOVE to USP	—	4(1/0)	—
NOP	—	4(1/0)	—
RESET	—	132(1/0)	—
RTE	—	20(5/0)	—
RTR	—	20(5/0)	—
RTS	—	16(4/0)	—
STOP	—	4(0/0)	—
SWAP	—	4(1/0)	—
TRAPV	—	4(1/0)	—
UNLK	—	12(3/0)	—

+ add effective address calculation time

TABLE D-13.

Move Peripheral Instruction Execution Times

Instruction	Size	Register → Memory	Memory → Register
MOVEP	Word	16(2/2)	16(4/0)
	Long	24(2/4)	24(6/0)

D-13 EXCEPTION PROCESSING EXECUTION TIMES

Table D-14 indicates the number of clock periods for exception processing. The number of clock periods includes the time for all stacking, the vector fetch, and the fetch of the first two instruction words of the handler routine. The number of bus read and write cycles is shown in parenthesis as (r/w).

TABLE D-14.

Exception Processing Execution Times

Exception	Periods
Address Error	50(4/7)
Bus Error	50(4/7)
CHK Instruction	44(5/4) +
Divide by Zero	42(5/4)
Illegal Instruction	34(4/3)
Interrupt	44(5/3) *
Privilege Violation	34(4/3)
$\overline{\text{RESET}}$ * *	40(6/0)
Trace	34(4/3)
TRAP Instruction	38(4/4)
TRAPV Instruction	34(4/3)

+ add effective address calculation time

* The interrupt acknowledge cycle is assumed to take four clock periods.

* * Indicates the time from when $\overline{\text{RESET}}$ and $\overline{\text{HALT}}$ are first sampled as negated to when instruction execution starts.

appendix B

complete summary of the MC68000 instruction set and addressing modes

Notation Used:

1. Direct Register Addressing
   - **(a)** Dn   Data Register Direct
   - **(b)** An   Address Register Direct
2. Direct Memory Addressing
   - **(a)** $s   Absolute Short
   - **(b)** $l   Absolute Long
3. Indirect Memory Addressing
   - **(a)** (An)   Register Indirect
   - **(b)** (An)+   Post-increment Register Indirect
   - **(c)** −(An)   Pre-decrement Register Indirect
   - **(d)** $w(An)   Register Indirect with Displacement
   - **(e)** $b(An,Xn)   Register Indirect with Index and Displacement
     Implied Register Addressing
     \<ea\> = SR, USP, SSP, PC
5. Program Counter Relative Addressing
   - **(a)** $w(PC)   PC relative with Displacement
   - **(b)** $b(PC,Xn)   PC relative with Index and Displacement
6. Immediate Data Addressing
   - **(a)** #$q   Immediate
   - **(b)** #p   Quick Immediate

Notes:

Dn	D0 to D7
An	A0 to A7
Xn	Contents of Data Register used as Index Register
b	8 bits (byte)
w	16 bits (word)
l	32 bits (longword) (upper byte not used on 68000)
q	8, 16, or 32 bits (b, w, or l)
p	A number between 1 and 8 (0 = 8, 1 = 1, 2 = 2, . . . , 7 = 7)
s	A short address in the range $0000–$7FFF
$	Denotes a hexadecimal number
#	Denotes data
\<Sea\>	Source Effective Address
\<Dea\>	Destination Effective Address
\<ea\>	Effective Address
*	Indicates a valid addressing mode
.B	Opcode suffix to denote byte-sized \<Sea\>
.W	Opcode suffix to denote word-sized \<Sea\>
.L	Opcode suffix to denote longword-sized \<Sea\>

ABCD   **Add Binary-Coded Decimal with Extend**

Assembler syntax: ABCD.B   \<Sea>,\<Dea>

\<Sea> Source effective address	\<Dea> Destination effective address												
	Dn An	(An)	(An) +	– (An)	$w (An)	$b (An, Xn)	$s	$l	$w (PC)	$b (PC, Xn)	#$q	SR	CCR
Dn – (An)	*			*									

Action: \<Sea> + \<Dea> + extend bit → \<Dea>

Status Register Condition Codes: X   N   Z   V   C
                                      *   U   *   U   *

X is set the same as the carry bit.
N is undefined.
Z is cleared if the result is nonzero. Unchanged otherwise.
V is undefined.
C is set if a BCD carry is generated from the most significant
   BCD nibble. Cleared otherwise.

Note: The low-order bytes of the \<Sea>, \<Dea>, and the extend bit are added, and
          the sum is stored in the \<Dea>.

Add Binary   *ADD*

Assembler syntax: ADD.B   Dn,<Dea>
.W
.L

<Sea> Source effective address	Dn	An	(An)	(An) +	− (An)	$w (An)	$b (An, Xn)	$s	$l	$w (PC)	$b (PC, Xn)	#$q	SR	CCR
							<Dea> Destination effective address							
Dn			*	*	*	*	*	*	*					

Assembler syntax: ADD.B   <Sea>,Dn
.W
.L

<Sea> Source effective address	Dn	<Dea> Destination effective address
Dn	*	(source is word and longword only)
An	*	
(An)	*	
(An) +	*	
− (An)	*	
$w (An)	*	
$b (An, Xn)	*	
$s	*	
$l	*	
$w(PC)	*	
$b (PC, Xn)	*	
#$q	*	
SR		
CCR		

Action: <Sea> + <Dea> → <Dea>

Status Register Condition Codes: X   N   Z   V   C
*   *   *   *   *

X is set the same as the carry bit.
N is set if the result is negative. Cleared otherwise.
Z is set if the result is zero. Cleared otherwise.
V is set if an overflow is generated. Cleared otherwise.
C is set if a carry is generated. Cleared otherwise.

Notes:

1. (a) Dn is added to the <Dea>, and the result is stored in the <Dea>.
   (b) The <Sea> is added to Dn, and the result is stored in Dn.
2. ADDA is used when the destination is an Address Register.
3. ADDI and ADDQ are used when the source is immediate data, and are selected automatically by some assemblers.

ADDA   Add Binary (To Address Register)

Assembler syntax: ADDA.W   <Sea>,An
                             .L

<Sea> Source effective address	<Dea> Destination effective address An	
Dn	*	
An	*	
(An)	*	
(An) +	*	
− (An)	*	
$w (An)	*	
$b (An, Xn)	*	
$s	*	
$l	*	
$w(PC)	*	
$b (PC, Xn)	*	
#$q	*	
SR		
CCR		

Action: <Sea> + <Dea> → <Dea>

Status Register Condition Codes: X N Z V C
                                     - - - - -

No bits are affected.

Notes:

1. The <Sea> is added to the <Dea>, and the result is stored in the <Dea>.
2. Word data is sign-extended to 32 bits.
3. The 32-bit <Dea> is used regardless of the operation size.
4. No Condition Code Register bits are affected.

ADDI   ADDI   Add Binary (Immediate Data)

Assembler syntax: ADDI.B   #$q,<Dea>
                             .W
                             .L

<Sea> Source effective address	<Dea> Destination effective address												
	Dn	An	(An)	(An) +	− (An)	$w (An)	$b (An, Xn)	$s	$l	$w (PC)	$b (PC, Xn)	#$q	SR	CCR
#$q	*		*	*	*	*	*	*	*					

Action: Immediate Data +<Dea> → <Dea>

Status Register Condition Codes: X N Z V C
                                     * * * * *

X is set the same as the carry bit.
N is set if the result is negative. Cleared otherwise.
Z is set if the result is zero. Cleared otherwise.
V is set if an overflow is generated. Cleared otherwise.
C is set if carry is generated. Cleared otherwise.

Notes: The immediate data is added to the <Dea>, and the result is stored in <Dea>.

Add Binary (Quick {1–8} Data)   *ADDQ*

Assembler syntax: **ADDQ.B   \<Sea>,\<Dea>**
                      .W
                      .L

\<Sea> Source effective address	\<Dea> Destination effective address													
	Dn	An	(An)	(An) +	− (An)	$w (An)	$b (An, Xn)	$s	$l	$w (PC)	$b (PC, Xn)	#$q	SR	CCR
#p	*	**	*	*	*	*	*	*	*					

Action: \<Sea> + \<Dea> → \<Dea>

Status Register Condition Codes: X N Z V C
                                     * * * * *

X is set the same as the carry bit.
N is set if the result is negative. Cleared otherwise.
Z is set if the result is zero. Cleared otherwise.
V is set if an overflow is generated. Cleared otherwise.
C is set if a carry is generated. Cleared otherwise.

Notes:

1. The immediate data must be between 1 and 8 ($ not required) (0 = 8, 1 = 1, 2 = 2, . . . , 7 = 7).
**2. Only word and longword sizes are used when an Address Register is the destination. The source data is sign-extended to longword, and the condition codes are NOT affected for this addressing mode.
3. The 32-bit \<Dea> is used regardless of operation size.

ADDX   Add Binary (Extended)   *ADDX*

Assembler syntax: **ADDX.B   \<Sea>,\<Dea>**
                      .W
                      .L

\<Sea> Source effective address	\<Dea> Destination effective address													
	Dn	An	(An)	(An) +	− (An)	$w (An)	$b (An, Xn)	$s	$l	$w (PC)	$b (PC, Xn)	#$q	SR	CCR
Dn − (An)	*				*									

Action: \<Sea> + \<Dea> + extend bit → \<Dea>

Status Register Condition Codes: X N Z V C
                                     * * * * *

X is set the same as the carry bit.
N is set if the result is negative. Cleared otherwise.
Z is cleared if result is nonzero. Unchanged otherwise.
V is set if an overflow is generated. Cleared otherwise.
C is set if a carry is generated. Cleared otherwise.

Notes:

1. The \<Sea>, the \<Dea>, and the extend bit are added, and the result is stored in the \<Dea>.
2. The Z bit is normally set with an instruction prior to execution of this instruction, allowing proper tests for a zero condition upon completion of multiple-precision operations.
3. Used for multiple precision on numbers larger than 32-bit longwords.

AND   **Bitwise Logical AND**

Assembler syntax: **AND.B   Dn,<Dea>**
                           .W
                           .L

<Sea> Source effective address	*<Dea>* Destination effective address													
	Dn	An	(An)	(An) +	– (An)	$w (An)	$b (An, Xn)	$s	$l	$w (PC)	$b (PC, Xn)	#$q	SR	CCR
Dn			*	*	*	*	*	*	*					

Assembler syntax: **AND.B   <Sea>,Dn**
                           .W
                           .L

<Sea> Source effective address	*<Dea>* Destination effective address Dn	
Dn	*	
An		
(An)	*	
(An) +	*	
– (An)	*	
$w (An)	*	
$b (An, Xn)	*	
$s	*	
$l	*	
$w (PC)	*	
$b (PC, Xn)	*	
#$q	*	
SR		
CCR		

Action: Logical AND each bit of <Sea> with <Dea> → <Dea>

Status Register Condition Codes: X   N   Z   V   C
                                                        -    *   *   0   0

X is not affected.
N is set if the result is negative. Cleared otherwise.
Z is set if result is zero. Cleared otherwise.
V is cleared.
C is cleared.

Notes:

1. Each bit of the <Sea> is AND'd with the <Dea>, and the result is stored in the <Dea>.
2. ANDI is used when the source is immediate data, and is selected automatically by some assemblers.

Bitwise Logical and (Immediate Data) *ANDI*

Assembler syntax: ANDI.B <Sea>,<Dea>
.W
.L

<Sea> Source effective address	Dn	An	(An)	(An) +	– (An)	$w (An)	$b (An, Xn)	$s	$l	$w (PC)	$b (PC, Xn)	#$q	SR	CCR
							<Dea> Destination effective address							
#$q	*		*	*	*	*	*	*	*				*	*

Action: Logical AND each bit of <Sea> with <Dea> → <Dea>

Status Register Condition Codes: X   N   Z   V   C
- * * 0 0

X is not affected.
N set if the result is negative. Cleared otherwise.
Z set if result is zero. Cleared otherwise.
V is cleared.
C is cleared.

Notes:

1. Each bit of the immediate data is AND'd with <Dea>, and the result is stored in <Dea>.
2. ANDI #$q,SR is a privileged instruction (supervisor mode only).
3. ANDI #$q,SR and ANDI #$q,CCR instructions affect the CCR bits as follows: Bits are cleared if the corresponding bits are low, unaffected otherwise.

<Sea> data bit: 4   3   2   1   0
CCR flags: X   N   Z   V   C

ASL   **Arithmetic Shift Left**

Assembler syntax: ASL.B   <Sea>,<Dea>
.W
.L

<Sea> Source effective address	<Dea> Destination effective address													
	Dn	An	(An)	(An) +	– (An)	$w (An)	$b (An, Xn)	$s	$l	$w (PC)	$b (PC, Xn)	#$q	SR	CCR
Dn #$q	* *													

Assembler syntax: ASL.W   <ea>

<ea> Effective address													
Dn	An	(An)	(An) +	– (An)	$w (An)	$b (An, Xn)	$s	$l	$w (PC)	$b (PC, Xn)	#$q	SR	CCR
		*	*	*	*	*	*	*					

Action: Left shift <Sea> by <Dea> → <Dea>
   or   Left shift <ea> by one bit → <ea>

Status Register Condition Codes: X   N   Z   V   C
                                      *   *   *   *   *

X is set according to the last bit shifted out. Unaffected for a shift count of zero.
N is set if the most significant bit of the result is set. Cleared otherwise.
Z is set if the result is zero. Cleared otherwise.
V is set if the most significant bit is changed at any time during the shift operation. Cleared otherwise.
C is set according to the last bit shifted out. Cleared for a shift count of zero.

Notes:

1. The <Sea> is shifted left by the count contained in the <Dea>, and the result is stored in the <Dea>.
2. The carry and extend bits receive the last bit shifted out.
3. A zero is shifted into the least significant bit of the operand.
4. The shift count is contained in Dn for Data Register Direct mode.
5. The shift count is between 1 and 8 for immediate data mode.
6. The ASL <ea> instruction shifts the memory word left by one bit only.

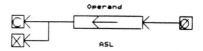

Arithmetic Shift Right   *ASR*

Assembler syntax: ASR.B <Sea>,<Dea>
              .W
              .L

<Sea> Source effective address	Dn	An	(An)	(An) +	– (An)	$w (An)	$b (An, Xn)	$s	$l	$w (PC)	$b (PC, Xn)	#$q	SR	CCR
					<Dea> Destination effective address									
Dn	*													
#$q	*													

Assembler syntax: ASR.W <ea>

	Dn	An	(An)	(An) +	– (An)	$w (An)	$b (An, Xn)	$s	$l	$w (PC)	$b (PC, Xn)	#$q	SR	CCR
						<ea> Effective address								
			*	*	*	*	*	*	*					

Action: Right shift <Sea> by <Dea>   → <Dea>
    or   Right shift <Dea> by <count> → <Dea>

Status Register Condition Codes: X   N   Z   V   C
                     *   *   *   *   *

X is set according to the last bit shifted out. Unaffected for a shift count of zero.
N is set if the most significant bit of the result is set. Cleared otherwise.
Z is set if the result is zero. Cleared otherwise.
V is set if the most significant bit is changed at any time during the shift operation.
  Cleared otherwise.
C is set according to the last bit shifted out. Cleared for a shift count of zero.

Notes:

1. The <Sea> is shifted right by the count contained in the <Dea>, and the result is stored in the <Dea>.
2. The carry and extend bits receive the last bit shifted out.
3. The sign bit is replicated into the high-order bit.
4. The shift count is contained in Dn for Data Register Direct mode.
5. The shift count is between 1 and 8 for Immediate Data mode.
6. The ASR <ea> instruction shifts the memory word right by one bit only.

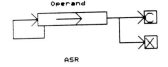

ASR

BCC   **Branch Conditionally**

> *Assembler syntax:*   Bcc.B <label>   *or*   Bcc.B displacement
>   .W   .W

cc is replaced by one of the conditions below:

Action: If (condition is true) then PC + displacement → PC

BCC	Branch if C bit is CLEAR
BCS	Branch if C bit is SET
BEQ	Branch if Z bit is SET
BGE	Branch if N and V bits are either both SET or both CLEAR
BGT	Branch if N and V bits are both SET and Z bit is CLEAR or
	if N, V, and Z bits are ALL CLEAR
BHI	Branch if C and Z bits are both CLEAR
BLE	Branch if the Z bit is SET or if the N bit is SET and the V bit is CLEAR or
	if the N bit is CLEAR and the V bit is SET
BLS	Branch if either the C or Z bit is SET
BLT	Branch if the N bit is SET and the V bit is CLEAR or
	if the N bit is CLEAR and the V bit is SET
BMI	Branch if the N bit is SET
BNE	Branch if the Z bit is CLEAR
BPL	Branch if the N bit is CLEAR
BVC	Branch if the V bit is CLEAR
BVS	Branch if the V bit is SET

Status Register Condition Codes: X   N   Z   V   C

             -   -   -   -   -

No bits are affected.

Notes:

1. If the specified condition is met, program execution continues at the location equal to the sum of the program counter's contents plus the displacement. Otherwise, execution resumes at the instruction following the branch instruction.
2. The displacement, which is sign-extended to 32 bits, allows branching −128 to +127 bytes away for a byte displacement or −32,768 to +32,766 bytes away for a word displacement.
3. The displacement must always be an even number.
4. The label corresponds to an instruction, all of which start on an even address boundary.

Bit Test and Change

BCHG

Assembler syntax: BCHG.B   <Sea>,<Dea>
.L

<Sea> Source effective address	<Dea> Destination effective address													
	Dn	An	(An)	(An) +	− (An)	$w (An)	$b (An, Xn)	$s	$l	$w (PC)	$b (PC, Xn)	#$q	SR	CCR
Dn	*		*	*	*	*	*	*	*					
#$q	*		*	*	*	*	*	*	*					

Action: The bit indicated by the <Sea> is tested, the Z bit is set or cleared accordingly, and the corresponding bit in the <Dea> is inverted.

Status Register Condition Codes: X N Z V C
\- - * - -

X is not affected.
N is not affected.
Z is set if the bit tested was zero before inversion. Cleared otherwise.
V is not affected.
C is not affected.

Notes:

1. When <Dea> is a Data Register, only longwords operations are allowed.
2. When <Dea> is a memory location, only byte operations are allowed.
3. Bits are numbered from 0, 0 being the least significant.

Bit Test and Clear

BCLR

Assembler syntax: BCLR.B   <Sea>,<Dea>
.L

<Sea> Source effective address	<Dea> Destination effective address													
	Dn	An	(An)	(An) +	− (An)	$w (An)	$b (An, Xn)	$s	$l	$w (PC)	$b (PC, Xn)	#$q	SR	CCR
Dn	*		*	*	*	*	*	*	*					
#$q	*		*	*	*	*	*	*	*					

Action: The bit indicated by the <Sea> is tested, the Z bit is set or cleared accordingly, and the corresponding bit in the <Dea> is cleared.

Status Register Condition Codes: X N Z V C
\- - * - -

X is not affected.
N is not affected.
Z is set if the bit tested was zero before being cleared. Cleared otherwise.
V is not affected.
C is not affected.

Notes:

1. When <Dea> is a Data Register, only longwords operations are allowed.
2. When <Dea> is a memory location, only byte operations are allowed.
3. Bits are numbered from 0, 0 being the least significant.

BRA   **Branch Always**

Assembler syntax: **BRA.B \<label\>   or   BRA.B displacement**
                                  .W   .W

Action: The displacement is added to the contents of the Program Counter and program execution resumes from that location

Status Register Condition Codes: X N Z V C
                                                            -   -   -   -   -

No bits are affected.

Notes:

1. The displacement is a two's-complement integer denoting the relative distance in bytes between the *beginning* of the instruction following the BRA instruction and the next instruction to be executed.
2. Byte displacements give a branch range of −128 to +126 bytes away from the BRA instruction.
3. Word displacements give a branch range of −32,768 to +32,766 bytes away from the BRA instruction.
4. The displacement must always be *even,* since instructions must begin on an even address.
5. Some assemblers use .S notation to denote .B (byte) displacements. Some do not require the .B or .W displacement size, calculating the proper size.

BSET   **Test a Bit and Set**

Assembler syntax: **BSET.B \<Sea\>,\<Dea\>**
                                    .L

\<Sea\> Source effective address	Dn	An	(An)	(An) +	− (An)	$w(An)	$b(An,Xn)	$s	$l	$w(PC)	$b(PC, Xn)	#$q	SR	CCR
Dn	*		*	*	*	*	*	*	*					
#$q	*		*	*	*	*	*	*	*					

Action: The bit indicated by the \<Sea\> is tested, the z bit is set or cleared accordingly, and the corresponding bit in the \<Dea\> is set

Status Register Condition Codes: X N Z V C
                                                            -   -   *   -   -

X is not affected.
N is not affected.
Z is set if the bit tested was zero before being set. Cleared otherwise.
V is not affected.
C is not affected.

Notes:

1. When \<Dea\> is a Data Register, only longwords operations are allowed.
2. When \<Dea\> is a memory location, only byte operations are allowed.
3. Bits are numbered from 0, 0 being the least significant.

Branch to Subroutine   *BSR*

> *Assembler syntax:* **BSR.B   <label>   or   BSR.B displacement**
>   .W   .W
>
> *Action:*   PC → −(SP); PC + displacement → PC
>
> *Status Register Condition Codes:*   X   N   Z   V   C
>   - - - - -

No bits are affected.

Notes:

1. The address of the next instruction to be executed is pushed on top of the stack. The displacement is added to the Program Counter contents, and execution continues from that location.
2. The displacement is a two's-complement integer denoting the relative distance in bytes between the *beginning* of the instruction following the BSR instruction and the next instruction to be executed.
3. Byte displacements give a branch range of −128 to +126 bytes away from the BSR instruction.
4. Word displacements give a branch range of −32,768 to +32,766 bytes away from the BSR instruction.
5. The displacement must always be *even,* since instructions must begin on an even address.
6. Some assemblers use .S notation to denote .B (byte) displacements. Some do not require the .B or .W displacement size, calculating the proper size.

Bit Test   *BTST*

> *Assembler syntax:*   **BTST.B   <Sea>,<Dea>**
>   .L

<Sea> Source effective address	\<Dea\> Destination effective address														
	Dn	An	(An)	(An) +	− (An)	$w(An)	$b(An,Xn)	$s	$l	$w(PC)	$b(PC,Xn)	#$q	SR	CCR
Dn	*		*	*	*	*	*	*	*					
#$q	*		*	*	*	*	*	*	*					

> *Action:*   The bit indicated by the <Sea> is tested, and the Z bit is set or cleared accordingly; if the tested bit is 0, the Z bit is set (TRUE); if the tested bit is 1, the Z bit is cleared (FALSE)
>
> *Status Register Condition Codes:*   X   N   Z   V   C
>   - - * - -

X is not affected.
N is not affected.
Z is set if the bit tested was zero. Cleared otherwise.
V is not affected.
C is not affected.

Notes:

1. When <Dea> is a Data Register, only longwords operations are allowed.
2. When <Dea> is a memory location, only byte operations are allowed.
3. Bits are numbered from 0, 0 being the least significant.

CHK

Check Register with Bounds

Assembler syntax: **CHK.W <Sea>,Dn**

<Sea> Source effective address	<Dea> Destination effective address	
	Dn	
Dn	*	
An		
(An)	*	
(An) +	*	
− (An)	*	
$w(An)	*	
$b(An,Xn)	*	
$s	*	
$l	*	
$w(PC)	*	
$b(PC,Xn)	*	
#$q	*	
SR		
CCR		

Action: If Dn<0 or Dn> (<Sea>), then TRAP

Status Register Condition Codes: X N Z V C
                                      - * U U U

X is not affected.
N is set if Dn<0; cleared if Dn> (<Sea>). Undefined otherwise.
Z is undefined.
V is undefined.
C is undefined.

Note: If the contents of Dn is less than zero or is greater than the contents of <Sea>, a trap is executed.

CLR

Clear Operand

Assembler syntax: **CLR.B <Dea>**
                              .W
                              .L

	<Dea> Destination effective address												
	Dn	An	(An)	(An) +	− (An)	$w(An)	$b(An,Xn)	$s	$l	$w(PC)	$b(PC,Xn) | #$q	SR	CCR
	*		*	*	*	*	*	*	*					

Action: 0 → <Dea>

Status Register Condition Codes: X N Z V C
                                      - 0 1 0 0

X is not affected.
N is cleared.
Z is set.
V is cleared.

Note: A memory destination is read before it is written to.

Compare   *CMP*

Assembler syntax: **CMP.B   \<Sea\>,Dn**
                          .W
                          .L

\<Sea\> Source effective address	Dn	\<Dea\> Destination effective address
Dn	*	(word and longword only)
An	*	
(An)	*	
(An) +	*	
− (An)	*	
$w(An)	*	
$b(An,Xn)	*	
$s	*	
$l	*	
$w(PC)	*	
$b(PC,Xn)	*	
#$q	*	
SR		
CCR		

Action: The \<Sea\> is subtracted from the \<Dea\>, and the condition codes are changed accordingly

Status Register Condition Codes: X   N   Z   V   C
                                     -   *   *   *   *

X is not affected.
N is set if the result is negative. Cleared otherwise.
Z is set if the result is negative. Cleared otherwise.
V is set if overflow occurs. Cleared otherwise.
C is set if a borrow occurs. Cleared otherwise.

Notes:

1. Neither the \<Sea\> nor \<Dea\> is changed.
2. CMPA is used when the \<Dea\> is an Address Register.
3. CMPI is used when the \<Sea\> is immediate data.
4. CMPM is used for memory-to-memory compares.

CMPA   **Compare Address**

Assembler syntax: CMPA.W   <Sea>,An
                                  .L

<Sea> Source effective address	<Dea> Destination effective address	
	An	
Dn	*	
An	*	
(An)	*	
(An) +	*	
− (An)	*	
$w (An)	*	
$b (An, Xn)	*	
$s	*	
$l	*	
$w(PC)	*	
$b (PC, Xn)	*	
#$q	*	
SR		
CCR		

Action: The <Sea> is subtracted from the <Dea>, and the condition codes are changed accordingly

Status Register Condition Codes: X   N   Z   V   C
                                      -   *   *   *   *

X is not affected.
N is set if the result is negative. Cleared otherwise.
Z is set if the result is zero. Cleared otherwise.
V is set if an overflow occurs. Cleared otherwise.
C is set if a borrow occurs. Cleared otherwise.

Notes:

1. Neither the <Sea> nor <Dea> is changed.
2. Word-length source operands are sign-extended to longwords before the compare.

Compare Immediate   *CMPI*

Assembler syntax: **CMPI.B   #$q,<Dea>**
                              .W
                              .L

<Sea> *Source* *effective* *address*	Dn	An	(An)	(An) +	– (An)	$w (An)	$b (An, Xn)	$s	$l	$w (PC)	$b (PC, Xn)	#$q	SR	CCR
#$q	*		*	*	*	*	*	*	*					

Action: The <Sea> is subtracted from the <Dea>, and the condition codes are changed accordingly

Status Register Condition Codes: X   N   Z   V   C
                                     -   *   *   *   *

X is not affected.
N is set if the result is negative. Cleared otherwise.
Z is set if the result is zero. Cleared otherwise.
V is set if an overflow occurs. Cleared otherwise.
C is set if a borrow occurs. Cleared otherwise.

Notes:

1. Neither the <Sea> nor <Dea> is changed.
2. The size of the <Sea> matches the operation size.

Compare Memory   *CMPM*

Assembler syntax: **CMPM.B   <Sea>,<Dea>**
                              .W
                              .L

<Sea> *Source* *effective* *address*	Dn	An	(An)	(An) +	– (An)	$w (An)	$b (An, Xn)	$s	$l	$w (PC)	$b (PC, Xn)	#$q	SR	CCR
(An) +				*										

Action: The <Sea> is subtracted from the <Dea>, and the condition codes are changed accordingly

Status Register Condition Codes: X   N   Z   V   C
                                     -   *   *   *   *

X is not affected.
N is set if the result is negative. Cleared otherwise.
Z is set if the result is zero. Cleared otherwise.
V is set if an overflow occurs. Cleared otherwise.
C is set if a borrow occurs. Cleared otherwise.

Note: Neither the <Sea> nor <Dea> is changed.

DBcc   Decrement and Branch Conditionally

Assembler syntax: **DBcc.W   Dn,<label>   or   DBcc.W   Dn, displacement**

cc is replaced by one of the conditions below.

Action: If (condition is false), then Dn − 1 → Dn; if Dn not equal −1, then PC + displacement → PC else PC + 2 → PC (resumes at next instruction)

DBCC	Terminate if C bit is CLEAR
DBCS	Terminate if C bit is SET
DBEQ	Terminate if Z bit is SET
DBGE	Terminate if N and V bits are either both SET or both CLEAR
DBGT	Terminate if the N and V bits are both SET and the Z bit is CLEAR or if the N, V, and Z bits are ALL CLEAR
DBHI	Terminate if the C and Z bits are both CLEAR
DBLE	Terminate if the Z bit is SET or if the N bit is SET and the V bit is CLEAR or if the N bit is CLEAR and the V bit is SET
DBLS	Terminate if EITHER the C or Z bit is SET
DBLT	Terminate if the N bit is SET and the V bit is CLEAR or if the N bit is CLEAR and the V bit is SET
DBMI	Terminate if the N bit is SET
DBNE	Terminate if the Z bit is CLEAR
DBPL	Terminate if the N bit is CLEAR
DBRA	Terminate on countdown of Dn (Decrement and Branch)
DBVC	Terminate if the V bit is CLEAR
DBVS	Terminate if the V bit is SET
DBF	Terminate on countdown of Dn only, equivalent to DBRA
DBT	Always terminate, no loop at all

Status Register Condition Codes: X   N   Z   V   C
                                      -   -   -   -   -

No bits are affected.

Note: If the specified condition is not met, the low-order 16 bits of Dn are decremented by one, and if it is not equal to −1, execution continues at <label>. Otherwise, execution resumes at the instruction following the branch instruction.

Signed Divide

DIVS

Assembler syntax: DIVS.W   <Sea>,Dn

<Sea> Source effective address	<Dea> Destination effective address	
	Dn	
Dn	*	
An		
(An)	*	
(An) +	*	
− (An)	*	
$w(An)	*	
$b(An,Xn)	*	
$s	*	
$l	*	
$w(PC)	*	
$b(PC,Xn)	*	
#$q	*	
SR		
CCR		

Action: <Dea>/<Sea> → <Dea>

Status Register Condition Codes: X   N   Z   V   C
$$- \quad * \quad * \quad * \quad 0$$

X is not affected.
N is set if the quotient is negative. Undefined if overflow.
  Cleared otherwise.
Z is set if the quotient is zero. Undefined if overflow.
  Cleared otherwise.
V is set if division overflow occurs. Cleared otherwise.
C is cleared.

Notes:

1. The longword-sized <Dea> is divided by a word-sized <Sea>, and the result is stored in the <Dea>.
2. The operation is performed using signed arithmetic.
3. The quotient is stored in the lower word of the <Dea>.
4. The remainder is stored in the upper word of the <Dea>.
5. The sign of the remainder is the same as the dividend unless it is equal to zero.
6. Division by zero causes a trap.
7. Overflow may occur before completion of the instruction. If this occurs, the condition is flagged and the operands are unaffected.
8. Overflow occurs if the quotient is larger than a 16-bit signed integer.

DIVU

Unsigned Divide

Assembler syntax: **DIVU.W** **<Sea>,Dn**

<Sea> Source effective address	<Dea> Destination effective address Dn	
Dn	*	
An		
(An)	*	
(An) +	*	
− (An)	*	
$w(An)	*	
$b(An,Xn)	*	
$s	*	
$l	*	
$w(PC)	*	
$b(PC,Xn)	*	
#$q	*	
SR		
CCR		

Action: $<Dea>/<Sea> \rightarrow <Dea>$

Status Register Condition Codes:
X	N	Z	V	C
-	*	*	*	0

X is not affected.
N is set if the quotient is negative. Undefined if overflow.
  Cleared otherwise.
Z is set if the quotient is zero. Undefined if overflow.
  Cleared otherwise.
V is set if division overflow occurs. Cleared otherwise.
C is cleared.

Notes:

1. The longword-sized <Dea> is divided by a word-sized <Sea>, and the result is stored in the <Dea>.
2. The operation is performed using unsigned arithmetic.
3. The quotient is stored in the lower word of the <Dea>.
4. The remainder is stored in the upper word of the <Dea>.
5. The sign of the remainder is the same as the dividend unless it is equal to zero.
6. Division by zero causes a TRAP.
7. Overflow may occur before completion of the instruction. If this occurs, the condition is flagged and the operands are unaffected.
8. Overflow occurs if the quotient is larger than a 16-bit unsigned integer.

Exclusive OR

EOR

Assembler syntax: EOR.B   Dn,<Dea>
                                    .W
                                    .L

<Sea> Source effective address	<Dea> Destination effective address													
	Dn	An	(An)	(An) +	– (An)	$w (An)	$b (An, Xn)	$s	$l	$w (PC)	$b (PC, Xn)	#$q	SR	CCR
Dn	*		*	*	*	*	*	*	*					

Action: The <Sea> and <Dea> are exclusive-OR'd, and the result is stored in the <Dea>

Status register condition codes: X   N   Z   V   C
                                                      -   *   *   0   0

X is not affected.
N is set if the most significant bit of the result is set.
   Cleared otherwise.
Z is set if the result is zero. Cleared otherwise.
V is cleared.
C is cleared.

Note: EORI is used when the <Sea> is immediate data.

Exclusive OR Immediate Data

EORI

Assembler syntax: EORI.B   #$q, the <Dea>
                                      .W
                                      .L

<Sea> Source effective address	<Dea> Destination effective address													
	Dn	An	(An)	(An) +	– (An)	$w (An)	$b (An, Xn)	$s	$l	$w (PC)	$b (PC, Xn)	#$q	SR	CCR
#$q	*		*	*	*	*	*	*	*					

Action: The <Sea> and <Dea> are exclusive-OR'd, and the result is stored in the <Dea>

Status Register Condition Codes: X   N   Z   V   C
                                                        -   *   *   0   0

X is not affected.
N is set if the most significant bit of the result is set.
   Cleared otherwise.
Z is set if the result is zero. Cleared otherwise.
V is cleared.
C is cleared.

Notes:

   1. The immediate data size matches the operation size.
   2. See the following instruction if the <Dea> is the CCR or SR.

EORI

Exclusive OR Immediate Data (<Dea> is SR or CCR)

Assembler syntax: EORI.B   <Sea>,CCR or
EORI.W   <Sea>,SR

<Sea> Source effective address	Dn	An	(An)	(An) +	− (An)	$w (An)	$b (An, Xn)	$s	$l	$w (PC)	$b (PC, Xn)	#$q	SR	CCR
#$q													*	*

<table>
<tr><td colspan="11"></td></tr>
</table>

Action: The <Sea> and <Dea> are exclusive-OR'd, and the result is stored in the <Dea>

Status Register Condition Codes: X   N   Z   V   C
* * * * *

X changed if bit 4 of <Sea> is one. Unchanged otherwise.
N changed if bit 3 of <Sea> is one. Unchanged otherwise.
Z changed if bit 2 of <Sea> is one. Unchanged otherwise.
V changed if bit 1 of <Sea> is one. Unchanged otherwise.
C changed if bit 0 of <Sea> is one. Unchanged otherwise.

Notes:

1. When the <Dea> is the CCR, byte size is used and the result is stored in the low-order byte of the status register.
2. When the <Dea> is the SR, word size is used, and all bits of the status register are affected. The processor must be in the supervisor mode, or a TRAP results.

EXG

Exchange Registers

Assembler syntax: EXG.L   <Sea>,<Dea>

<Sea> Source effective address	Dn	An	(An)	(An) +	− (An)	$w (An)	$b (An, Xn)	$s	$l	$w (PC)	$b (PC, Xn)	#$q	SR	CCR
Dn	*	*												
An	*	*												

Action: The contents of the two registers are exchanged

Status Register Condition Codes: X   N   Z   V   C
-   -   -   -   -

No bits are affected.

Note: The exchange is always a longword.

Sign Extend   *EXT*

Assembler syntax: **EXT.W   Dn**
                          .L

	Dn	An	(An)	(An) +	– (An)	$w(An)	<Dea> Destination effective address $b(An,Xn)	$s	$l	$w(PC)	$b(PC,Xn)	#$q	SR	CCR
	*													

Action: The sign bit is extended from a byte to a word or from a word to a longword, depending on the size selected

Status Register Condition Codes: X   N   Z   V   C
                                      -   *   *   0   0

X is not affected.
N is set if the result is negative. Cleared otherwise.
Z is set if the result is zero. Cleared otherwise.
V is cleared.
C is cleared.

Notes:

1. If the operation is word sized, bit 7 of the data register is copied to bits 8 to 15.
2. If the operation is longword sized, bit 15 of the data register is copied to bits 16 to 31.

Illegal Instruction   *ILLEGAL*

Assembler syntax: **No mnemonic, hex code is 4AFC**

Action: PC → −(SSP); SR → −(SSP); (illegal instruction vector) → PC

Status Register Condition Codes: X   N   Z   V   C
                                      -   -   -   -   -

No bits are affected.

Note: This is not really an instruction, but is guaranteed to cause an illegal instruction TRAP on all 68000 family processors.

Jump to Effective Address   *JMP*

Assembler syntax: **JMP   <label>   or   JMP   <ea>**

	Dn	An	(An)	(An) +	– (An)	$w (An)	<ea> Effective address $b (An, Xn)	$s	$l	$w (PC)	$b (PC, Xn)	#$q	SR	CCR
			*				*	*	*	*	*			

Action: <ea> → Program Counter

Status Register Condition Codes: X   N   Z   V   C
                                      -   -   -   -   -

No bits are affected.

Note: Program execution resumes at the address specified by the <ea>.

JSR   Jump to Subroutine

Assembler syntax: JSR   <label>   or   JSR   <ea>

						<ea> *Effective address*								
Dn	An	(An)	(An) +	– (An)	$w (An)	$b (An, Xn)	$s	$l	$w (PC)	$b (PC, Xn)	#$q	SR	CCR
		*			*	*	*	*	*	*			

Action: PC → –(SP); <ea> → Program Counter

Status Register Condition Codes: X   N   Z   V   C
                                        -   -   -   -   -

No bits are affected.

Notes:

1. The address of the next instruction following the JSR instruction is saved on the stack, and then program execution resumes at the <ea>.
2. Normal resumption of a program is done with a RTS at the end of the subroutine.

LEA   Load Effective Address

Assembler syntax: LEA.L   <Sea>,An

<Sea> Source effective address	*<Dea>* Destination effective address	
	An	
Dn		
An		
(An)	*	
(An) +		
– (An)		
$w (An)	*	
$b (An, Xn)	*	
$s		
$l	*	
$w(PC)	*	
$b (PC, Xn)	*	
#$q		
SR		
CCR		

Action: <Sea> → An

Status Register Condition Codes: X   N   Z   V   C
                                        -   -   -   -   -

No bits are affected.

Notes:

1. The <Sea> is loaded into An. All 32 bits of An are affected.
2. This instruction is used to write position-independent programs.
3. A constant can be added to An without altering the CCR flags using the $w(An) <Sea>.

Link and Allocate

LINK

Assembler syntax: **LINK An,#<displacement>**

Action: An → −(SP); SP → An; SP + displacement → SP

Status Register Condition Codes: X   N   Z   V   C
                                     -   -   -   -   -

No bits are affected.

Notes:

1. The content of the Address Register is pushed onto the stack, and is then loaded from the updated Stack Pointer.
2. A 16-bit displacement is added to the Stack Pointer. A negative displacement is specified to allocate stack area.

Logical Shift Left

LSL

Assembler syntax: **LSL.B   <Sea>,Dn**
                      .W
                      .L

<Sea> Source effective address	<Dea> Destination effective address														
	Dn	An	(An)	(An) +	− (An)	$w (An)	$b (An, Xn)	$s	$l	$w (PC)	$b (PC, Xn)	#$q	SR	CCR
Dn	*													
#$q	*													

Action: Left shift <Sea> by <count stored in Dn> → <Dea>

Assembler syntax: **LSL.W   <ea>**

<Dea> Destination effective address														
Dn	An	(An)	(An) +	− (An)	$w (An)	$b (An, Xn)	$s	$l	$w (PC)	$b (PC, Xn)	#$q	SR	CCR
		*	*	*	*	*	*	*					

Action: Left shift <ea> by one bit → <ea>

Status Register Condition Codes: X   N   Z   V   C
                                     *   *   *   0   *

X is set according to the last bit shifted out.
    Unaffected for a shift count of zero.
N is set if the result is negative. Cleared otherwise.
Z is set if the result is zero. Cleared otherwise.
V is cleared.
C is set according to the last bit shifted out.
    Cleared for a shift count of zero.

Notes:

1. The carry and extend bits receive the last bit shifted out.
2. The shift count is contained in Dn for Data Register Direct mode.
3. The shift count is between 1 and 8 for Immediate Data mode.

LSR   Logical Shift Right

Assembler syntax: **LSR.B <Sea>,Dn**
                                            .W
                                            .L

<Sea> Source effective address	Dn	An	(An)	(An) +	– (An)	$w (An)	$b (An, Xn)	$s	$l	$w (PC)	$b (PC, Xn)	#$q	SR	CCR
Dn	*													
#$q	*													

<Dea> Destination effective address

Assembler syntax: **LSR.W <ea>**

Dn	An	(An)	(An) +	– (An)	$w (An)	$b (An, Xn)	$s	$l	$w (PC)	$b (PC, Xn)	#$q	SR	CCR
		*	*	*	*	*	*	*					

<Dea> Destination effective address

Action: Right shift <Dea> by <count> → <Dea>
     or   Right shift <ea> by one bit → <ea>

Status Register Condition Codes: X   N   Z   V   C
                                                        *   *   *   0   *

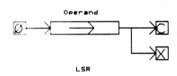

LSR

X is set according to the last bit shifted out. Unaffected for a shift count of zero.
N is set if the result is negative. Cleared otherwise.
Z is set if the result is zero. Cleared otherwise.
V is cleared.
C is set according to the last bit shifted out. Cleared for a shift count of zero.

Notes:

1. The carry and extend bits receive the last bit shifted out.
2. The sign bit is replicated into the high-order bit.
3. The shift count is contained in Dn for Data Register Direct mode.
4. The shift count is between 1 and 8 for Immediate Data mode.

Move Source to Destination   *MOVE*

Assembler syntax: MOVE.B   <Sea>,<Dea>
                            .W
                            .L

<Sea> Source effective address	<Dea> Destination effective address													
	Dn	An	(An)	(An) +	− (An)	$w(An)	$b(An,Xn)	$s	$l	$w(PC)	$b(PC,Xn)	#$q	SR	CCR
Dn	*		*	*	*	*	*	*	*					
An	*		*	*	*	*	*	*	*					
(An)	*		*	*	*	*	*	*	*					
(An) +	*		*	*	*	*	*	*	*					
− (An)	*		*	*	*	*	*	*	*					
$w(An)	*		*	*	*	*	*	*	*					
$b(An,Xn)	*		*	*	*	*	*	*	*					
$s	*		*	*	*	*	*	*	*					
$l	*		*	*	*	*	*	*	*					
$w(PC)	*		*	*	*	*	*	*	*					
$b(PC,Xn)	*		*	*	*	*	*	*	*					
#$q	*		*	*	*	*	*	*	*					
SR														
CCR														

Action: <Sea> → <Dea>

Status Register Condition Codes: X   N   Z   V   C
                                      -   *   *   0   0

X is not affected.
N is set if the result is negative. Cleared otherwise.
Z is set if the result is zero. Cleared otherwise.
V always cleared.
C always cleared.

Notes:

1. MOVEA is used when the destination is an Address Register.
2. MOVEM allows moving of a group of registers to/from memory.
3. MOVEP is used for moving data to 8-bit peripheral devices.
4. MOVEQ is used for moving a BYTE, which gets sign extended to 32 bits.
5. The CCR bits ARE affected by this instruction, unlike some 8-bit microprocessors.
6. See the next instruction if the <Sea> or <Dea> is the CCR or SR.

MOVE   Move Source to Destination (CCR or SR)

Assembler Syntax: MOVE.W   <Sea>,<Dea>

<Sea> Source effective address	Dn	An	(An)	(An)+	-(An)	$w (An)	$b (An, Xn)	$s	$l	$w (PC)	$b (PC, Xn)	#$q	SR	CCR
Dn													*	*
An														
(An)													*	*
(An) +													*	*
- (An)													*	*
$w (An)													*	*
$b (An, Xn)													*	*
$s													*	*
$l													*	*
$w (PC)													*	*
$b (PC, Xn)													*	*
#$q													*	*
SR (1)	*		*	*	*	*	*	*	*					
CCR (2)	*		*	*	*	*	*	*	*					

(3) (4)

Action: (1) SR → <Dea>
(2) CCR → <Dea> (MC68010 only)
(3) If in supervisor state <Sea> → SR, else TRAP
(4) <Sea> → CCR

Status Register Condition Codes: X   N   Z   V   C
(1) No bits are affected.   -   -   -   -   -
(2) No bits are affected.   -   -   -   -   -
(3) Set according to the source operand.
(4)   *   *   *   *   *
X is set the same as bit 4 of the source operand.
N is set the same as bit 3 of the source operand.
Z is set the same as bit 2 of the source operand.
V is set the same as bit 1 of the source operand.
C is set the same as bit 0 of the source operand.

MOVE   Move User Stack Pointer (Privileged Instruction)

Assembler syntax: MOVE.L   <Sea>,<Dea>

<Sea> Source effective address	<Dea> Destination effective address	
	An	USP
An		*
USP	*	

Action: If in supervisor state <Sea> → <Dea>, else TRAP

Status Register Condition Codes: X   N   Z   V   C
                                        -    -   -   -   -

No bits are affected.

Note: The contents of the user stack pointer are transferred to or from An.

MOVEA (To Address Register)   *MOVEA*

Assembler Syntax: MOVEA.W   <Sea>, An
                           .L

<Sea> Source effective address	<Dea> Destination effective address An	
Dn	*	
An	*	
(An)	*	
(An) +	*	
− (An)	*	
$w (An)	*	
$b (An, Xn)	*	
$s	*	
$l	*	
$w(PC)	*	
$b (PC, Xn)	*	
#$q	*	
SR		
CCR		

Action: <Sea> → An

Status Register Condition Codes: X   N   Z   V   C
                                      -   -   -   -   -

No bits are affected.

Notes:

1. The <Sea> is stored in An.
2. Only word and longword data sizes are allowed.
3. All 32 bits of the An are affected.
4. Word-sized data are sign-extended to 32 bits.
5. The CCR bits are not affected, unlike the MOVE instruction.

MOVEM

Move Multiple Registers

Assembler syntax: **MOVEM.W** **<register list>,<Dea>**
.L

	Dn	An	(An)	(An) +	– (An)	$w (An)	$b (An, Xn)	$s	$l	$w (PC)	$b (PC, Xn)	#$q	SR	CCR
			*		*	*	*	*	*					

Column group header: <Dea> Destination effective address (spans Dn through CCR)

Action: Selected registers → consecutive memory locations starting at <Dea>

Assembler syntax: **MOVEM.W** **<Sea>,<register list>**
.L

<Sea> Source effective address	<Dea> Destination effective address	
	Dn	An
Dn		
An		
(An)		*
(An) +		*
– (An)		
$w(An)		*
$b(An,Xn)		*
$s		*
$l		*
$w(PC)		*
$b(PC,Xn)		*
#$q		
SR		
CCR		

Action: Selected registers are loaded from consecutive memory locations starting at <Sea>

Status Register Condition Codes: X N Z V C
                                     - - - - -

Notes:

1. Only word and longword data sizes are allowed.
2. All 32 bits of the <Dea> are affected.
3. Word-sized data are sign-extended to 32 bits.
4. <register list> format examples:
   D0-DN   is D0 to Dn
   A2-An   is A2 to An
   D1-D3/A3 is D1 to D3, and A3

Move (To 8-Bit Peripheral)   *MOVEP*

Assembler syntax: MOVEP.W   <Sea>,<Dea>
                                           .L

<Sea> Source effective address	<Dea> Destination effective address														
	Dn	An	(An)	(An) +	− (An)	$w (An)	$b (An, Xn)	$s	$l	$w (PC)	$b (PC, Xn)	#$q	SR	CCR
Dn An (An) (An) + − (An) $w (An)	*													

Action: Data is transferred between a Data Register and alternate bytes of memory, starting at <Dea> and incrementing by two

Status Register Condition Codes: X N Z V C
                                                          - - - - -

No bits are affected.

Notes:

1. Only word and longword data sizes are allowed.
2. If the <Dea> is even, transfers are made on the high-order half of the data bus.
3. If the <Dea> is odd, transfers are made on the low-order half of the data bus.
4. This instruction simplifies data transfers to/from 8-bit devices.
5. If a longword operation is specified, four 8-bit bytes will be transferred.

Move (Quick Data)   *MOVEQ*

Assembler syntax: MOVEQ.L   #$q,Dn

<Sea> Source effective address	Dn	<Dea> Destination effective address
#$q	*	

Action: #$q → Dn

Status Register Condition Codes: X N Z V C
                                                          - - - - -

No bits are affected.

Notes:

1. The immediate data is limited to byte size, allowing constants in the range −128 to +127 decimal.
2. The data is sign-extended to a longword and all 32 bits stored in the Dn.
3. Dn will contain a 32-bit representation of the signed number between −128 and +127.

MULS   Signed Multiply

Assembler syntax: MULS.W   <Sea>,Dn

<Sea> Source effective address	<Dea> Destination effective address Dn	
Dn	*	
An		
(An)	*	
(An) +	*	
− (An)	*	
$w (An)	*	
$b (An, Xn)	*	
$s	*	
$l	*	
$w(PC)	*	
$b (PC, Xn)	*	
#$q	*	
SR		
CCR		

Action: <Sea>*Dn → Dn

Status Register Condition Codes:

X	N	Z	V	C
-	*	*	0	0

X is not affected.
N is set if result is negative. Cleared otherwise.
Z is set if result is zero. Cleared otherwise.
V is cleared.
C is cleared.

Notes:

1. The two 16-bit signed operands are multiplied using signed arithmetic, and 32 bits of the product are saved in Dn.
2. Signed arithmetic is used.

Unsigned Multiply   *MULU*

Assembler syntax: **MULU.W   <Sea>,Dn**

<Sea> Source effective address	<Dea> Destination effective address	
	Dn	
Dn	*	
An		
(An)	*	
(An) +	*	
− (An)	*	
$w (An)	*	
$b (An, Xn)	*	
$s	*	
$l	*	
$w(PC)	*	
$b (PC, Xn)	*	
#$q	*	
SR		
CCR		

Action: <Sea>*Dn → Dn

Status Register Condition Codes:
X   N   Z   V   C
-   *   *   0   0

Z is not affected.
N is set if the most significant bit of the result is negative.
  Cleared otherwise.
Z is set if result is zero. Cleared otherwise.
V is cleared.
C is cleared.

Note: The two 16-bit unsigned operands are multiplied using unsigned arithmetic,
and 32 bits of the product are saved in Dn.

NBCD   **Negate Decimal with Extend**

Assembler syntax: **NBCD.B   \<Dea>**

					\<Dea>								
					Destination effective address								
Dn	An	(An)	(An) +	− (An)	$w (An)	$b (An, Xn)	$s	$l	$w (PC)	$b (PC, Xn)	#$q	SR	CCR
*		*	*	*	*	*	*						

Action: 0−\<Dea>−extend bit → \<Dea>

Status Register Condition Codes: X   N   Z   V   C
                                      *   U   *   U   *

X is set the same as the carry bit.
N is undefined.
Z cleared if the result is not zero. Unchanged otherwise.
V is undefined.
C is set if a borrow (decimal) occurs. Cleared otherwise.

Notes:

1. The \<Dea> and the extend bit are subtracted from zero using decimal arithmetic.
2. This instruction produces the ten's complement of the \<Dea> if the extend bit is clear, or the nine's complement if the extend bit is set.
3. The Z bit is normally set by another instruction before this instruction, to allow successful tests for zero results upon completion of multiple-precision operations.

NEG   **Negate**

Assembler syntax: **NEG.B   \<Dea>**
                           .W
                           .L

					\<Dea>								
					Destination effective address								
Dn	An	(An)	(An) +	− (An)	$w (An)	$b (An, Xn)	$s	$l	$w (PC)	$b (PC, Xn)	#$q	SR	CCR
*		*	*	*	*	*	*						

Action: 0−\<Dea> → \<Dea>

Status Register Condition Codes: X   N   Z   V   C
                                      *   *   *   *   *

X is set the same as the carry bit.
N is set if the result is negative. Cleared otherwise.
Z is set if the result is zero. Cleared otherwise.
V is set if an overflow occurs. Cleared otherwise.
C is cleared if the result is zero. Set otherwise.

Note: The \<Dea> is subtracted from zero, and the result is stored in the \<Dea>.

Negate (with Extend)   *NEGX*

Assembler syntax: NEG.B   \<Dea\>
.W
.L

							\<Dea\> Destination effective address								
Dn	An	(An)	(An) +	– (An)	$w (An)	$b (An, Xn)	$s	$l	$w (PC)	$b (PC, Xn)	#$q	SR	CCR
*		*	*	*	*	*	*	*					

Action: 0–\<Dea\>–extend bit → \<Dea\>

Status Register Condition Codes: X   N   Z   V   C
*   *   *   *   *

X is set the same as the carry bit.
N is set if the result is negative. Cleared otherwise.
Z is set if the result is zero. Cleared otherwise.
V is set if an overflow occurs. Cleared otherwise.
C is cleared if the result is zero. Set otherwise.

Notes:

1. The \<Dea\> and the extend bit are subtracted from zero, and the result is stored in the \<Dea\>.
2. The Z bit is normally set by another instruction before this instruction, to allow successful tests for zero results upon completion of multiple-precision operations.

No operation   *NOP*

Assembler syntax: NOP

Action: No operation occurs; execution continues with the next instruction

Status Register Condition Codes: X   N   Z   V   C
-   -   -   -   -

No bits are affected.

NOT   **One's Complement**

Assembler syntax: NOT.B   \<Dea>
                          .W
                          .L

						\<Dea> Destination effective address								
Dn	An	(An)	(An) +	− (An)	$w (An)	$b (An, Xn)	$s	$l	$w (PC)	$b (PC, Xn)	#$q	SR	CCR
*		*	*	*	*	*	*	*					

Action: One's complement of the \<Dea> → \<Dea>

Status Register Condition Codes: X   N   Z   V   C
                                      -   -   *   -   -

X is not affected.
N is not affected.
Z is set if the bit tested is zero. Cleared otherwise.
V is not affected.
C is not affected.

Logical OR

OR

Assembler syntax: OR.B Dn,<Dea>
.W
.L

<Sea> Source effective address	Dn	An	(An)	(An) +	− (An)	$w (An)	$b (An, Xn)	$s	$l	$w (PC)	$b (PC, Xn)	#$q	SR	CCR
Dn			*	*	*	*	*	*	*					

Assembler syntax: OR.B <Sea>,Dn
.W
.L

<Sea> Source effective address	Dn	<Dea> Destination effective address
Dn	*	
An		
(An)	*	
(An) +	*	
− (An)	*	
$w (An)	*	
$b (An, Xn)	*	
$s	*	
$l	*	
$w(PC)	*	
$b (PC, Xn)	*	
#$q	*	
SR		
CCR		

Action: Logical-OR each bit of <Sea> with <Dea> → <Dea>

Status Register Condition Codes: X N Z V C
- * * 0 0

X is not affected.
N is set if the most significant bit of the result is negative.
   Cleared otherwise.
Z is set if the result is zero. Cleared otherwise.
V is cleared.
C is cleared.

Note: ORI is used when the source is immediate data.

ORI   **Logical OR (Immediate Data)**

Assembler syntax: ORI.B   #$q,<Dea>
.W
.L

<Sea> Source effective address	<Dea> Destination effective address													
	Dn	*An*	*(An)*	*(An)* +	– *(An)*	*$w (An)*	*$b (An, Xn)*	*$s*	*$l*	*$w (PC)*	*$b (PC, Xn)*	*#$q*	*SR*	*CCR*
#$q	*		*	*	*	*	*	*	*					

Action: Logical-OR each bit of <Sea> with <Dea> → <Dea>

Status Register Condition Codes: X   N   Z   V   C
-   *   *   0   0

X is not affected.
N is set if the most significant bit of the result is negative.
  Cleared otherwise.
Z is set if the result is zero. Cleared otherwise.
V is cleared.
C is cleared.

Note: The size of the immediate data matches the operation size.

Assembler syntax: ORI.B   #$q,CCR (SR or CCR)
or   ORI.W   #$q,SR

<Sea> Source effective address	<Dea> Destination effective address													
	Dn	*An*	*(An)*	*(An)* +	– *(An)*	*$w (An)*	*$b (An, Xn)*	*$s*	*$l*	*$w (PC)*	*$b (PC, Xn)*	*#$q*	*SR*	*CCR*
#$q													*	*

Action: Logical-OR each bit of immediate data with CCR → CCR; or if in supervisor state, logical-OR each bit of immediate data with SR → SR, else TRAP

Status Register Condition Codes: X   N   Z   V   C
*   *   *   *   *

X is set if bit 4 of the <Sea> is 1. Unchanged otherwise.
N is set if bit 3 of the <Sea> is 1. Unchanged otherwise.
Z is set if bit 2 of the <Sea> is 1. Unchanged otherwise.
V is set if bit 1 of the <Sea> is 1. Unchanged otherwise.
C is set if bit 0 of the <Sea> is 1. Unchanged otherwise.

Push Effective Address   *PEA*

Assembler syntax: **PEA.L   <Dea>**

| | | <Dea> Destination effective address | | | | | | | | | | | |
Dn	An	(An)	(An) +	– (An)	$w (An)	$b (An, Xn)	$s	$l	$w (PC)	$b (PC, Xn)	#$q	SR	CCR
		*			*	*	*	*	*	*			

Action: <Dea> → – (SP)

Status Register Condition Codes: X   N   Z   V   C
                                       -   -   -   -   -

No bits are affected.
Note: The longword effective address is pushed onto the stack.

Reset External Devices (Privileged Instruction)   *RESET*

Assembler syntax: **RESET**

Action: All external devices are reset.

Status Register Condition Codes: X   N   Z   V   C
                                       -   -   -   -   -

No bits are affected.

Notes:

1. If in the supervisor state, the RESET line is asserted, else TRAP is executed.
2. The processor state is not affected and execution resumes with the next instruction.

ROL   **Rotate Left**

Assembler syntax: **ROL.B   <Sea>,Dn**
                                      .W
                                      .L

<Sea> Source effective address	Dn	An	(An)	(An) +	− (An)	$w (An)	$b (An, Xn)	$s	$l	$w (PC)	$b (PC, Xn)	#$q	SR	CCR
							<Dea> Destination effective address							
Dn	*													
#$q	*													

Assembler syntax: **ROL.W   <Dea>**

Dn	An	(An)	(An) +	− (An)	$w (An)	$b (An, Xn)	$s	$l	$w (PC)	$b (PC, Xn)	#$q	SR	CCR
						<Dea> Destination effective address							
		*	*	*	*	*	*	*					

Action: <Dea> rotated left by <count> → <Dea>

Status Register Condition Codes: X   N   Z   V   C
                                                            -    *   *   0   *

X is not affected.
N is set if the most significant bit of the result is negative.
   Cleared otherwise.
Z is cleared if the result is zero. Cleared otherwise.
V is cleared.
C is set according to the last bit shifted out.
   Cleared for a shift count of zero.

ROL

Notes:

   1. The carry and least significant bit of the operand receive the last bit shifted out.
   2. The shift count is contained in Dn for Data Register Direct mode.
   3. The shift count is between 1 and 8 for Immediate Data mode.

Rotate Right   *ROR*

Assembler syntax: ROR.B   <Sea>,Dn
                              .W
                              .L

<Sea> Source effective address	Dn	An	(An)	(An) +	– (An)	<Dea> Destination effective address $w (An)	$b (An, Xn)	$s	$l	$w (PC)	$b (PC, Xn)	#$q	SR	CCR
Dn	*													
#$q	*													

Assembler syntax: ROR.W   <Dea>

	Dn	An	(An)	(An) +	– (An)	<Dea> Destination effective address $w (An)	$b (An,Xn)	$s	$l	$w (PC)	$b (PC,Xn)	#$q	SR	CCR
			*	*	*	*	*	*	*					

Action: <Dea> rotated right by <count> → <Dea>

Status Register Condition Codes: X   N   Z   V   C
                                     *   *   *   0   *

X is not affected.
N is set if the most significant bit of the result is negative.
  Cleared otherwise.
Z is set if the result is zero. Cleared otherwise.
V is cleared.
C is set according to the last bit shifted out.
  Cleared for a shift count of zero.

Notes:

1. The carry bit and the most significant bit receive the last bit shifted out.
2. The shift count is contained in Dn for Data Register Direct mode.
3. The shift count is between 1 and 8 for Immediate Data mode.

ROXL   **Rotate Left with Extend**

Assembler syntax: ROXL.B   <Sea>,Dn
.W
.L

<Sea> Source effective address	<Dea> Destination effective address												
	Dn	An	(An)	(An) +	– (An)	$w (An)	$b (An, Xn)	$s	$l	$w (PC)	$b (PC, Xn)	#$q	SR	CCR
Dn	*													
#$q	*													

Assembler syntax: ROXL.W   <Dea>

<Dea> Destination effective address												
	Dn	An	(An)	(An) +	– (An)	$w (An)	$b (An,Xn)	$s	$l	$w (PC)	$b (PC,Xn)	#$q	SR	CCR
			*	*	*	*	*	*	*					

Action: Rotate left <Sea> by <count> → <Dea>

Status Register Condition Codes: X   N   Z   V   C
 *   *   *   0   *

X is set according to the last bit shifted out.
   Unaffected for a shift count of zero.
N is set if the most significant bit of the result is negative.
   Cleared otherwise.
Z is set if the result is zero. Cleared otherwise.
V is cleared.
C is set according to the last bit shifted out.
   Cleared for a shift count of zero.

ROXL

Notes:

1. The carry and extend bits receive the last bit shifted out, and the extend bit is shifted to the least significant bit of the operand.
2. The shift count is contained in Dn for Data Register Direct mode.
3. The shift count is between 1 and 8 for Immediate Data mode.

Rotate Right with Extend

ROXR

Assembler syntax: ROXR.B   <Sea>,Dn
.W
.L

<Sea> Source effective address	<Dea> Destination effective address													
	Dn	An	(An)	(An) +	– (An)	$w (An)	$b (An, Xn)	$s	$l	$w (PC)	$b (PC, Xn)	#$q	SR	CCR
Dn #$q	* *													

Assembler syntax: ROXR.W   <Dea>

<Dea> Destination effective address													
Dn	An	(An)	(An) +	– (An)	$w (An)	$b (An,Xn)	$s	$l	$w (PC)	$b (PC,Xn)	#$q	SR	CCR
		*	*	*	*	*	*	*					

Action: <Dea> rotated right by <count> → <Dea>

Status Register Condition Codes: X N Z V C
* * * 0 *

X is set according to the last bit shifted out.
  Unaffected for a shift count of zero.
N is set if the most significant bit of the result is negative.
  Cleared otherwise.
Z is set if the result is zero. Cleared otherwise.
V is cleared.
C is set according to the last bit shifted out.
  Cleared for a shift count of zero.

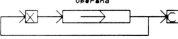

ROXR

Notes:

1. The carry and extend bits receive the last bit shifted out, and the preceding extend bit is shifted to the most significant bit of the operand.
2. The shift count is contained in Dn for Data Register Direct mode.
3. The shift count is between 1 and 8 for Immediate Data mode.

Return and Deallocate

RTD

Assembler syntax: RTD   #<displacement>

Action: (SP) + → PC; SP + displacement → SP

Status Register Condition Codes: X N Z V C
- - - - -

No bits are affected.

Note: The Program Counter is pulled from the stack. The displacement value is sign extended to 32 bits and added to the Stack Pointer.

RTE   ReTurn from Exception (Privileged Instruction)

> *Assembler syntax:*   **RTE**
>
> *Action:* $(SP) + \rightarrow SR; (SP) + \rightarrow PC$
>
> *Status Register Condition Codes:*   X   N   Z   V   C
>   *   *   *   *   *
>
> Bits are set according to the contents of the word on the stack.
>
> *Note:*   A TRAP is executed if this instruction is executed in the user mode.

RTR   Return and Restore Condition Codes

> *Assembler syntax:*   **RTR**
>
> *Action:* $(SP) + \rightarrow SR; (SP) + \rightarrow PC$
>
> *Status Register Condition Codes:*   X   N   Z   V   C
>   *   *   *   *   *
>
> Bits are set according to the contents of the word on the stack.

RTS   Return from Subroutine

> *Assembler syntax:*   **RTS**
>
> *Action:* $(SP) + \rightarrow PC$
>
> *Status Register Condition Codes:*   X   N   Z   V   C
>   -   -   -   -   -
>
> No bits are affected.
>
> *Note:*   The Program Counter is pulled from the stack, and program execution resumes at that address.

Subtract Decimal with Extend   *SBCD*

Assembler syntax: SBCD.B \<Sea>,Dn

\<Sea> Source effective address	\<Dea> Destination effective address													
	Dn	An	(An)	(An) +	− (An)	$w (An)	$b (An, Xn)	$s	$l	$w (PC)	$b (PC, Xn)	#$q	SR	CCR
Dn − (An)	*				*									

Action: Dn − \<Sea> − extend bit → Dn

Status Register Condition Codes: X N Z V C
                                               * U * U *

X is set the same as the carry bit.
N is undefined.
Z is cleared if the result is nonzero. Unchanged otherwise.
V is undefined.
C is set if a BCD borrow is generated from the most significant BCD digit. Cleared otherwise.

Notes:

1. The LOW-ORDER bytes of the Data Registers are subtracted, and the result is stored in the destination Data Register.
2. The pre-decrement addressing mode is used for subtracting multiple bytes in memory. Start at the highest address, which contains the most significant BCD digit.
3. The Z bit is normally set with an instruction prior to execution of this instruction, allowing proper tests for a zero condition upon completion of multiple-precision operations.
4. The subtraction is done using binary-coded-decimal arithmetic.

Set Conditionally   *Scc*

Assembler syntax:	**Action:**
SCC \<ea>	Set \<ea> if the C bit is CLEAR
SCS \<ea>	Set \<ea> if the C bit is SET
SEQ \<ea>	Set \<ea> if the Z bit is SET
SGE \<ea>	Set \<ea> if the N and V bits are either both SET or both CLEAR
SGT \<ea>	Set \<ea> if the N and V bits are both set and the Z bit is CLEAR or if the N, V, and Z bits are ALL CLEAR
SHI \<ea>	Set \<ea> if the C and Z bits are both CLEAR

SLE \<ea>	Set \<ea> if the Z bit is set or
	        if the N bit is set and the V bit is CLEAR or
	        if the N bit is CLEAR AND the V bit is SET

SLS \<ea>	Set \<ea> if either the C or Z bits are SET
SLT \<ea>	Set \<ea> if the N bit is SET and the V bit is CLEAR or if the N bit is CLEAR and the V bit is SET
SMI \<ea>	Set \<ea> if the N bit is SET
SNE \<ea>	Set \<ea> if the Z bit is CLEAR
SPL \<ea>	Set \<ea> if the N bit is CLEAR
SVC \<ea>	Set \<ea> if the V bit is CLEAR
SVS \<ea>	Set \<ea> if the V bit is SET
SF \<ea>	Never set \<ea>
ST \<ea>	Always set \<ea>

Status Register Condition Codes: X N Z V C
                                              - - - - -

No bits are affected.

Note: All bits of the byte size \<ea> are set if the condition is true, or the byte is cleared if the condition is false.

STOP

Stop (Privileged Instruction)

Assembler syntax: **STOP   #<16 bit data>**

Action: <data> → SR; STOP

Status Register Condition Codes: X   N   Z   V   C
                                      *   *   *   *

Set according to bits 5 to 0 of the immediate operand.

Notes:

1. This instruction provides a way to both enable interrupts and wait for an interrupt to occur.
2. Bit 13 (corresponding to the S bit) must be set to ensure supervisor state operation. A TRAP occurs otherwise.

SUB

Subtract Binary

Assembler syntax: **SUB.B   Dn,<Dea>**
                          .W
                          .L

<Sea> Source effective address	Dn	An	(An)	(An)+	−(An)	$w (An)	$b (An, Xn)	$s	$l	$w (PC)	$b (PC, Xn)	#$q	SR	CCR
							Dea Destination effective address							
Dn			*	*	*	*	*	*	*					

Assembler syntax: **SUB.B   <Sea>,Dn**
                          .W
                          .L

<Sea> Source effective address	Dn	<Dea> Destination effective address
Dn	*	
An	*	(word and longword only)
(An)	*	
(An)+	*	
−(An)	*	
$w (An)	*	
$b (An,Xn)	*	
$s	*	
$l	*	
$w (PC)	*	
$b (PC,Xn)	*	
#$q	*	
SR		
CCR		

Action: <Dea> − <Sea> → <Dea>

Status Register Condition Codes: X   N   Z   V   C
                                      *   *   *   *   *

X is set the same as the carry bit.
N is set if the result is negative. Cleared otherwise.
Z is set if the result is zero. Cleared otherwise.
V is set if an overflow is generated. Cleared otherwise.
C is set if a borrow is generated. Cleared otherwise.

Notes:

1. SUBA is used when the destination is an Address Register.
2. SUBI and SUBQ are used when the source is immediate data.

Subtract Binary (to Address Register)   *SUBA*

Assembler syntax: SUBA.W   <Sea>,An
                            .L

<Sea> Source effective address	<Dea> Destination effective address	
	An	
Dn	*	
An	*	
(An)	*	
(An) +	*	
− (An)	*	
$w (An)	*	
$b (An, Xn)	*	
$s	*	
$l	*	
$w(PC)	*	
$b (PC, Xn)	*	
#$q	*	
SR		
CCR		

Action: <Dea> − <Sea> → <Dea>

Status Register Condition Codes: X N Z V C
                                      - - - - -

No bits are affected.

Notes:

1. Word-sized source operands are sign-extended to 32 bits.
2. The 32-bit destination address register is used for both the .W & .L formats.
3. The condition codes are NOT affected.

SUBI   Subtract Binary (Immediate Data)

Assembler syntax: SUBI.B   #$q,<Dea>
                                   .W
                                   .L

<Sea> Source effective address	<Dea> Destination effective address														
	Dn	An	(An)	(An) +	− (An)	$w (An)	$b (An, Xn)	$s	$l	$w (PC)	$b (PC, Xn)	#$q	SR	CCR
#$q	*		*	*	*	*	*	*	*					

Action: <Dea> − <Sea> → <Dea>

Status Register Condition Codes: X   N   Z   V   C
                                                       *   *   *   *   *

X is set the same as the carry bit.
N is set if the result is negative. Cleared otherwise.
Z is set if the result is zero. Cleared otherwise.
V is set if an overflow is generated. Cleared otherwise.
C is set if a borrow is generated. Cleared otherwise.

Note: The size of the immediate data matches the operation size.

SUBQ   Subtract Binary (Quick {1–8} Data)

Assembler syntax: SUBQ.B   #$q,<Dea>
                                   .W
                                   .L

<Sea> Source effective address	<Dea> Destination effective address														
	Dn	An	(An)	(An) +	− (An)	$w (An)	$b (An, Xn)	$s	$l	$w (PC)	$b (PC, Xn)	#$q	SR	CCR
#$q	*	*	*	*	*	*	*	*	*					

Action: <Dea> − <Sea> → <Dea>

Status Register Condition Codes: X   N   Z   V   C
                                                       *   *   *   *   *

X is set the same as the carry bit.
N is set if the result is negative. Cleared otherwise.
Z is set if the result is zero. Cleared otherwise.
V is set if an overflow is generated. Cleared otherwise.
C is set if a borrow is generated. Cleared otherwise.

Notes:

1. The immediate data must be between 1 and 8 ($ not required) (0 − 8, 1 = 1, 2 = 2, ..., 7 = 7).
2. Only word and longword sizes are used for the Address Register Direct (An) mode, the source data is sign-extended to longword, and the condition codes are not affected for this addressing mode.
3. The 32-bit destination Address Register is used regardless of the operation size.

Subtract Binary (Extended)   *SUBX*

Assembler syntax: SUBX.B   <Sea>,<Dea>
                           .W
                           .L

<Sea> Source effective address	Dn	An	(An)	(An) +	– (An)	$w (An)	$b (An, Xn)	$s	$l	$w (PC)	$b (PC, Xn)	#$q	SR	CCR
							<Dea> Destination effective address							
Dn – (An)	*				*									

Action: <Dea> – <Sea> – extend bit → <Dea>

Status Register Condition Codes: X   N   Z   V   C
                                                  *   *   *   *   *

X is set the same as the carry bit.
N is set if the result is negative. Cleared otherwise.
Z is set if result is zero. Cleared otherwise.
V is set if an overflow is generated. Cleared otherwise.
C is set if a borrow is generated. Cleared otherwise.

Note: The Z bit is normally set with an instruction prior to execution of this instruction, allowing proper tests for a zero condition upon completion of multiple-precision operations.

Swap Data Register Halves   *SWAP*

Assembler syntax: SWAP.W   Dn

	Dn	An	(An)	(An) +	– (An)	$w (An)	$b (An, Xn)	$s	$l	$w (PC)	$b (PC, Xn)	#$q	SR	CCR
<ea> Effective address														
	*													

Action: Bits 31 to 16 and bits 15 to 0 of the Data Register are swapped

Status Register Condition Codes: X   N   Z   V   C
                                                  -   *   *   0   0

X is not affected.
N is set if the most significant bit (bit 31) is set. Cleared otherwise.
Z is set if the result is zero. Cleared otherwise.
V is cleared.
C is cleared.

TAS   Test and Set Operand

Assembler syntax: TAS.B <ea>

						<ea> Effective address							
Dn	An	(An)	(An) +	– (An)	$w (An)	$b (An, Xn)	$s	$l	$w (PC)	$b (PC, Xn)	#$q	SR	CCR
*		*		*	*	*	*	*					

Action: The <ea> is tested, setting or clearing the CCR N and Z bits;
bit 7 of the <ea> is set

Status Register Condition Codes: X N Z V C
                                                - * * 0 0

X is not affected.
N is set if the most significant bit of the operand is negative. Cleared otherwise.
Z is set if the operand was zero. Cleared otherwise.
V is cleared.
C is cleared.

Note: This operation is indivisible, to allow synchronization of several processors.

TRAP   Trap (To Exception Processing)

Assem- bler syntax:	Action:
TRAP #0	Trap to address 00080
TRAP #1	Trap to address 00084
TRAP #2	Trap to address 00088
TRAP #3	Trap to address 0008C
TRAP #4	Trap to address 00090
TRAP #5	Trap to address 00094
TRAP #6	Trap to address 00098
TRAP #7	Trap to address 0009C
TRAP #8	Trap to address 000A0
TRAP #9	Trap to address 000A4
TRAP #10	Trap to address 000A8
TRAP #11	Trap to address 000AC
TRAP #12	Trap to address 000B0
TRAP #13	Trap to address 000B4
TRAP #14	Trap to address 000B8
TRAP #15	Trap to address 000BC

Action: PC → – (SSP); SR → – (SSP); (vector) → PC

Status Register Condition Codes: X N Z V C
                                                - - - - -

No bits are affected.

Notes:

1. Normally used in user mode to call supervisor mode routines.
2. The USER Program Counter and Status Register are saved on the SUPERVISOR stack, the Program Counter is loaded with the corresponding trap address, and program execution resumes from that location.
3. Return to main program execution is done with the RTE instruction.

Trap on Overflow   *TRAPV*

Assembler syntax: **TRAPV**

Action: PC → − (SSP); SR → − (SSP); (vector) → PC

Status Register Condition Codes: X   N   Z   V   C
                                     -   -   -   -   -

No bits are affected.

Notes:

1. Trap to address 0001C if V bit is set, resume processing otherwise.
2. The CPU switches to the supervisor mode.
3. Return to main program execution is done with the RTE instruction.

TST

Assembler syntax: **TST.B   <ea>**
                       .W
                       .L

| | | | | | | | <ea>
Effective address | | | | | | | |
Dn	An	(An)	(An) +	− (An)	$w (An)	$b (An, Xn)	$s	$l	$w (PC)	$b (PC, Xn)	#$q	SR	CCR
*		*	*	*	*	*	*	*					

Action: The <ea> is compared with zero, setting or clearing the CCR N and Z bits

Status Register Condition Codes: X   N   Z   V   C
                                     -   *   *   0   0

X is not affected.
N is set if the operand is negative. Cleared otherwise.
Z is set if the operand is zero. Cleared otherwise.
V is cleared.
C is cleared.

Note: The <ea> is not changed.

UNLK

Assembler syntax: **UNLK   An**

| | | | | | | | <ea>
Effective address | | | | | | | |
Dn	An	(An)	(An) +	− (An)	$w (An)	$b (An, Xn)	$s	$l	$w (PC)	$b (PC, Xn)	#$q	SR	CCR
	*												

Action: An → SP; (SP) + → An

Status Register Condition Codes: X   N   Z   V   C
                                     -   -   -   -   -

No bits are affected.

Notes:

1. The Stack Pointer is loaded from the Address Register. The Address Register is then loaded with the longword pulled from the top of the stack.
2. This instruction performs the opposite function as the LINK instruction.
3. This instruction functions properly regardless of the number of PUSH and POP instructions done between the LINK and UNLK instructions.

reprint of *radio-electronics* article

PC's aren't just for word processing, spreadsheet analysis, and database management. In fact, when a PC can collect data from remote locations it can make decisions based on that data, so it becomes a powerful tool for controlling the environment. The problem is that special I/O cards might be required, and they typically cost hundreds of dollars. Also, I/O boards usually require installation within the PC, taking up yet another slot.

In this article we will describe the hardware and software of an I/O control system that can be implemented for less than $50.00, will interface to any personal computer through an RS-232 port, is modular, and has full duplex operation for both input and output.

The heart of the system is a little-known special-purpose IC made by Motorola, the MC14469. The MC14469 is an addressable asynchronous receiver/transmitter that is especially well-suited for remote data collection and control.

The control software is written in Microsoft "C" for the IBM PC and compatibles. Adapting the software to other compilers and computers should be easy.

System overview

Figure 1 shows an overview of the system. It's composed of a PC, control software, a combination RS-232 interface and power-distribution center, and one or more control nodes connected in parallel over a four-conductor bus. The conductors carry power and ground, and the transmit and receive signals.

A control node is shown in Fig. 2. Each node has a unique 7-bit address that is set via DIP-switch S1, which connects to the seven address lines (A0–A6) of IC3.

To communicate with a node, the software on the host PC must first transmit an address byte, over the common receive line (RI). Each node on the bus then compares the received address against its own address, which is set by the DIP switch. If the values match, then that node will accept the control byte that follows

The control byte is latched until a new address and control byte are received by the node. The control and address bytes are distinguished by the value of the most significant bit.

The control data may be used in conjunction with two other MC14469 control signals to direct the activity of the node. The other control signals are Valid Address Pulse (VAP), which is generated after a valid address is detected, and Command Strobe (CS), which is generated after a valid control byte, has been received.

Data transmission back to the host PC is initiated by toggling the SEND input (pin 30) from low to high. The

RS-232

MONITOR/CONTROL SYSTEM

Control the environment with your PC and our simple interface.

STEVEN J. FRICKEY

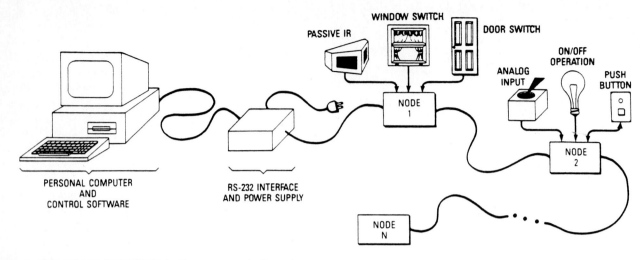

FIG. 1—THE RS-232 INTERFACE buffers communications between the host PC and all nodes.

data that is present on the sixteen input pins (ID0–ID7 and S0–S7) will be transmitted back to the host after SEND is toggled. Data is sent one byte at a time; we'll discuss the details shortly.

After receiving the data from the selected node, the host software could compare that value against the previous value from the same node, perform some action based on the comparison, and then continue on, polling the next node.

By creating an appropriate interface between external devices and any given node, the software can be tailored to a number of monitor and control situations. For example, a number of inputs could be connected to door and window switches. If one of those switches were opened before a master switch, an alarm might be sounded.

Node circuit

Connector J3 provides eight pulled-up input lines (S0–S7) that may be driven by reed switches, pushbuttons, mercury switches, tilt switches, relays, and other mechanical-switching devices. That connector also provides eight ground lines for attaching lead wires.

Connector J4 provides access to the seven output-control lines (C0–C6) of the MC14469, eight more input lines (ID0–ID7), and various control signals. To use the node in its basic configuration, jumpers should be installed across pins 9 and 10, 11 and 12, and so on, through pins 25 and 26. Later on, we'll show how J4 can be used to interface an 8-bit A/D converter to a node.

As shown in Fig. 2, the four-conductor bus runs straight through each node from J1 to J2. One line is for ground, another for +12-volts DC, one for the common transmit line (TRO), and another for the common receive line (RI). The overall length of the bus (from the RS-232 interface to the last node on the bus) depends on the degree of electrical noise in the operating environment. The author has successfully operated three nodes, using 20-gauge unshielded cable, at a cumulative length of 200 feet.

With a seven-bit address, the possible number of nodes in a system is 127, but that is not a practical limit. Realistically, the number of nodes is limited by the amount of current supplied by the +12-volt power source. Each node (with no expansion circuitry) draws 50 mA.

IC4, a 7805 voltage regulator, drops the +12-volt bus

voltage to +5 volts for powering the logic circuitry. Because the +12-volt line is also available at J4, you can attach off-the-shelf alarm-system components, such as passive infrared detectors, buzzers, beepers, etc.; all of which typically operate at +12 volts.

The MC14469's baud-rate clock can be generated internally across pins 1 and 2, or an external clock can be fed directly to pin 1. The maximum baud rate (4800) is re-

NODE-CIRCUIT PARTS LIST
IC1—74ALS161, synchronous 4-bit counter
IC2—74ALS05 or open-collector hex inverter
IC3—MC14469, addressable asynchronous receiver/ transmitter
IC4—7805, 5-volt regulator
IC5—4-MHz TTL crystal oscillator
RP1, RP2—4700 ohms, 10-pin SIP
S1—8-position DIP switch.
C1—0.33 μF, 12 volts, tantalum
C2—22 μF, 25 volts, electrolytic
C3—0.1 μF, monolithic ceramic
J1—9-pin D, female
J2—9-pin D, male
J3—16-pin, PC-mount, screw terminal block
J4—26-pin dual-row header strip

RS-232 INTERFACE PARTS LIST
IC1—1488, quad RS-232 line driver
IC2—1489, quad RS-232 line receiver
R1—4700 ohms, ¼ watt, 10%
J1—25-pin D, male
J2—9-pin D, male
J3—4-pin power connector
Miscellaneous: Power supply with ±12- and +5-volt outputs, cases, interconnecting cables, etc.
The following are available from Steven J. Frickey, 3661 North Lena Ave., Boise, ID 83704 (FAX: 208-377-9410); MC14469P with spec sheet and ap note, $15.47; Node PC board, Rev 2.0 with assembly notes, $19.40; Interface PC board, Rev 2.0 with assembly notes, $19,40; Monitor V2.0 software with source code for IBM PC's and clones, $10.00; Node kit Version 2, $56.37; Interface kit, Version 2, original option, $28.29; Interface kit Version 2, MAX232 option, $34.19; Prototyping board Kit, $25.56; DC-DC converter kit, $15.97. All orders should include $3.50 for shipping and handling.

stricted by the +5-volt supply. The required clock rate is 64 times the baud rate, or in this case, 307.2 kHz.

Because that's a non-standard frequency, the circuit uses a readily available 4-MHz TTL clock oscillator (IC5), a 74ALS05 (IC2) open-collector inverter, and a 74ALS161 (IC1) four-bit counter to divide the 4-MHz signal by 13, thereby providing a 307.69-kHz signal. The communications protocol is fixed at one start bit, eight data bits, an even parity bit, and one stop bit. So at 307.69 kHz, the maximum sampling time error over the entire 11 bits is 35.7 μs, well within one-half of a data bit period, which is 104 μs at 4800 baud.

A second gate on the 74ALS05 (IC2-d) inverts the serial data from IC3 and drives the common transmit line (TRO). The pull-up resistor for IC2-d is actually located in the RS-232 interface circuit (shown in Fig. 3 as R1). The open-collector outputs of all nodes are pulled up by that resistor, which makes it a wired-OR circuit.

A local reset is generated by each node at power up by an RC circuit consisting of 22-μF capacitor C2 and a 4.7K resistor inside RP1. The reset signal is also provided at J4, should your expansion circuitry require access to that signal.

The 7-bit address for each node is set on pins 4 through 10 (A0–A6) of IC3. Table 1 shows the relationship between switch settings and node numbers.

The voltage supplied to IC3 can range from 4.5–18 volts. At five volts, the output drive current of each pin (I_{OH}) is typically 0.35 mA, providing a fan-out capacity of 17 ALS devices. The output-high voltage (V_{OH}) is typically 5.0; the low voltage (V_{OL}) is typically 0.0. The input high voltage is typically 2.75; the input-low voltage is typically 2.25. For more information on the MC14469, consult *CMOS/NMOS Special Functions Data*, Motorola Inc., 1984, and Application Note AN806A, *Operation Of The MC14469*, Motorola Inc., 1984.

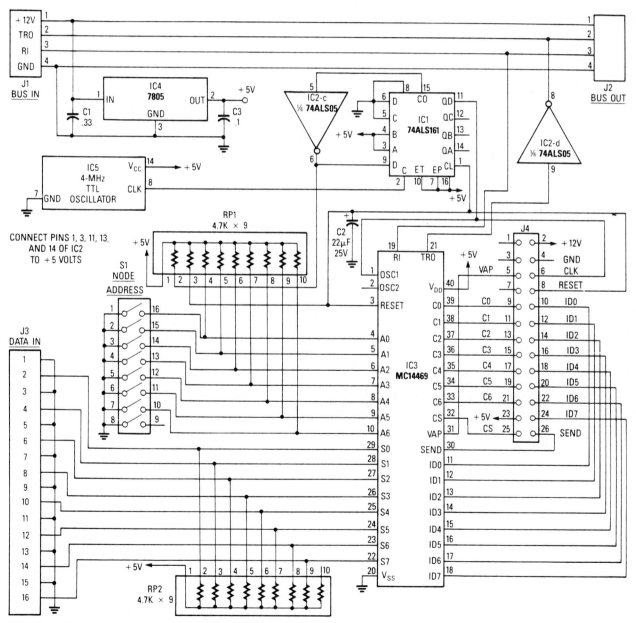

FIG. 2—THE HEART OF A NODE is the MC14469, an addressable UART. When a node is addressed, data present on pins 11–18 and 22–29 is transmitted to the host. The address is set on pins 4–10.

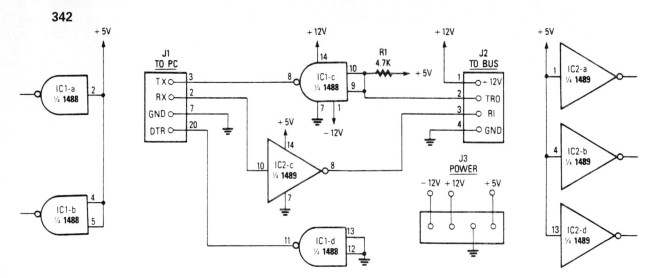

FIG. 3—THE RS-232 INTERFACE routes 12-volt power to the nodes, and buffers data between the nodes and the host PC.

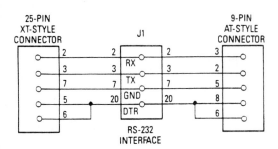

FIG. 4—CONNECT AN XT TO J1 of the RS-232 interface as shown at left, and to an AT as shown at right.

Node operation

The communication software first transmits a seven-bit address that is received simultaneously on pin 19 (RI) of all MC14469's in the system. Each node then checks the state of the most significant bit. If it's high, then the remaining seven bits are compared against the address set on A0–A6. If the values are identical, then VAP is generated on pin 31. VAP is not used in the node circuit shown in Fig. 2, but it is used internally by the MC14496 to latch a control byte on output pins 33–39 (C0–C6). Control-byte data is latched only after a valid address has been received, and it remains latched until another address byte is received.

Transmitting data back to the host PC is accomplished by toggling pin 30 (SEND) high. After receiving the SEND pulse, the MC14469 will transmit, via pin 21 (TRO), the data present on pins 11–18 (ID0–ID7), followed by the data on pins 22–29 (S7–S0). The only stipulation is that the rising edge of the SEND pulse must

TABLE 1—NODE ADDRESSES

Node	A6	A5	A4	A3	A2	A1	A0
127	H	H	H	H	H	H	H
126	H	H	H	H	H	H	L
125	H	H	H	H	H	L	H
124	H	H	H	H	H	L	L
...							
2	L	L	L	L	L	H	L
1	L	L	L	L	L	L	H
0	L	L	L	L	L	L	L

occur within eight bit times after the generation of either VAP or CS. At 4800 baud, eight bit times provides a maximum of 1.667 ms.

Receipt of a control byte generates a Control Strobe (CS) pulse on pin 32. In our circuit, CS is normally connected to SEND through J4. In this configuration, data will be transmitted to the host as soon as a control byte has been received.

What is the minimum interval between events that this system can detect? The time it takes to transmit and receive data from the same node twice, which works out to (1/4800) × 11 bits/byte × 8 bytes, or about 18 ms.

Realistically, the minimum time is much longer, at least on the order of hundreds of milliseconds, because of the amount of time the software processing takes, especially when relatively slow I/O devices (disk, BIOS video routines, printer) are being accessed. Just don't try to detect more than three events per second.

RS-232 interface and power supply

Figure 3 is the schematic for the RS-232 interface, which uses a 1489 (IC2) for the line receiver and a 1488 (IC1) for the line driver. Pin 2 of J2 is the common transmit line (TRO) that receives data from the open-collector output of each node, and R1 is the pull-up resistor.

Power is supplied to the system via four-pin connector J3. As stated earlier, a single node draws about 50 mA from the +12-volt supply. Low-current sources of +5 and −12 volts are also required.

Figure 4 shows the cable wiring required to connect the RS-232 interface to a 25-pin XT-style port (on the left); and to a 9-pin AT-style port (on the right).

Assembly and testing

Figures 5-*a* and 5-*b* show how to mount the components on the PC boards. The Node board, shown in Fig. 5-*a*, is a double-sided board. You can use the patterns shown in PC Service to build your own, or you can purchase the board from the source mentioned in the Parts List. The pattern for the RS-232 board is also shown in PC Service, but because it is so simple, a commercial product has not been made available.

After you assemble the system, test it using the sample

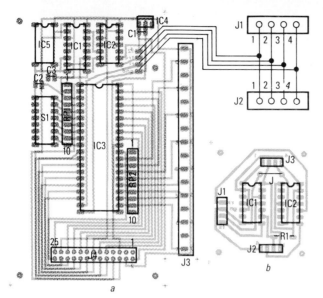

FIG. 5—MOUNT ALL COMPONENTS on the node circuit board as shown in 5-a and mount all components on the RS-232 circuit board as shown in 5-b.

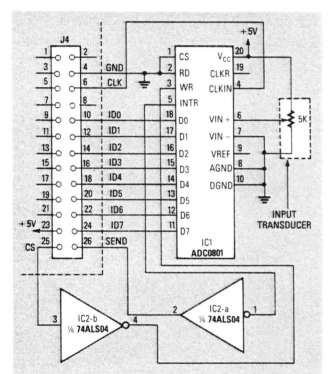

FIG. 6—THE OPTIONAL A/D CONVERTER is shown here. CS from the Node board starts the conversion process; INTR from IC1 here informs the Node that the process is complete.

program that will be discussed shortly. Apply ±12- and +5-volts DC to the RS-232 module, and connect it to your PC and to a single node configured as address 0. Then run the test program. If you receive any error messages (especially a time-out error), check your cabling carefully—the chances are that the RS-232 module hs not been connected to your PC properly.

When the software seems to be running correctly, temporarily short several of J3's even-numbered pins to ground, one at a time. Then terminate the test program according to the directions given on the screen. An ASCII text file called MONITOR.LOG should be present in your current directory. That file should contain a number of

messages corresponding to the state of the input lines of J3 at startup, and it should also include messages indicating that it sensed the shorts.

A/D expansion example

Figure 6 shows how to interface an eight-bit analog-to-digital converter (the ADC0801) to a node via connector J4. The component labeled Input Transducer is shown as a 5K potentiometer, but in real life it might be a temperature sensor, a pressure sensor, etc.

In this circuit, CS initiates the analog-to-digital conversion (WR), and the end-of-conversion (INTR) pulse from the ADC initiates data transmission to the host by toggling the MC14469's SEND input.

The ADC uses the 307-kHz node clock. At that rate, a single conversion will take at most 240 μs, which is well within the 1.667-ms time limit between the CS and the SEND pulses.

The software

Because of space limitations, we are unable to print the 600-line C source listing here. However, we will give an overview of how the software works. In addition, both executable files and the full source code have been posted on the RE-BBS (516-293-2283). Download file RS232MON.ARC at 300 or 1200 baud, eight data bits, one stop bit, and no parity. (Source and executable files of an additional program that demonstrates use of the A/D converter is also included.)

The program is a simple event-logging system that continually polls a single node, logs the date, the time, and the input device(s) that changed state since the last time that node was polled. Execute it by typing the name of the program followed by the number of the serial port being used (0 = COM1, 1 = COM2, etc.).

The program communicates directly with the serial port through BIOS interrupt 14h. That means the program can reconfigure the port-communications protocol, read a byte, write a byte, and check the status of the port. Several error conditions can also be determined when using the interrupt. If an error does occur during execution, the program stops and a message is displayed on the screen indicating the type of error.

In the program, each node is represented by a data structure that contains the node address, the initial value of the control byte, a mask value indicating which bit values to respond to, a copy of the last data values returned from the node, and sixteen other fields that correspond to the bit values returned from a node. The sixteen fields contain names that identify what a bit represents, what its *on* state is, and what its *off* state is.

When the program starts, each record is accessed sequentially, and the corresponding node address and control byte are sent. For each node, required functions are initialized, communications checked, and initial conditions logged in a disk file called MONITOR.LOG.

After initialization the program begins to loop, sequentially polling each node and checking the return values against the previous values from the same node. If a new value is different from a previous value, and if those particular bits that indicate a difference are not masked, then the event is logged in the log file with the date and time. Polling continues in that way until the user terminates the program by pressing a key.◆CD◆

ordering information for data acquisition project

Inner Sanctum Software's

DATA AQUISITION & CONTROL
CATALOG

---

featuring:

ERBERUS

the Experimenter's

Serial Monitor/Control System

Table of contents

INTRODUCTION

The kits in this catalog are intended for assembly and use by students, electronic hobbyists, experimenters and engineers in the development of low speed data acquisition and control systems.

RS-232 MONITOR/CONTROL SYSTEM

The ISS Monitor/Control System provides the basis for a low cost, low speed solution for data acquisition and control.

As shown in figure 1, the system is comprised of one or more node circuits, one RS-232 interface circuit, a power supply, control software, and a host computer. Each node is a nexus for I/O with 16 digital input lines and 7 digital output lines. As standard, each node kit comes with PC board screw terminals with pull up resistors on 8 ot the input lines. This configuration allows those 8 lines to be used immediately with switch type inputs such as contact switchs, door switches,

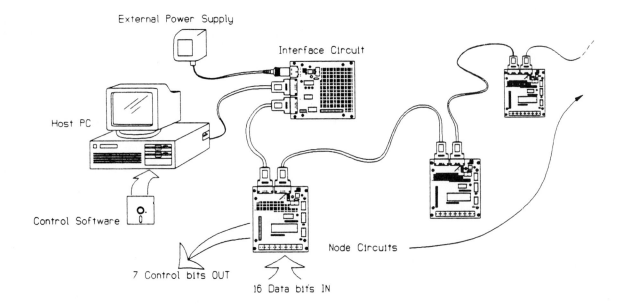

External Power Supply
Interface Circuit
Host PC
Control Software
7 Control bits OUT
16 Data bits IN
Node Circuits

relays, IR detectors, etc. Each node can also be customized with the addition of user designed control hardware, sensors, transducers to utilize all input and control lines.

The system interfaces to the host computer through one of the host's RS-232 serial ports. Each node in the system is built around an Addressable Asynchronous Receiver/Transmitter.

To operate, the control software first transmits the address of the desired node. The corresponding node "wakes up" to receive one byte of control data, invoke any custom hardware that may be designed onto the node by the end user, and transmit two bytes of data back to the host computer.

In a rapid polling mode over 100 transactions can take place every second at 4800 baud.

Floppy disks containing executable and source code listings of the control software is available in GW-BASIC, 80x86 Assembly language, and 'C'. This source code aids the designer in incorporating real world control and data collection into their projects.

Kit documentation includes hardware interface examples, and control source code listings in GW-BASIC, Assembly, and 'C'.

Documentation for connecting to RS-422 is also included in the kit documentation.

A complete functional description of the system and its design is featured in an article titled, *RS-232 Monitor/Control System*, which appeared in Radio-Electronics Magazine, August 1988, page 83.

- Assembly time for each kit is approximately 20 minutes to 1 hour.
- 4800 baud operation.
- Node to Node link is a "wired or" using TTL logic levels.
- Interface to PC link uses RS-232 logic levels.
- Nodes are addressable via DIP switch. The address range is 0 to 127 inclusive.
- Each board has a prototyping area for adding your own circuit modifications.

- One or more nodes can be linked together. Maximum circuit configuration is dependent upon power requirements, length of wire run, and environment noise.
- Source code interfacing examples are available for the IBM PC/XT/AT and compatibles in GW-BASIC, assembly, and 'C'.
- DC-DC converter kit is optional for lower power consumption.
- As designed, each node draws approximately 40ma at +5Vdc.

Prototyping breadboards are also available with the same form factor, allowing custom logic to be added to a node and stacked via the 34 pin connector.

Enclosures

The PC boards in each kit are designed to fit in readily available enclosures from the following manufacturers:

Pactec
Enterprise and Executive Avenues
Philadelphia, PA 19153
    Pactec enclosures: CM5-200 or CM5-125

BUD West, Inc.
7733 West Olive Ave.
Peoria, Arizona 85345-0350
    BUD enclosures: PC-11402

(check for distributors in your area)

Enclosures CM5-125 and PC-11420 are 5" x 5.25" x 1.25" and will accommodate one PC board.

Enclosure CM5-200 is 2" high and it will accommodate two PC boards, one stacked on the other.

Power Supplies

An external power source is required to operate the system. The source voltages required are dependent upon the configuration options chosen. In general, the range of source voltages required are a +12V source, or a +12V, +5V, -12V source, or a +5V source. Refer to the specifications of each kit to determine the specific power requirements.

KITS

INTERFACE CIRCUIT Kit Rev. 2.0

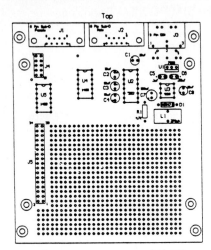

INTERFACE BOARD, REV 2.0, COMPONENT SIDE

- Board Size - 4.1" x 4.7"

- Breadboard area - 6.5 square inches.

- Two configuration options are available:
  The MAX232 configuration requires +5Vdc for operation
  The LM1488/LM1489 configuration requires +5Vdc, +12Vdc
  and -12Vdc for operation. This is the configuration originally
  described in the Radio-Electronics article.

- The MAX232 configuration uses an LM7805 to regulate a higher voltage
  down to the +5Vdc required. Optionally the DC-DC converter components
  (part #DC-001) can be installed rather than the LM7805.

- In all cases the interface circuit requires less than 10ma of current.

- Connectors are provided for both 9-Pin and 25-Pin serial communication
  ports (COM1. COM2, etc.).

- Assembly time is approximately 30 minutes (31 components with the
  MAX232 configuration, and 27 components with the 1488/89 configuration).

Interface kit, MAX232 configuration, **KIT-004A** $34.19
Interface kit, 1488/89 configuration, **KIT-004B** $28.29

NODE CIRCUIT Kit Rev. 2.0

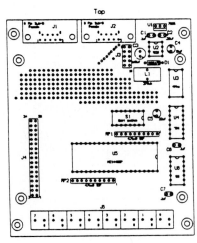

NODE BOARD, REV 2.0, COMPONENT SIDE

- Board Size - 4.1" x 4.7"

- Breadboard area - 1.5 square inches.

- 4800 baud operation can be changed to 2400 or 1200 by replacing the 4Mhz
  oscillator with either a 2Mhz or 1Mhz crystal respectively.

- 16 digital input lines (8 with pull up resistors).

- 8 of the digital input line are provided with pull up resistors and can be accessed via screw terminals installed on the board.

- 7 digital output control lines.

- Board operation requires +5Vdc. In the standard configuration the +5Vdc is
  provided via an LM7805. The LM7805 is used to regulate down to +5Vdc
  from a higher input voltage. Optionally the DC-DC converter components
  (part #DC-001) can be installed rather than the LM7805.

- During operation the Node circuit will draw approximately 40ma of current.

- Assembly time is approximately one hour. The kit is comprised of 50
  separate components.

Node Kit, **KIT-002** $56.73

PROTOTYPING Board Rev. 1.0

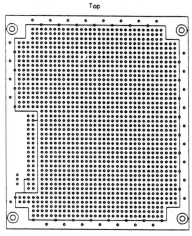

PROTOTYPE BOARD, REV 1.0, COMPONENT SIDE

- Board Size - 4.1" x 4.7"

- Breadboard area - 8.9 square inches.

- Provided with 1" standoffs to piggyback directly on top of the Interface or Node PC boards with enough clearance for wire-wrap prototypes.

- Connects to the Node or Interface PCB via the 34 pin IDC connector.

- 6" ribbon connector is provided for board to board connection.

- 15 components total in the kit.

Proto Kit, **KIT-003**................................ Available in Nov. '89.

MONITOR source code, PC Version 2.0

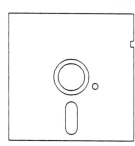

- Disk contains example monitor and control programs which interface to the monitor/control hardware via serial ports.

- Example programs are provided in:
    GW-BASIC
    80x86 Assembly Language
    'C'

- Both executable and source code is provided.

- System requirements:
    IBM PC, XT, AT, PS/2 or compatible.
    At least one serial port (Com1, Com2, ...)
    256K memory
    DOS 3.1 or higher.

IBM 360K, 5¼" format, **DSK-001**................................ $10.00
IBM 760K, 3½" format, **DSK-002**................................ $10.00

DC-DC converter option

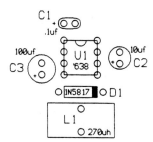

- A DC-DC step-down switching regulator circuit based on the MAX638.

- 75% to 80% typical efficiency.

- Input voltage range of + 10.5Vdc to + 16Vdc.

- Output voltage +5Vdc.

- Maximum output current is 95ma.

- 6 components including 3 capacitors, 1 inductor, 1 diode rectifier, 1 IC, and 1 DIP socket.

DC-DC converter components, **DC-001** $15.79

INDIVIDUAL COMPONENTS

NODE CIRCUIT BOARD, Rev. 1.0 (original version)

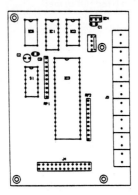

- Board Size - 4.1" x 2.9"
- This is the original board design as described in the RS-232 Monitor/Control article, Radio-Electronics Magazine, August 1988.
- 4800 baud operation.
- 16 digital input lines (8 with pull up resistors).
- 8 of the digital input line are provided with pull up resistors and can be accessed via screw terminals installed on the board.
- 7 digital output control lines.
- Board operation requires +5Vdc. In the standard configuration the +5Vdc is provided via an LM7805. The LM7805 is used to regulate down to +5Vdc from a higher input voltage.

Node Circuit Board, **PCB-001** (While Supply Lasts)...... $16.00

POWER SUPPLY

- Plug in, class 2 wall transformer
- Input: 120V 60Hz 25W
- Output: 12Vdc, 800ma, unregulated
- Indoor use only

Power Supply, +12Vdc, 800ma, **PS-12-800** $11.12

KIT COMPONENTS

┌─those parts and quantities included with the interface kits. Those parts included only with **KIT-004A** are prefixed with an **A**. Those parts included only with **KIT-004B** are prefixed with a **B**.

　┌─those parts and quantities included with the node kit, **KIT-002**.

　　┌─those parts and quantities included with the Prototying board kit, **KIT-003**.

　　　┌─those parts and quantities for the DC-DC converter, **DC-001**.

A	B	C	D	Part #	Description	PRICE each
				PCB-001	Node PCB, Rev1.0, PCB00100 (while supply lasts)	16.00
	1			PCB-202	Node PCB, Rev2.0, PCB00202	19.40
1				PCB-402	Interface PCB, Rev. 2.0, PCB00402	19.40
		1		PCB-301	Prototype PCB, Rev. 1.0, PCB00301 (Available in Nov. '89)	29.40
				PS-12-800	12Vdc, 800ma unregulated plug in wall power supply	11.12
			1	1N5817	1N5817, Schottky Barrier rectifier	.64
			1	IND270	270μh, Toroid Inductor	5.76

A	B	C	D	Description (cont.)		PRICE each
			1	CPE100-25	100μf, Electrolytic capacitor, radial leads, 25Vdc	.29
	1			CPE022-25	22μf, Electrolytic capacitor, radial leads, 25Vdc	.20
A4			1	CPE010-25	10μf, Electrolytic capacitor, radial leads, 25Vdc	.19
	2		1	CPM.100-50	0.1μf, Monolythic capacitor, 50Vdc	.36
A1	1			CPT.300-35	0.33μf, Tantalium capacitor, 35Vdc	.20
1				R4.7KE	4.7kΩ, resistor, 1/4W 5%	.03
	2			RP4.7K10	4.7kΩ, resistor SIP, 9 element, common terminal	.45
	1			MC14469P	MC14469P, Addressable Async receiver/transmitter	15.47
A1	1			LM7805T	LM7805T, Voltage regulator, 5Vdc	.42
	1			74LS05	74LS05, Inverter w/open collector outputs	.26
	1			74LS161	74LS161, 4 bit counter	.46
B1				LM1488N	LM1488N, RS-232 Line Driver	.90
B1				LM1489N	LM1489N, RS-232 Line Receiver	.90
A1				MAX232EPE	MAX232EPE, RS-232 Line Driver/Receiver, 5Vdc	6.10
			1	MAX638ACPA	MAX638ACPA, Step down switching regulator	7.90
	1			OSC4	4MHZ crystal clock oscillator	8.40
			1	SKT8DP	8 Pin DIP Socket, Tin PCB	.15
B2	2			SKT14DP	14 Pin DIP Socket, Tin PCB	.24
A1	1			SKT16DP	16 Pin DIP Socket, Tin PCB	.27
	1			SKT40DP	40 Pin DIP Socket, Tin PCB	.67
	1			SWD8	8 Ckt, DIP Switch, SPST, Tin	2.10
	8			TMS2	2 Position PCB screw terminal block	.44
1				DB09PM	9 Pin Sub-D, right angle, male PCB connector	.74
1	2			DB09PF	9 Pin Sub-D, right angle, female PCB connector	.64
	2			DB09SM	9 Pin Sub-D, male solder cup connector	.36
2				DB09SF	9 Pin Sub-D, female solder cup connector	.40
1				DB25SF	25 Pin Sub-D, female solder cup connector	.62
1				DIN5PF	5 Pin, right angle, female, circular DIN PCB connector	.78
1				DIN5SM	5 Pin, circular DIN connector and housing assembly	1.30
	1	1		IDC34DH	34 Pin, dual row, IDC header	1.38
		1		IDCASY6	6" ribbon cable assembly with two, 34 Pin , dual row, IDC connectors	6.04
	9			SJ2	Socket Jumpers	.30
2	2			HD09	Hood, 9 Pin Sub-D	.34
1				HD25	Hood, 25 Pin Sub-D	.48
1				STR-1	Strain relief tubing, 1"	.10
2	1			WD4-4	Cable 22AWG, 4 conductor, 4ft.	1.32
		4		HWSP440-100	4-40x1" round threaded spacer	.38
4	4			HWSP440-050	4-40x.5" round threaded spacer	.26
4	4	8		HWSC440-025	4-40x.25" machine screw	.02
1	1			DOC-ASMV2	Assembly Manual for Node & Interface KIT	4.00

TERMS & CONDITIONS of SALE

Inner Sanctum Software
Terms and Conditions of Sale

All orders are subject to (ISS) Inner Sanctum Softwares' standard terms and conditions of sale.

ISS sells it products under the following terms and conditions:

1 - Specifications

The specifications listed the ISS catalog are subject to change without notice. ISS reserves the right to make product improvements or changes at any time.

2 - Misprints

While Inner Sanctum Software has made every effort to ensure the accuracy of information contained in this catalog, ISS is not responsible for printing errors, or omissions.

3 - Software License Agreements

Unless otherwise stated on the software package or documentation all executable code may be reproduced and distributed to demonstrate the use of the RS-232 Monitor/Control system. All source code may be reproduced and distributed for educational purposes, demonstrating how a programmer can write code to interface a host computer with the RS-232 Monitor/Control system. While programmers are encouraged to utilize any portion of the code for their own use, no part of the code may be used in any way, for any commercial product without the written consent of Inner Sanctum Software.

4 - Warranty

All of the kits and parts listed in the ISS catalog have been carefully chosen for their price, performance, function and for the manufacturers that stand behind them. The kits have been designed for assembly by students, electronic hobbyists, experimenters, engineers and those experienced in electronics assembly.

The manufacturers of the products and kit components contained herein warrant that their products will be free from defects in materials and workmanship for a period of time and under conditions specified. A copy of the separate manufacturers' warranties is available by writing to ISS at the address listed below.

Replacements for factory-defective parts will be supplied free for 90 days from the date of purchase. Replacement parts are warranted for the remaining portion of the original warranty period. You can obtain warranty parts directly from ISS by writing to the address listed below.

You may receive free consultation on any problem you might encounter in the assembly or use of your ISS kit by writing to the address below or FAXing your questions to the number listed below. We will respond as soon as possible. Sorry, we cannot accept collect calls.

Correction of assembly errors, adjustments, calibration, and damage due to misuse, abuse, or negligence are not covered by this warranty.

Use of corrosive solder and/or modification of the kits voids this warranty in its entirety. In other words, this warranty applies only to the kits when assembled and tested as per assembly instructions. Claims against this warranty must be made before user modifications are made to the kits.

Other than the express warranties set forth above, Inner Sanctum Software makes no other warranty, express or implied, regarding its products, including, without limitation, the implied warranty of merchantability and fitness for a particular purpose.

This warranty covers only ISS products and is not extended to other equipment or components that a customer uses in conjunction with our products.

Some states do not allow the exclusion or limitation of incidental or consequential damages, so the above limitations or exclusions may not apply to you.

5 - Limitation of Liability

In no event shall ISS, Inner Sanctum Software, be liable for any defect in the software, for loss of or inadequacy of data of any kind, or for any direct, indirect, incidental, or consequential damages in connection with or arising out of the performance or use of any product furnished hereunder. Inner Sanctum Software liability shall in no event exceed the purchase price of the product furnished hereunder.

6 - Prices

All prices are net, FOB Boise, Idaho, USA. All transportation charges and insurance shall be paid by the customer. The prices listed are subject to change without notice. Exact prices are determined by the prices in effect the day the order is accepted by ISS.

7 - Ordering

Orders may be placed by mail or by fax. For Idaho residents add 5% sales tax. If your order is non-taxable, you must submit proof of tax exempt status and your tax number if applicable.

8 - Taxes

All prices listed are exclusive of all sales, use, and like taxes and are the responsibility of the buyer in their own state. ISS will collect sales tax on all applicable orders from within the state of Idaho.

9 - Returns and Exchanges

Requests for returns or exchanges must be made within 15 days of receipt of merchandise. A copy of your invoice and a written memo with the reason for return is required. All returned merchandise must be shipped freight prepaid in a new unused and resalable condition, complete with all accessories and documentation.

We reserve the right to refuse returns or adjustments on partially assembled kits, soldered components, opened magnetic recording media (software), or special order items.

10 - Claims

Inspect all shipments immediately upon receipt. Missing cartons or obvious damage to a carton should be noted on the delivery receipt before signing. Concealed damage or loss should be reported at once to the carrier and an inspection requested. All claims for shortage or damage must be made within TEN (10) days of receipt of the shipment. You must save damaged or pilfered items until the claim is settled. Claims for lost shipments must be made within 20 days of receipt of invoice or other notification of shipment.

11 - General Provisions

These terms and conditions are governed by the laws of the State of Idaho and will become binding only when an order is accepted by Inner Sanctum Software. The terms constitute the entire agreement between parties with respect to the subject matter hereof. These terms and conditions will prevail notwithstanding any different, conflicting, or additional terms and conditions which may appear on any order submitted by the purchaser. Deviations from these terms and condition are not valid unless confirmed in writing by an authorized agent for Inner Sanctum Software.

Address:
Inner Sanctum Software
3661 N. Lena Ave.
Boise, Idaho, 83704.

FAX & Voice Message number:
(208) 377-9410

Motorola MC14469 data sheets

ADDRESSABLE ASYNCHRONOUS RECEIVER/TRANSMITTER

The MC14469 Addressable Asynchronous Receiver Transmitter is constructed with MOS P-channel and N-channel enhancement devices in a single monolithic structure (CMOS). The MC14469 receives one or two eleven-bit words in a serial data stream. One of the incoming words contains the address and when the address matches, the MC14469 will then transmit its information in two eleven-bit-word data streams. Each of the transmitted words contains eight data bits, even parity bit, start and stop bit.

The received word contains seven address bits and the address of the MC14469 is set on seven pins. Thus 2^7 or 128 units can be interconnected in simplex or full duplex data transmission. In addition to the address received, seven command bits may be received for data or control use.

The MC14469 finds application in transmitting data from remote A-to-D converters, remote MPUs or remote digital transducers to the master computer or MPU.

- Supply Voltage Range — 4.5 Vdc to 18 Vdc
- Low Quiescent Current — 75 μAdc maximum @ 5 Vdc
- Data Rates to 4800 Baud @ 5 V, to 9600 Baud @ 12 V
- Receive — Serial to Parallel
  Transmit — Parallel to Serial
- Transmit and Receive Simultaneously in Full Duplex
- Crystal or Resonator Operation for On-Chip Oscillator
- See also Application Note AN-806

CMOS LSI
(LOW-POWER COMPLEMENTARY MOS)

ADDRESSABLE ASYNCHRONOUS RECEIVER/TRANSMITTER

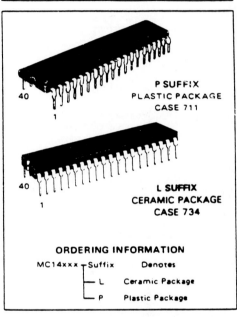

P SUFFIX
PLASTIC PACKAGE
CASE 711

L SUFFIX
CERAMIC PACKAGE
CASE 734

ORDERING INFORMATION

MC14xxx ⌐Suffix   Denotes
        ├─ L   Ceramic Package
        └─ P   Plastic Package

BLOCK DIAGRAMS

PIN ASSIGNMENTS

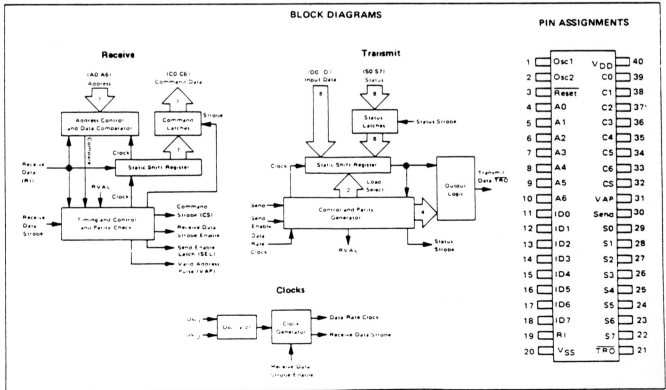

MAXIMUM RATINGS (Voltages referenced to V_{SS}, Pin 20.

	Symbol	Value	Unit
DC Supply Voltage	V_{DD}	-0.5 to +18	Vdc
Input Voltage, All Inputs	V_{in}	-0.5 to V_{DD} + 0.5	Vdc
DC Current Drain per Pin	I	10	mAdc
Operating Temperature Range	T_A	-40 to +85	°C
Storage Temperature Range	T_{stg}	-65 to +150	°C

This device contains circuitry to protect the inputs against damage due to high static voltages or electric fields; however, it is advised that normal precautions be taken to avoid application of any voltage higher than maximum rated voltages to this high impedance circuit. For proper operation it is recommended that V_{in} and V_{out} be constrained to the range $V_{SS} \leq (V_{in}$ or $V_{out}) \leq V_{DD}$.

Unused inputs must always be tied to an appropriate logic voltage level (e.g., either V_{SS} or V_{DD}).

ELECTRICAL CHARACTERISTICS

Characteristic	Symbol	V_{DD} Vdc	-40°C Min	-40°C Max	25°C Min	25°C Typ	25°C Max	+85°C Min	+85°C Max	Unit
Output Voltage "0" Level	V_{OL}	5.0	—	0.05	—	0	0.05	—	0.05	Vdc
V_{in} = V_{DD} or 0		10	—	0.05	—	0	0.05	—	0.05	
		15	—	0.05	—	0	0.05	—	0.05	
"1" Level	V_{OH}	5.0	4.95	—	4.95	5.0	—	4.95	—	Vdc
V_{in} = 0 or V_{DD}		10	9.95	—	9.95	10	—	9.95	—	
		15	14.95		14.95	15	—	14.95	—	
Input Voltage # "0" Level	V_{IL}									Vdc
(V_O = 4.5 or 0.5 Vdc)		5.0	—	1.5	—	2.25	1.5	—	1.5	
(V_O = 9.0 or 1.0 Vdc)		10	—	3.0	—	4.50	3.0	—	3.0	
(V_O = 13.5 or 1.5 Vdc)		15	—	4.0	—	6.75	4.0	—	4.0	
"1" Level	V_{IH}									Vdc
(V_O = 0.5 or 4.5 Vdc)		5.0	3.5	—	3.5	2.75	—	3.5	—	
(V_O = 1.0 or 9.0 Vdc)		10	7.0	—	7.0	5.50	—	7.0	—	
(V_O = 1.5 or 13.5 Vdc)		15	11.0	—	11.0	8.25	—	11.0	—	
Output Drive Current (Except Pin 2)	I_{OH}									mAdc
(V_{OH} = 2.5 Vdc) Source		5.0	-1.0	—	-0.8	-1.7	—	-0.6	—	
(V_{OH} = 4.6 Vdc)		5.0	-0.2	—	-0.16	-0.35	—	-0.12	—	
(V_{OH} = 9.5 Vdc)		10	-0.5	—	-0.4	-0.9	—	-0.3	—	
(V_{OH} = 13.5 Vdc)		15	-1.4	—	-1.2	-3.5	—	-1.0	—	
(V_{OL} = 0.4 Vdc) Sink	I_{OL}	5.0	0.52	—	0.44	0.88	—	0.36	—	mAdc
(V_{OL} = 0.5 Vdc)		10	1.3	—	1.1	2.25	—	0.9	—	
(V_{OL} = 1.5 Vdc)		15	3.6	—	3.0	8.8	—	2.4	—	
Output Drive Current (Pin 2 Only)	I_{OH}									mAdc
(V_{OH} = 2.5 Vdc) Source		5.0	-0.19	—	-0.16	-0.32	—	-0.13	—	
(V_{OH} = 4.6 Vdc)		5.0	-0.04	—	-0.035	-0.07	—	-0.03	—	
(V_{OH} = 9.5 Vdc)		10	-0.09	—	-0.08	-0.16	—	-0.06	—	
(V_{OH} = 13.5 Vdc)		15	-0.29	—	-0.27	-0.48	—	-0.2	—	
(V_{OL} = 0.4 Vdc) Sink	I_{OL}	5.0	0.1	—	0.085	0.17	—	0.07	—	mAdc
(V_{OL} = 0.5 Vdc)		10	0.17	—	0.14	0.28	—	0.1	—	
(V_{OL} = 1.5 Vdc)		15	0.50	—	0.42	0.84	—	0.3	—	
Maximum Frequency	f_{max}	4.5	400	—	365	550	—	310	—	kHz
Input Current	I_{in}	15	—	±0.3	—	±0.00001	±0.3	—	±1.0	µAdc
Pull-Up Current (Pins 4-18)	I_{UP}	15	12	120	10	50	100	8.0	85	µAdc
Input Capacitance (V_{in} = 0)	C_{in}	—	—	—	—	5.0	7.5	—	—	pF
Quiescent Current	I_{DD}	5.0	—	75	—	0.010	75	—	565	µAdc
(Per Package)		10	—	150	—	0.020	150	—	1125	
		15	—	300	—	0.030	300	—	2250	
Supply Voltage	V_{DD}	—	+4.5	+18.0	+4.5	—	+18.0	+4.5	+18.0	Vdc

Noise immunity specified for worst-case input combination.

Noise Margin both "1" and "0" level = 1.0 Vdc min @ V_{DD} = 5.0 Vdc
                        2.0 Vdc min @ V_{DD} = 10 Vdc
                        2.5 Vdc min @ V_{DD} = 15 Vdc

MC14469

DATA FORMAT AND CORRESPONDING DATA POSITION AND PINS FOR MC14469 AND MC6850

RECEIVE DATA (RI; Pin 19)

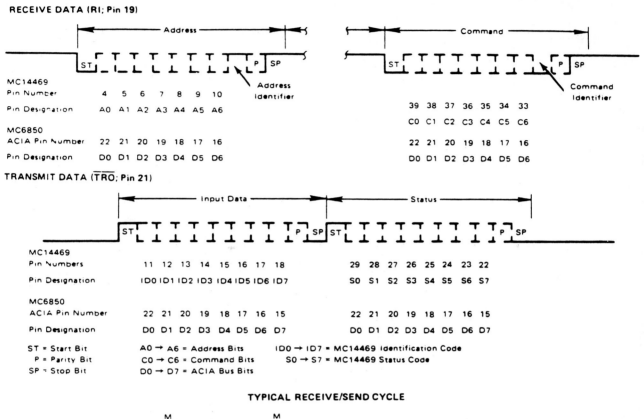

MC14469 Pin Number	4	5	6	7	8	9	10		39	38	37	36	35	34	33
Pin Designation	A0	A1	A2	A3	A4	A5	A6		C0	C1	C2	C3	C4	C5	C6
MC6850 ACIA Pin Number	22	21	20	19	18	17	16		22	21	20	19	18	17	16
Pin Designation	D0	D1	D2	D3	D4	D5	D6		D0	D1	D2	D3	D4	D5	D6

TRANSMIT DATA (TRO; Pin 21)

MC14469 Pin Numbers	11	12	13	14	15	16	17	18		29	28	27	26	25	24	23	22
Pin Designation	ID0	ID1	ID2	ID3	ID4	ID5	ID6	ID7		S0	S1	S2	S3	S4	S5	S6	S7
MC6850 ACIA Pin Number	22	21	20	19	18	17	16	15		22	21	20	19	18	17	16	15
Pin Designation	D0	D1	D2	D3	D4	D5	D6	D7		D0	D1	D2	D3	D4	D5	D6	D7

ST = Start Bit   A0 → A6 = Address Bits   ID0 → ID7 = MC14469 Identification Code
P = Parity Bit   C0 → C6 = Command Bits   S0 → S7 = MC14469 Status Code
SP = Stop Bit   D0 → D7 = ACIA Bus Bits

TYPICAL RECEIVE/SEND CYCLE

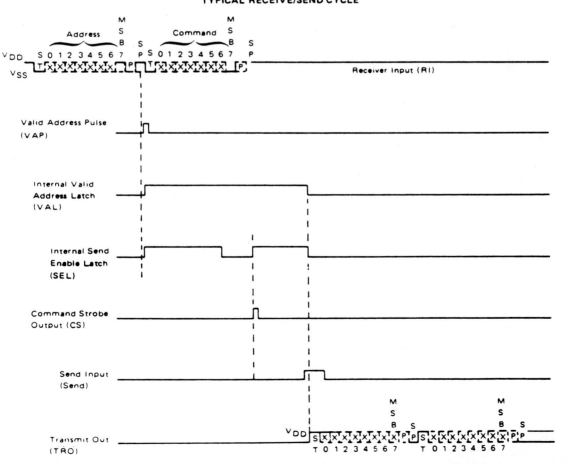

DEVICE OPERATION

OSCILLATOR (Osc1, Osc2; Pins 1, 2) — These pins are the oscillator input and output. (See Figure 1.)

RESET ($\overline{\text{Reset}}$; Pin 3) — When this pin is pulled low, the circuit is reset and ready for operation.

ADDRESS (A0-A6; Pin 4, 5, 6, 7, 8, 9, 10) — These are the address setting pins which contain the address match for the received signal.

INPUT DATA (ID0-ID7; Pins 11, 12, 13, 14, 15, 16, 17, 18) — These pins contain the input data for the first eight bits of data to be transmitted.

RECEIVE INPUT (RI; Pin 19) — This is the receive input pin.

NEGATIVE POWER SUPPLY (V_{SS}; Pin 20) — This pin is the negative power supply connection. Normally this pin is system ground.

TRANSMIT REGISTER OUTPUT SIGNAL ($\overline{\text{TRO}}$; Pin 21) — This pin transmits the outgoing signal. Note that it is inverted from the incoming signal. It must go through one stage of inversion if it is to drive another MC14469.

SECOND or STATUS INPUT DATA (S0-S7; Pins 22, 23, 24, 25, 26, 27, 28, 29) — These pins contain the input data for the second eight bits of data to be transmitted.

SEND (Send; Pin 30) — This pin accepts the send command after receipt of an address.

VALID ADDRESS PULSE (VAP; Pin 31) — This is the output for the valid address pulse upon receipt of a matched incoming address.

COMMAND STROBE (CS; Pin 32) — This is the output for the command strobe signifying a valid set of command data on pins 33-39.

COMMAND WORD (C0-C6; Pins 33, 34, 35, 36, 37, 38, 39) — These pins are the readout of the command word which is the second word of the received signal.

POSITIVE POWER SUPPLY (V_{DD}; Pin 40) — This pin is the package positive power supply pin.

OPERATING CHARACTERISTICS

The receipt of a start bit on the Receive Input (RI) line causes the receive clock to start at a frequency equal to that of the oscillator divided by 64. All received data is strobed in at the center of a receive clock period. The start bit is followed by eight data bits. Seven of the bits are compared against states of the address of the particular circuit (A0-A6). Address is latched 31 clock cycles after the end of the start bit of the incoming address. The eighth bit signifies an address word "1" or a command word "0". Next, a parity bit is received and checked by the internal logic for even parity. Finally a stop bit is received. At the completion of the cycle if the address compared, a Valid Address Pulse (VAP) occurs. Immediately following the address word, a command word is received. It also contains a start bit, eight data bits, even parity bit and a stop bit. The eight data bits are composed of a seven-bit command, and a "0" which indicates a command word. At the end of the command word a Command Strobe Pulse (CS) occurs.

A positive transition on the Send input initiates the transmit sequence. Send must occur within 7 bit times of CS. Again the transmitted data is made up of two eleven-bit words, i.e., address and command words. The data portion of the first word is made up from Input Data inputs (ID0-ID7), and the data for the second word from Second Input Data (S0-S7) inputs. The data on inputs ID0-ID7 is latched one clock before the falling edge of the start bit. The data on inputs S0-S7 is latched on the rising edge of the start bit. The transmitted signal is the inversion of the received signal, which allows the use of an inverting amplifier to drive the lines. TRO begins either ½ or 1½ bit times after Send, depending where Send occurs.

The oscillator can be crystal controlled or ceramic resonator controlled for required accuracy. Pin 1 may be driven from an external oscillator. See Figure 1.

MC14469

FIGURE 1 — OSCILLATOR CIRCUIT

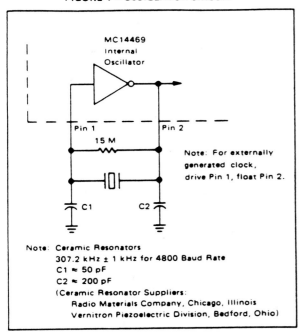

FIGURE 2 — RECTIFIED POWER FROM DATA LINES CIRCUIT

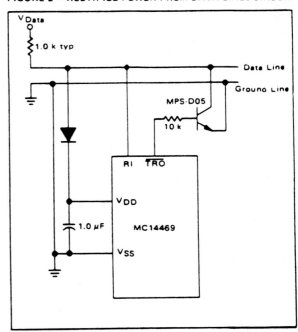

FIGURE 3 — A-D CONVERTER INTERFACE

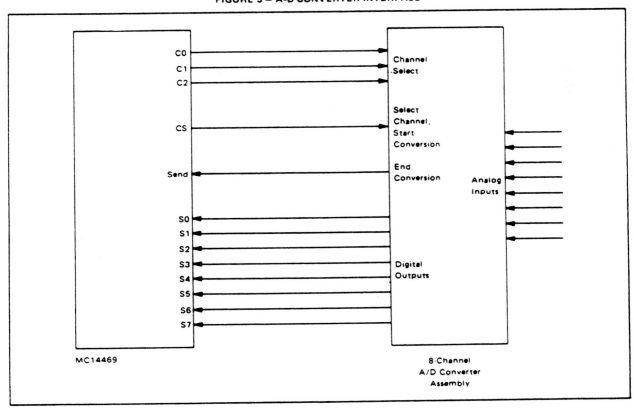

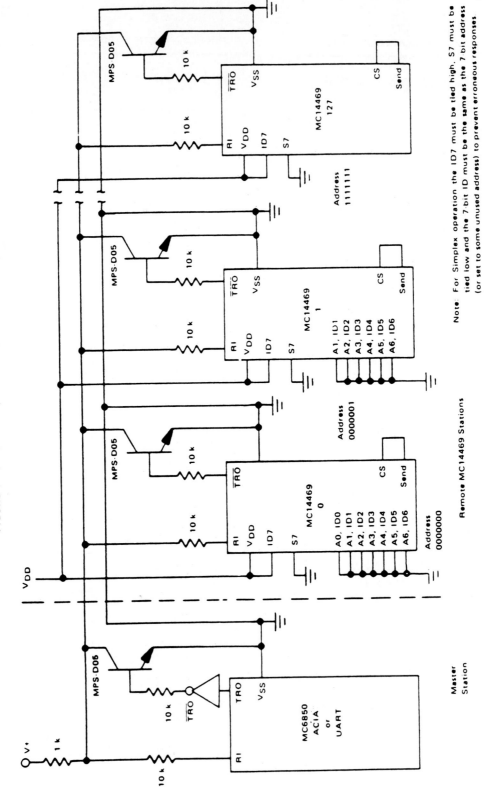

FIGURE 4 – SINGLE LINE, SIMPLEX DATA TRANSMISSION

Note: For Simplex operation the ID7 must be tied high, S7 must be
tied low and the 7-bit ID must be the same as the 7 bit address
(or set to some unused address) to prevent erroneous responses.

FIGURE 5 — DOUBLE LINE, FULL DUPLEX DATA TRANSMISSION

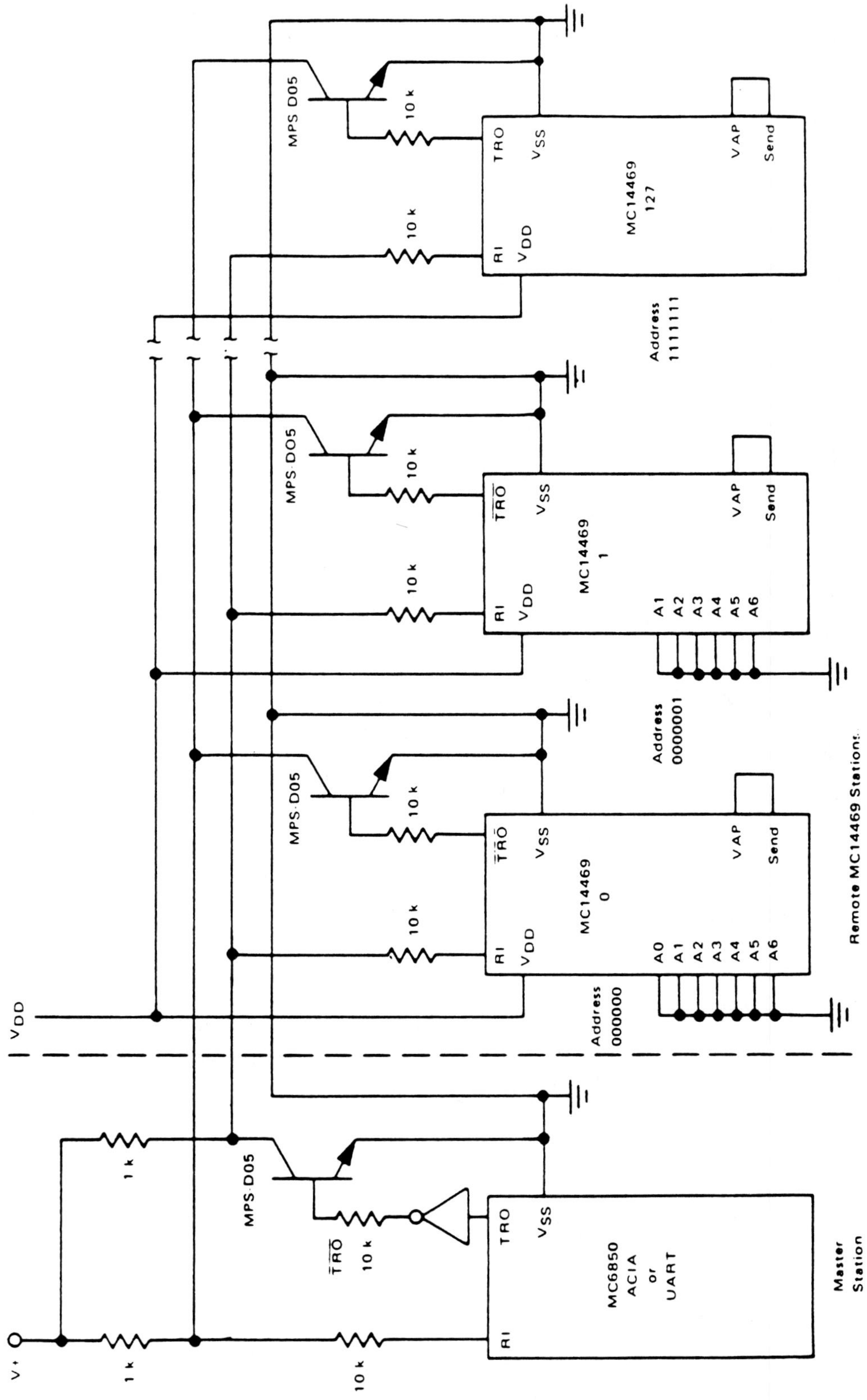

FIGURE 6 – FLOW CHART OF MC14469 OPERATION

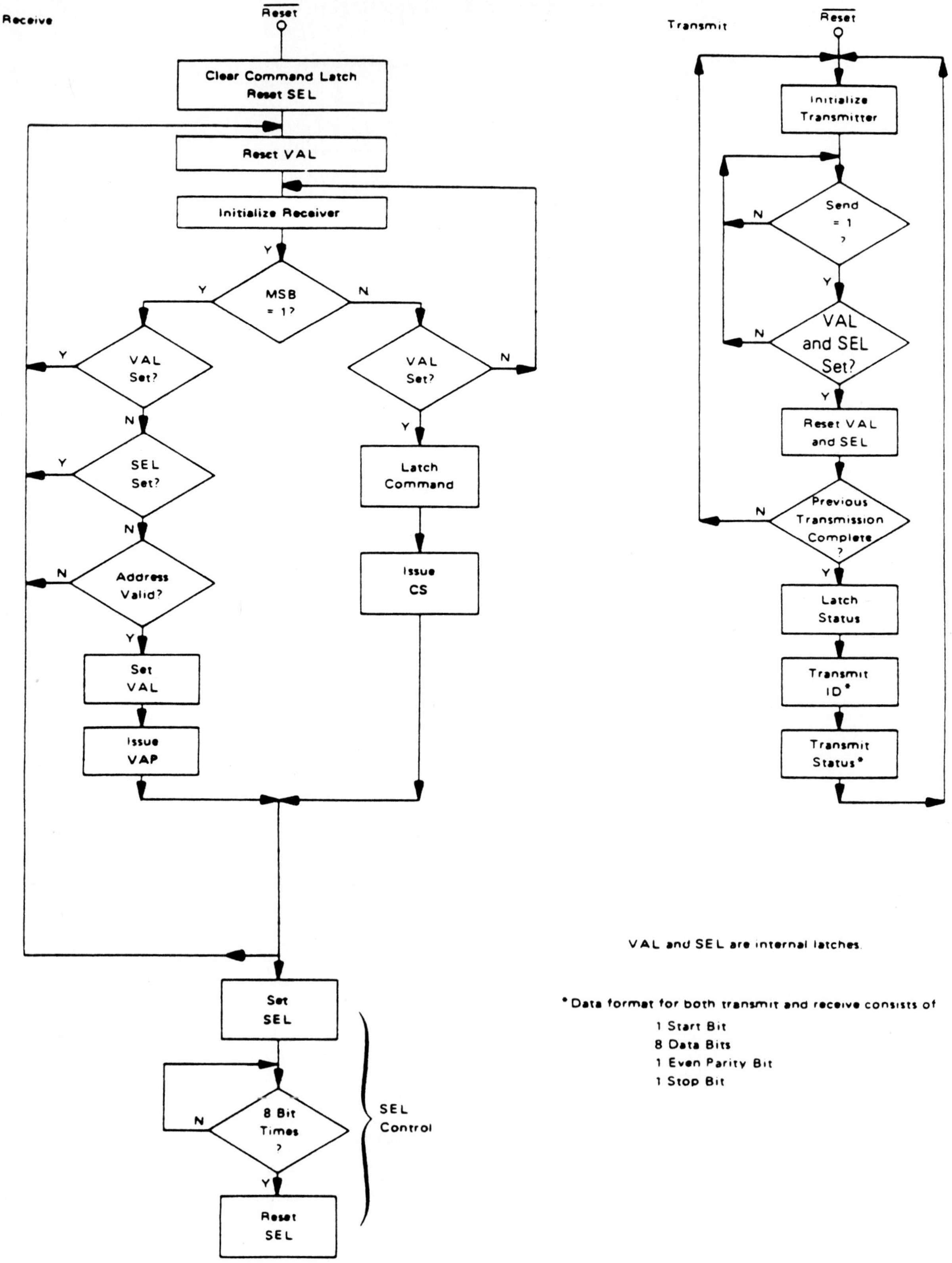
Receive
Reset
Clear Command Latch
Reset SEL
Reset VAL
Initialize Receiver
MSB = 1?
VAL Set?
SEL Set?
Address Valid?
Set VAL
Issue VAP
Latch Command
Issue CS
Set SEL
8 Bit Times ?
Reset SEL
SEL Control
Transmit
Reset
Initialize Transmitter
Send = 1 ?
VAL and SEL Set?
Reset VAL and SEL
Previous Transmission Complete ?
Latch Status
Transmit ID*
Transmit Status*
VAL and SEL are internal latches.
*Data format for both transmit and receive consists of
1 Start Bit
8 Data Bits
1 Even Parity Bit
1 Stop Bit

appendix F

Motorola MC14469 application note

OPERATION OF THE MC14469

Prepared By:
Len Bogle and Bill Cravy
Logic and Special Functions Applications
Austin, Texas

The MC14469 is an addressable asynchronous receiver transmitter that finds applications in control of remote devices, transfer of data to and from remote locations on a shared wire and as an interface from remote sensors to a central processor.

OPERATION OF THE MC14469

The MC14469 is an asynchronous receiver/transmitter fabricated in metal-gate CMOS technology. The asynchronous data format consists of a serial stream of data bits, preceded by a start bit and followed by one or more stop bits. The asynchronous data format is used to eliminate the need to transmit the system clock along with the data bit stream. The fact that the MC14469 is made in CMOS technology means that it offers the high noise immunity and low power consumption characteristic of this technology.

The MC14469 can receive one or two eleven-bit words in a serial data stream. The first received word contains a seven-bit address and if it matches the programmed address of the receiver, the transmitter can be enabled to transmit its two data words. The 7 bits of the received address word must correlate bit by bit with the 7 address pins of the MC14469. A second word may optionally be received for data or control use. This word will contain seven data bits which will be latched onto the command data outputs if it has a valid command format. With 7 address lines, 2^7 or 128 separate units may be interconnected for simplex or full duplex data transmission. The MC14469 is capable of operation at data rates in excess of 30,000 baud controlled by an on chip oscillator. Applications include transmitting data from remote A/D converters, temperature sensors, or remote digital transducers as well as single line control of remote devices such as motors, lights or security devices.

DEVICE OPERATION

As shown in the block diagram of Figure 1, the MC14469 consists of three different sections: the receiver, the transmitter, and the oscillator. The receiver must receive (at least) a valid address on its receive data input (pin 19) in order to set up the necessary internal conditions to allow the transmitter to transmit its two data words. The address word consists of

a start bit, seven address bits, the address identifier, an even parity bit and a stop bit. The address will be valid only if: a) the seven address bits match the address that is programmed on input pins A0 through A6, b) if the address identifier is high, and c) if the state of the parity bit causes the total number of ones in the address word, including the address identifier and parity bit, to be even. After reception of a valid address, the MC14469 can optionally receive a command word. Similar to the address word, the command consists of a start bit, seven data bits, a command identifier, an even parity bit and a stop bit. The command will be valid if the command identifier is low and the total number of ones, including the parity bit, is even. The reception of either a valid address or both valid address and a valid command can be used to set up the necessary internal conditions for transmission. The format of address and command words is shown in Figure 2.

Upon receipt of a valid address data stream, the MC14469 generates a valid address pulse (VAP) which in turn sets the internal valid address latch (VAL) and the internal send enable latch (SEL). See Figure 3 for a timing diagram. SEL remains high for eight data bit times or until the send input (pin 30) is taken high. If SEL is allowed to time out and a valid command word is subsequently received, a command strobe (CS) is generated which sets SEL high again. It again remains high for eight data bit times after being set. However, once the valid address latch (VAL) is set high, it will remain high until SEND goes high and resets it.

In order for the MC14469 to transmit its two data words, SEND must receive a rising edge while the valid address latch and send enable latch are both set high. Therefore, a send input must occur within eight bit times after the generation of either a valid address pulse or a command strobe, depending on the system configuration. After eight bit times, SEL will time out and transmission will be inhibited.

361

Figure 1A. MC14469 Block Diagram

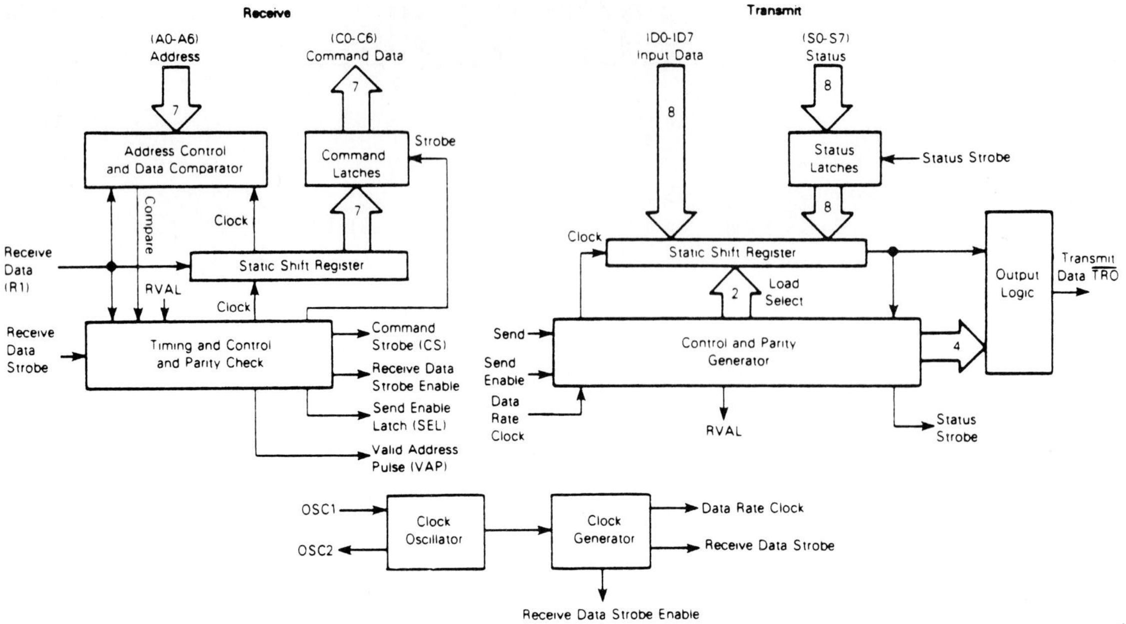

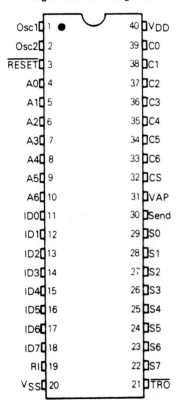

Figure 1B. Pin Assignments

SEND going high resets VAL and SEL, and initiates the transmission of the data defined by input pins 11-18 and the status word defined by input pins 22-29. The transmitted words each contain a start bit, eight data bits, an even parity bit and a stop bit, all in UART compatible format. The transmitted data has the format shown in Figure 3. Note that the transmitted data must be inverted before being presented to the receiving device. This is usually accomplished by the line driver or transistor used to drive the common transmit wire.

OSCILLATOR OPERATION

The oscillator can be controlled by a ceramic resonator, a crystal or by an externally generated clock, and will typically operate at frequencies up to 2 MHz at a V_{DD} of 12 V. The oscillator frequency is divided by 64 to derive the receive data strobe and the data rate clock. Thus, the data bit period is 64 times the oscillator period. To allow for maximum phase jitter, the receive data strobe is centered at the middle of each data bit. The receipt of a start bit initiates the receive data strobe and synchronizes the strobe to the receive data bit stream.

Since data is sent asynchronously, the transmit oscillator and receive oscillator must be the same frequency to ensure that the receive data strobe occurs at the middle of the bit period. The maximum permissible variation in oscillator frequency between a transmitting unit and a receiving unit can be such that over the entire receive data word time the total error is plus or minus one-half data bit period.

Figure 2A. Data Format

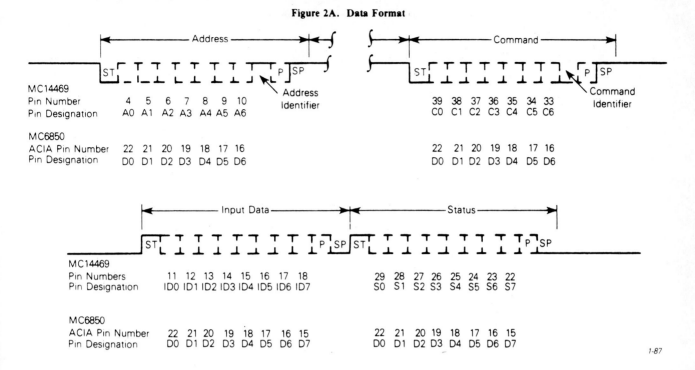

Figure 2B. Example Data Words

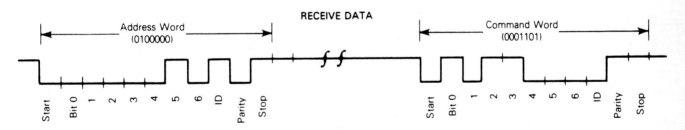

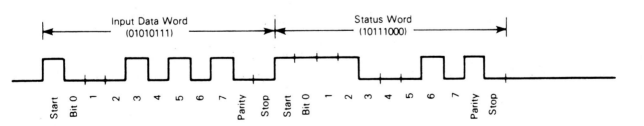

Figure 3. Typical Receive/Send Cycle

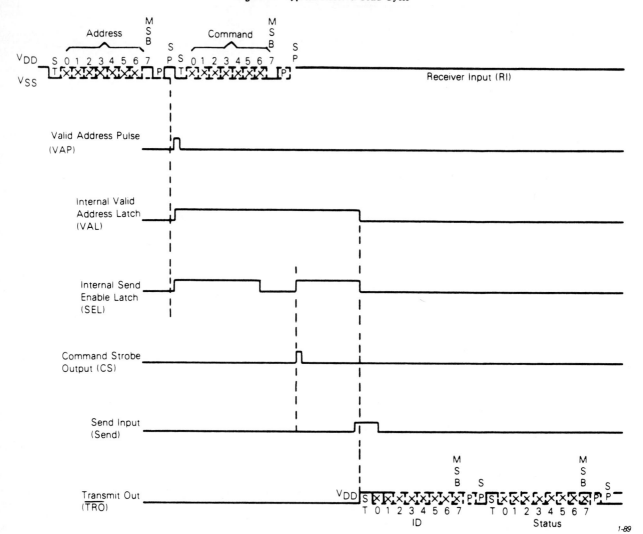

Each received data word consists of 11 bits, and thus the variation in oscillators cannot be more than half a bit time divided by 11 bit times or 4.5%.

The internal oscillator active circuitry consists of a normal CMOS inverter. When a high value resistor is used to provide DC feedback, the inverter is biased into its linear region and acts as an AC simplifier. The size of the feedback resistor is unimportant but needs to be small enough to overcome leakages and large enough to not load the oscillator output. Values in the range of 1 MΩ to 22 MΩ are common.

With the inverter biased as an AC amplifier the usual oscillator design is the Pierce type oscillator using a parallel resonant crystal. See Figure 4. Two capacitors, one from input to ground and one from output to ground, present the required capacitive load to the crystal. The series connection of the capacitors through ground avoids feedback of signal through the parallel capacitive path. An inductor or ceramic resonator can be substituted for the crystal to form a Colpitts

oscillator ususally at less cost than a crystal but at the expense of frequency stability.

MODES OF OPERATION

The various modes of operation of the MC14469 are discussed below. For most applications, the send input is tied to either VAP or CS for fully automatic operations. If this is not done, the send input must receive a rising edge within eight bit times after VAP or CS in order for a transmission to occur.

It is possible to operate the MC14469 in a receive only mode by tying SEND to V_{SS}. The device can receive valid address words only or both address and command words. Three different modes of operation of the MC14469 are possible depending on the signal used to drive the SEND input. These are RECEIVE ONLY MODE, SEND EQUALS VAP and SEND EQUALS CS.

Figure 4. Oscillator Circuit

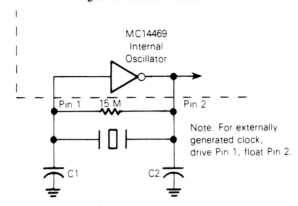

Note: Ceramic Resonators
307.2 kHz ± 1 kHz for 4800 Baud Rate
C1 ≈ 50 pF
C2 ≈ 200 pF
(Ceramic Resonator Suppliers:
    Radio Materials Company, Chicago, Illinois
    Vernitron Piezoelectric Division, Bedford, Ohio)

1-90

RECEIVE ONLY MODE

If the MC14469 is in the receive only mode (i.e., if SEND is tied to V_{SS}) and if it is receiving valid address words only, it will respond with a valid address pulse after every other valid address. The intervening addresses will cause no output. The reason for this can be seen by examining the flow chart (Figure 5).

Assume the MC14469 has been reset, the receiver is initialized and is ready to be addressed. If the MSB of the first word received is a one (signifying an address), the device checks to see if the valid address and send enable latches are set. If neither is set, and if the word is a valid address, VAL and SEL are set and a valid address pulse is generated. If SEND is not taken high within eight bit times, SEL is reset and the device is re-initialized and ready to receive a command. If the next word received is an address rather than a command, the MC14469 will find that VAL is still set. It will then reset VAL, initialize the receiver, and wait for another address to be sent. As a result, the second consecutive address to be received will not result in the generation of a VAP. This problem does not arise when the device is enabled to transmit every time it is addressed, since VAL is reset during the transmission cycle. Notice that once VAL has been set by the reception of a valid address, the only way it can be reset without rejecting an address is by going through the transmission cycle.

A similar situation arises when the MC14469 is in the receive only mode and valid addresses and valid commands are alternately received. On the reception of the first valid address, VAL and SEL are set and a VAP is generated. After eight bit times, SEL is reset and the receiver is re-initialized. If the next received word is a command, the MSB will be zero, and when the device checks the valid address latch, it will find that it is set. If the command word has a valid format, it will be latched onto the command data outputs (C_0-C_6). A command strobe will be generated and the send enable latch will be set. Once again SEL will be reset after eight bit times and the receiver will be re-initialized. Thus, the reception of the first valid address and command words

will result in the generation of a valid address pulse and a command strobe respectively, as expected. However, since data has not been transmitted, the next incoming address word will be rejected because the valid address latch has not been reset. The MC14469 will then reset VAL and re-initialize the receiver. The following word is a command word and because the valid address latch is not set, the command is also rejected and the receiver is re-initialized.

The next address and command words received will result in the generation of a VAP and CS. Thus, in the receive only mode, every other address and command words will be rejected.

SEND EQUALS VAP

If only addresses are being received and if VAP is tied to SEND, a VAP is generated and data is transmitted when a valid address is received. Normally the transmit cycle is completed before a new address word is received. It is possible, however, for the reception of a new address to overlap the transmission of data. If this occurs, the current transmission is completed before the transmitter is allowed to start another transmission (see Figure 5).

If address and command words are alternately received while SEND is tied to VAP, a VAP is generated every time an address is received. The transmission of data begins as soon as a VAP is generated. This results in VAL and SEL being reset before completion of the received command word and causes the command word to be ignored.

SEND EQUALS CS

If SEND is tied to CS and if address and command words are alternately received, a VAP and CS are generated every time an address and command are received. Data is transmitted every time a CS is generated. Once again, data transmission can overlap the reception of a new address and command word. However, the current transmission will be allowed to finish transmitting (see Figure 5).

Figure 5. Flow Chart of MC14469 Operation

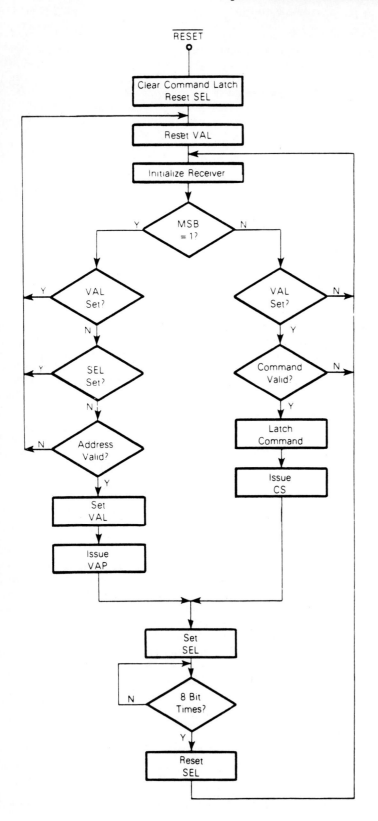

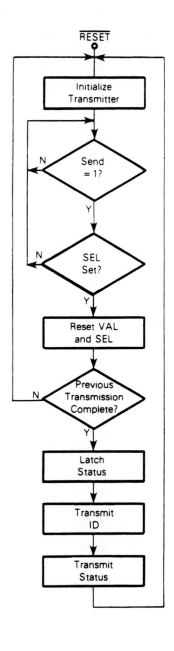

1-91

THE MC14469 AS A MASTER TRANSMITTER

The MC14469 can transmit only after it has received a valid address. For this reason it is usually considered to be a remote or slave device controlled by a UART, MPU or similar control system. However, it is possible to use the MC14469 as a master transmitter by giving it a start pulse on its receive input that has the format of a valid address. The idea is to set the address of the MC14469 that is to be used as a master transmitter in such a way that a valid address will consist of a single pulse which goes low for a certain number of bit times and then goes back high and remains high. This will allow the use of a one-shot or RC network to generate a start pulse which will look like a valid address to the MC14469. On receiving the start pulse, the MC14469 will generate a valid address pulse which can be tied to the SEND input in order to initiate a transmission.

As shown in Figure 2B, if the address of the MC14469 has an even number of ones, the parity bit will be high. The address identifier and stop bit are also high. Therefore, if the address begins with any odd number of zeros and ends with an even number of ones, the address word (start pulse) will need to go low for the start bit, stay low for an odd number of address bits, go high for the rest of the address bits, the address identifier, parity and stop bits. For example, if the address of the MC14469 is set to hex 00, a valid address will consist of a signal which goes low for eight bit times and then goes back high (see Figures 6 and 7). The other addresses for which the scheme will work are hex 60, hex 78 and hex 7E.

Figure 6.

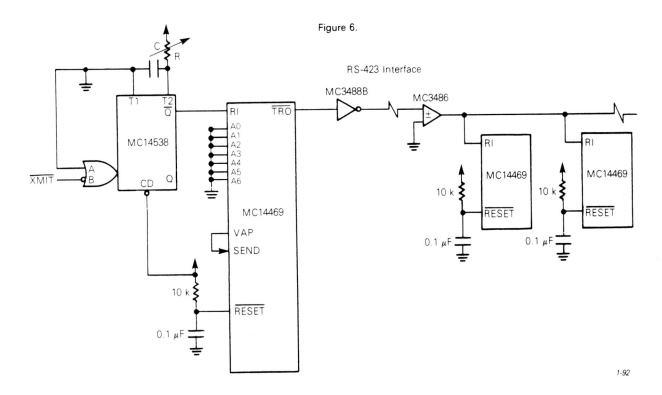

1-92

Figure 7. Transmitter Timing Diagram

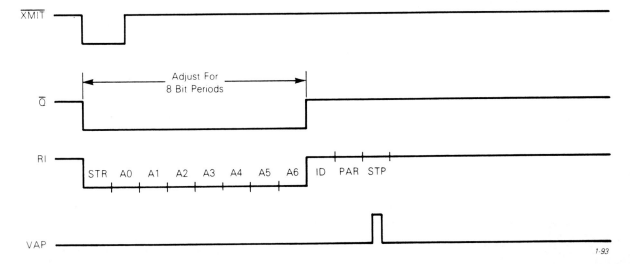

1-93

flowcharting example using shareware flowcharter program

Before introduction of computer-aided drafting programs, the vast majority of students (this one included) always created the required flowchart for a computer programming assignment AFTER the program was created, debugged, and ready for submission. The manual torture using the T-square, triangles, and templates often took as long as the program creation. However, when the programming assignments approach the difficult stage, it is a known fact that considerable time will be saved by FIRST creating the flowchart.

The computer-aided approach is best described by some informative words from the instruction manual for ***Flowcharter***, a shareware program worthy of serious consideration. The shareware version is available on computer bulletin boards.

Flowcharter Overview

The purpose of a flowchart is to provide a method of displaying complex procedures or events in a graphical manner using standard symbols which are easy to understand. The most common use of flowcharts is in the design of complex computer programs. ***These flowcharts show the basic design of the program independent of the computer language in which they will later be written.*** In this way, other programmers can translate the design into the actual program using the flowcharts.

FLOWCHARTER is designed to provide a programmer with the capability to produce flowcharts easily using simple commands which are then translated into actual flowchart symbols and output to a dot matrix printer or laser printer.

This method allows flowcharts to be updated easily as the program design changes and eliminates the tedious job of drawing and redrawing flowcharts while maintaining up-to-date drawings. Updates or corrections are made to the flowchart using any standard text editor, such as Word Star, PC Write, Brief, etc. and then reprinted with minimal effort.

Registered copies of the FLOWCHARTER program can be obtained from the author (JJ Enterprises) at the address shown below. Registration costs $25 ($35 if you want a printed copy of the manual) and provides you with the following benefits:

Current FLOWCHARTER diskette (5¼-inch DSDD format only)
Current FLOWCHARTER manual on diskette
FLOWCHARTER support for a year
Telephone support for questions

A printed copy of the FLOWCHARTER manual can be obtained for an additional $10. Contact: JJ Enterprises, 13133 Thomas Circle, Burnsville, MN 55337, (612) 890-5405.

Following is the text file used for creation of the flowchart for Example 6.22, shown as Figure 6-25:

```
        .$XM   .50,0,.20,9.40
        .$XH   F2,1.0,9.6
        .$XU   N
Start   .$H    Figure 6-6-0 – Bubble Sort Flowchart
        .$P    Point To First Number (Fixed Pointer)
        .$C    See comment 1
Again   .$P    Clear SWAP Flag
        .$C    See comment 2
        .$P    Point To Last Number (Variable
               Pointer)
        .$C    See comment 3
Read    .$P    Get First Number To Compare
        .$C    See comment 4
        .$P    Decrement Variable Pointer
        .$C    See comment 5
        .$T    Pointing Past Top Number?
        .$A    Yes,Check No
        .$C    See comment 6
        .$P    Get Second Number To Compare
        .$C    See comment 7
        .$T    Is 1st > 2nd?
        .$A    Yes,Read No Swap
        .$C    See comment 8
Swap    .$P    Swap Numbers
        .$P    Set Swap Flag
        .$C    See comment 9
        .$J    Read
Check   .$C    See comment 10
        .$T    Swap Flag Set?
        .$AE   No,EXIT Yes,Again
```

In summary, FLOWCHARTER is a powerful and versatile program, but only a relatively few of its features are needed in order to get started drawing useful flowcharts.

problems

Problem 1

Write a program that adds the two 16-bit numbers in memory locations $7000 and $7002 and stores the sum in location $7500 and $7501.

Problem 2

Write a program that adds the two 64-bit numbers in memory locations starting at $7000 and stores the sum starting at location $7500. (*Note:* The sum could be 65 bits long.)

Problem 3

Write a program that converts a string of 10 ASCII numbers to BCD format. The string starts at location $7000. Store the BCD numbers starting at location $7500.

Problem 4

Write a program that converts two 8-bit **BCD** numbers (starting at location $7000) to a single 8-bit **binary** number (stored at location $7500). The BCD numbers can range between 00 and 09.

Problem 5

Write a program that converts four 8-bit BCD numbers (starting at location $7000) to a single binary number (stored in locations $7500 and $7501). The BCD numbers can range between 00 and 09.

Problem 6

Write a program that divides a 32-bit number (stored in locations $7000 to $7003) by a 16-bit number (stored in locations $7004 and $7005). Store the **quotient** in locations $7500 and $7501 and the **remainder** in locations $7502 and $7503.

Problem 7

Write a program that finds the smaller of two 8-bit numbers, one located at $7000 and the other at $7004. Store the answer in location $7500. (Preload the locations with some known values.)

Problem 8

Write a program that finds the smallest 8-bit number stored in locations $7000 to $7400. Store the answer in location $7500. (Preload the locations with some known values.)

Problem 9

Write a program that finds the number of times a (8-bit) ZERO is found in locations $7000 to $7400. Store the answer in location $7500. (Preload the memory with some known values.)

Problem 10

Write a program that finds the number of times a (8-bit) zero is found in the ECB or MAX 68000 ROM/EPROM monitor program that starts at location $0000. Store the answer in location $7500.

Problem 11

Find the larger of two 32-bit unsigned integers stored in locations $7000 and $7004. Store it in location $7500.

Problem 12

Move a block of data. Move the contents of memory locations $7000 to $7100 to locations $7500 to $7600.

Problem 13

Find the maximum value contained in a region of memory. Write a program that will find the largest 16-bit unsigned integer stored in locations $7000 to $71FE. The data is all word-sized, starting at location $7000. Store the answer in location $7500.

Problem 14

Write a program that blinks an LED as long as an input switch is closed, basically combining the techniques of Examples 7.1 and 7.2.

Problem 15

Assume a memory location contains a decimal value (e.g., the sum of some numbers) that is between 00 and 99 (00 to 63 hex). Write the program to convert and display this sum as two BCD digits connected as in Section 7.4.

Problem 16

Write the program necessary to display eight digits for the circuit described in Figure 7-4-2.

Problem 17

Write the program necessary to read the A/D thermistor setup in Example 7.12 and to display the measured temperature.

Problem 18

Write the program necessary to control the dc motor setup in Example 7.13.

Problem 19

Write the program necessary to control the dc motor with tach feedback described in Example 7.14.

Problem 20

Write the program necessary to control the tank described in Example 7.15.

Problem 21

Write the program necessary to perform the data acquisition project described in Example 7.10.

Problem 22

Write the program outlined in Example 8.1.

Problem 23

Connect two 68000 computer systems via their RS-232 interface and establish communications, using Example 8.2 as a guide.

Problem 24

Construct a simple MC68008 computer system and develop a program to perform input and output, using Example 8.3 as a guide.

references

1. *The 68000 Microprocessor: Architecture, Software, and Interfacing Techniques* by Walter A. Triebel and Avtar Singh. Prentice Hall, ISBN 0-13-811357-2

2. *68000 Microcomputer Systems: Design and Troubleshooting* by Alan D. Wilcox. Prentice Hall, ISBN 0-13-811399-8

3. *Introduction to 6800/68000 Microprocessors* by Frederick F. Driscoll. Breton Publishers, ISBN 0-534-07692-0

4. *Programming the 68000* by Steve Williams. SYBEX Inc., ISBN 0-89588-133-0

5. *Dr. Dobb's Toolbox of 68000 Programming* by The Editors of *Dr. Dodd's Journal of Software Tools*. A Brady Book, ISBN 0-13-216557-0

6. *Using and Troubleshooting the MC68000* by James W. Coffron. Reston, ISBN 0-8359-8158-4

7. *The Motorola MC68000 Microprocessor Family: Assembly Language, Interface Design, and System Design* by Thomas L. Harman. Prentice Hall, ISBN 0-13-603960-X

8. *16-Bit Microprocessors: Architecture, Software, and Interfacing Techniques* by Walter A. Triebel and Avtar Singh. Prentice Hall, ISBN 0-13-811407-2

9. *Microprocessor/Hardware Interfacing and Applications* by Barry B. Brey. Charles E. Merrill, ISBN 0-675-20158-6

10. *Assembly and Assemblers: The Motorola MC68000 Family* by George W. Gorsline. Prentice Hall, ISBN 0-13-049982-X

11. *68000 Microprocessor Handbook* by Gerry Kane. Osborne/McGraw-Hill, ISBN 0-931988-41-1

12. *Microcomputer Experimentation with the Motorola MC68000ECB* by Lance A. Leventhal. Holt, Rinehart & Winston, ISBN 0-03-011782-8

13. *Introduction to Microprocessors: Software, Hardware, Programming* by Lance A. Leventhal. Prentice Hall, ISBN 0-13-487868-X

14. *Microcomputer Electronics* by Daniel L. Metzger. Prentice Hall, ISBN 0-13-579871-X

15. *The 68000 Microprocessor: Hardware and Software Principles and Applications* by James L. Antonakos. Charles E. Merrill, ISBN 0-675-21043-7

16. *The M68000 Family: Architecture, Addressing Modes, and Instruction Set* by Werner Hilf and Anton Nausch. Prentice Hall, ISBN 0-13-541533-0

17. *The 68000 Microprocessor: Architecture, Programming, and Applications* by Michael A. Miller. Charles E. Merrill, ISBN 0-675-20522-0

18. *MC68000 Assembly Language and Systems Programming* by William Ford and William Topp. D.C. Heath.

Index

THE SECRETS OF SCHOOL BOARD SUCCESS

Practical Tips for Board Members

Lynn Hamilton

Rowman & Littlefield Education
Lanham, Maryland • Toronto • Plymouth, UK
2008

Published in the United States of America
by Rowman & Littlefield Education
A Division of Rowman & Littlefield Publishers, Inc.
A wholly owned subsidary of
The Rowman & Littlefield Publishing Group, Inc.
4501 Forbes Boulevard, Suite 200, Lanham, Maryland 20706
www.rowmaneducation.com

Estover Road
Plymouth PL6 7PY
United Kingdom

British Library Cataloguing in Publication Information Available

Library of Congress Cataloging-in-Publication Data

Hamilton, Lynn, 1948–
    The secrets of school board success : practical tips for board members /
Lynn Hamilton.
        p. cm.
    ISBN-13: 978-1-57886-714-1 (hardcover : alk. paper)
    ISBN-13: 978-1-57886-715-8 (pbk. : alk. paper)
    ISBN-10: 1-57886-714-2 (hardcover : alk. paper)
    ISBN-10: 1-57886-715-0 (pbk. : alk. paper)
    1. School boards. 2. School board members—Professional
relationships. I. Title.
    LB2831.H295 2008
    379.1'531—dc22
                                                                2007030025

CONTENTS

7 Creating a Positive Relationship with the Public 33
8 Effective School Board Meetings 38
9 Continual Growth and Learning 44
10 The Top Ten Secrets of a Successful
School Board Member 47

**PART II LEARNING THE SECRETS OF SCHOOL
BOARD SUCCESS THE HARD WAY**

11 My First Political Campaign, My First Political Defeat
(1986) 53
12 Redemption! A Winning School Board Campaign
(1988) 59
13 A Tumultuous First Year (1988–1989) 68
14 Finishing My First Term with a Fizzle (1989–1990 and
1990–1991) 85
15 My School Board Dark Ages (1991–1994) 95
16 My School Board Renaissance (1994–1997) 102
17 A Resolution to Conflict with the Teachers' Union
(1997–2000) 109
18 Ending on a High Note (2000–2003) 117

Appendix: Historical Note and Timeline 121
About the Author 129

INTRODUCTION

Something clicked about midway through my 15-year career on the Board of Education in North Little Rock, Arkansas. I realized our board had gone from struggling as individuals to working as a cohesive unit for the good of our schools, and that we had the overwhelming support of our community.

Initially, though, our six-member board had bickered and fought like children in our elementary schools. Only half of the board routinely supported the superintendent. In my first meeting, we repeatedly voted three against three on multiple motions attempting to elect a board president for the coming year. We finally selected as our leader the other rookie member who, like me, never had served on a school board.

Our meetings always seemed to run long that year, often grinding argumentatively and unproductively for hours into the evening. Once, we finished a particularly grueling session after midnight, and on another occasion, two board members ended up screaming, calling each other names. The newspapers routinely reported our squabbles, and editorials criticized our contentious behavior. We were an embarrassment to ourselves and to the community.

I wasn't exactly innocent in the discord. Two years earlier in my first school board campaign, I had been defeated by an incumbent who remained on the board when I finally was elected. I believe we both still held some bitter feelings left over from our strenuously contested election. I certainly remembered the sting of my defeat, and I think my former opponent must have resented my having put her through a reelection campaign that she shouldn't have had to run.

She had done a good job and didn't deserve an opponent. Board members aren't compensated monetarily in Arkansas. We simply serve as volunteers because we're interested in the schools. Regardless, no love was lost when it came time for the two of us to work together.

Eventually, our board made a complete reversal. We became such a cooperative team and held so many of the same values that we seldom disagreed. Our meetings became highly efficient, rarely lasting longer than an hour or two and sometimes less than an hour. Often, because all members were doing their homework and trusted one another, we had little disagreement or even discussion.

Then, the editor of the local weekly newspaper routinely began criticizing us for not having *enough* debate. She thought we must be violating the state's Freedom of Information Act by holding illegal private meetings to work out our differences. But, in fact, we weren't meeting privately. Individual board members simply had learned to support one another, ask questions before meetings, and serve the interests of students.

Our accomplishments over the years included quietly and compassionately removing an abrasive, divisive superintendent. We spared our community the turmoil of an open fight. The failed superintendent was replaced by a highly competent successor who productively led our district for 15 years. We passed six out of seven tax initiatives at the polls, and we learned from our one failure. Perhaps our most unique success was dramatically gaining the

support of the teaching staff. Their trust enabled us to drop recognition of the teachers' union and eliminate collective bargaining.

One of many rewarding personal experiences came nine years into my tenure when I was supported in a reelection bid by that same woman who had defeated me so many years earlier. I learned firsthand that those whom we think are our enemies today can become our allies tomorrow.

Our school board made a journey from chaos to cohesion, and I was along for a sometimes wild ride, growing mentally, emotionally, and spiritually all the while. Part I of this book tells the secrets of success I learned along the way. Part II relates the story of the trip.

The secrets summarized in the initial section provide a quick examination of the keys to effective board service. However, major improvements in performance don't occur quickly or easily. Our board's successful change in superintendents and termination of collective bargaining took place through a series of events and good decisions made over a period of years. Part II explores that success in detail.

My growth as a board member largely came through introspection. I spent time consciously considering my role in the schools, so it was natural for me to reflect back on all I had learned at the end of my service. Over the years, I had arduously—and sometimes painfully—acquired some specialized skills I would no longer be able to use after I left the board. I valued those hard-earned skills, and I hated to see the knowledge simply fade away into the recesses of my memory.

I somehow felt compelled to put what I had learned into written form, if for no other reason than to preserve it for my grandchildren or perhaps their children. In a way, this book almost began writing itself as a collection of random thoughts soon after I left the school board. In the first year after my departure, I made a presentation at the Arkansas School Boards Association's annual meeting for new board members, and my notes from that speech

became part of this book, too. Then, at some point I realized that perhaps my reflections might be of interest to other people who share my passion for the public schools.

My sincere gratitude goes to the education professionals, other employees, students, parents, fellow board members, friends, and family who traveled parts of my school board journey with me. It was through them that my growth and this book were made possible. One such person is Ann Marshall, a longtime friend of Arkansas' public schools who in recent years has become my writing coach. Her thorough yet gracious guidance was invaluable in the early drafting of this work.

I also especially appreciate the love and support of my wife, Betti, and children, Logan, Wyatt, Kayce, Kelly, and Meggie, who have so fully enlivened and enriched my life.

PART I

THE SECRETS TO BECOMING A SUCCESSFUL SCHOOL BOARD MEMBER

I

THE IMPORTANCE OF INTENTIONALLY BUILDING POSITIVE RELATIONSHIPS

So, what did I learn in 15 years on a local public school board? In short, I learned that a board member can function effectively only through other people. Building positive relationships is the crux of serving the schools successfully.

I felt a sense of acquired competence in the last few years of my board tenure. Having survived a myriad of mistakes, I had eventually become comfortable in my ability to perform at a consistently high level.

Parts of my skill set came from professional training sessions and instructional materials. These concepts are familiar to anyone who has been around a school board at all. Most of my ideas, though, came from my own experience and may be different, or perhaps at odds, with what you've heard before. The common factor among all these secrets of success is that they worked in actual practice. Keep an open mind as you read. Take what you like and leave the rest.

The recommendations within these pages include a few procedural and technical suggestions, but my ideas primarily concern the manner in which board members relate to other people. This

book is intended to be read by board members and by the education professionals who work with them. My words, though, are addressed directly to those admirable individuals who actively demonstrate commitment to their communities by serving on local school boards. I've experienced their struggle firsthand. My text comes from the trenches of public education. The "secrets" contained within these pages are my answers to common dilemmas faced every day by local boards of education across America.

As a school board member, you have two sources of power to help students and families: (1) the limited *formal* authority that comes from being elected (or appointed) and (2) the potentially much more effective *informal* authority freely given by other board members, employees of the school district, and citizens in the community.

Formal authority mandates your right to vote at meetings but provides little else. You're one small component of a larger body, so your formal power is limited to a fraction of the authority required to make a decision. For example, your opinions and your vote on a seven-member board are worth one-seventh of the whole. When you vote alone (the times you cast the only dissenting vote), you can't accomplish anything. Your formal power is worthless. In fact, repeatedly voting alone is worse than worthless because the practice causes people to lose respect for you.

Formal authority also gives you the right to speak at board meetings; however, no one is forced to listen to you. You don't have the authority to tell the superintendent what to do, and you don't have the power to give orders to a principal, a teacher, or even a student. If you rely solely on formal authority, your words will be ignored like meaningless background noise or derided as uninformed nonsense.

Can you as an individual school board member really expect to help students? The answer is yes, definitely yes, but not just because you've been elected. Your election gives you the opportunity to assist your schools only if you earn the *informal* authority that

comes when you inspire the respect and trust of the people around you. You acquire the power to help students by inspiring confidence. Real leaders are inspirational.

Most (if not all) board members want to play a leading role in their school districts, and they want to help students. However, serving successfully means more than just filling a seat on a school board. Anyone can exercise the right to talk and vote. Some members stay on their boards for years and never really make a significant contribution. In fact, some board members do great injury to their schools with their talking and voting.

Gaining the informal authority to positively influence the education of students takes time, but the process can be accelerated by a school board member who understands the concept. People know you're interested in the schools when you first go on the board, and they'll be watching you. Your reputation, either good or bad, soon will be known throughout the schools and the community.

A board member who consistently builds informal authority by inspiring respect and trust can have a significant impact on the quality of education in his or her district. Administrators, teachers, fellow board members, and citizens all will respond to the member's ideas and encouragement.

When an entire board (or most of the board) is composed of members who inspire confidence, good things will happen in the schools, because teachers and administrators consistently will feel true leadership coming from the top. The atmosphere the board creates will spread through the schools; staff creativity will soar; and ultimately, students will benefit. Any organization will reflect its leadership, whether good or bad.

So how do you go about inspiring respect and trust, and acquiring the power of informal authority? You intentionally and consciously build productive relationships throughout your school district. The quality of those relationships will determine your success. You have the ability to create a large reserve of influence, or informal authority, through your relationships with others.

Here's an analogy. Think of having an influence account the same way you have a bank account. You can build the value of your account by making deposits or reduce the account by making withdrawals. You make deposits to your influence account by building positive relationships with other people in the school district. Similarly, your account will be diminished through interactions that create ill will.

Sometimes your account may be decreased through no real fault of your own. For instance, a delay in salary increases may be unfairly blamed on the school board. A board member, or entire board, who has built little or no influence reserve may be devastated by such a crisis. The key is to make sure your deposits far outweigh your withdrawals.

Over a long period of time, a board member who consistently creates positive relationships will build a large balance in his or her influence account, thereby establishing a significant asset for the entire school district. Likewise, a board member who consistently overdraws his or her account will be a serious drain on the schools.

The chapters that follow discuss a board member's various relationships with specific groups of people in the school district. However, we'll first explore the primary motivation of truly successful school board members and the personal characteristics they display in all relationships.

2

WHY WOULD ANYONE WANT TO BE A SCHOOL BOARD MEMBER?

Throughout my board service, the comment I heard most often was, "I don't know why anyone would want to be on a school board. It's such a thankless job." So why *would* anyone want to be a school board member? And what are the common sources of motivation that successful members share?

In late 1985, I was 37 years old and busy with my family and profession, but I wanted a challenge and more personal growth in my life. Community involvement was becoming appealing, too. All my adult life, part of my nature has been to stay busy seeking challenges and growth and always preferring to be a participant rather than a spectator.

I had earned an undergraduate degree in mathematics and a teaching certificate in four years while working full-time at night and marrying at age 20. College was followed immediately by a job as an inner-city high school teacher in Memphis, Tennessee. A low military draft number changed my life forever, though, causing me to quit teaching a few months into the position and join the U.S. Navy.

Three years as a navy officer preceded a brief fling in a computer business before settling into my career with one of two statewide newspapers, the *Arkansas Democrat*, in 1974. Shortly after joining the paper, I also began taking night classes at the local university, eventually completing an MBA and enough undergraduate accounting courses to pass the CPA exam.

By 1985, I had been out of school four years and missed the stimulation of working toward academic goals. Also, my last promotion at work had come three years earlier. At a fairly young age, my newspaper career had topped out. The organization had no place for me to move upward. Relocating to take another job wasn't an appealing option. My family had shuffled residences too much in my childhood, and I loved the Little Rock area. My life was full but lacked immediate goals or challenges.

One day driving home from work, a thought struck me, "Why not consider running for a political office?" Statewide and national elections always had captured my enthusiasm although local politics had never been of interest. I hadn't made a political contribution and certainly hadn't worked in any candidate's campaign. The idea excited me, though.

Soon afterward, I dropped by city hall and started asking questions. I couldn't name the local city or county offices, the people who held them, the length of their terms, or the timing of elections. The city clerk furnished all this information and at some point she mentioned the board of education. I'd heard of school boards before but hadn't given them much thought. In Arkansas, school elections are held separately from the general elections for other political offices, so I had voted infrequently in school elections, if at all.

The city clerk's words caught my attention. Public education was important to me, and the news media had been vigorously reporting that our county's three local school districts were embroiled in lawsuits surrounding their possible consolidation. My teaching career had ended prematurely but not my interest in education. In fact, my

aborted teaching experience had left me with feelings of unfinished business. My brief time at the head of a classroom had given me particular insight into the problems confronting urban school districts, so the idea of becoming a school board member fit with what I knew and what I needed. Perhaps, with a background in both education and business, I had something to offer the schools.

My intentions weren't solely altruistic, though. Most people who first consider running for political office surely stop for at least a few moments to think about what it would be like to see their name on a ballot, on yard signs, and in the newspaper. A desire to serve the public and the belief that you can do a better job than someone else go hand in hand with anticipating recognition for involvement.

Recognition and a sense of fulfillment from meeting a challenge are intrinsic rewards successful public officials receive for serving. Perhaps a need for these rewards is even a defining characteristic of people who are willing to seek public offices of all kinds, especially school boards where other more tangible rewards are small or nonexistent. Further, perhaps no volunteer in any unpaid job serves for purely selfless reasons. For an elective officeholder to claim otherwise is probably disingenuous.

Once elected, my motives for serving began to evolve, or more accurately, they began to expand. As the years went by, I continued to enjoy meeting the challenge, lending my personal skills to problem solving, and being recognized in the community for my role in the schools. But board service also became gratifying in unexpected ways.

I thoroughly enjoyed being more involved in my kids' lives through the board. I knew more about their activities and problems and got to know their teachers much better than most fathers. Also, my role set an example of community service for my children to follow.

The high school in our community graduates about 500 students each year. The board members attend the graduation ceremony,

and the president confers diplomas to the graduating seniors, while individual members present diplomas to their children. It was a great thrill for me to hand diplomas to my six kids (who included three children, two stepchildren, and a girl who later became my daughter-in-law) as well as many of their friends. Hanging prominently in my office today is one of my proudest possessions: a large, framed collage of pictures that shows me giving diplomas to each of them.

As the years passed, profound admiration and a special affection developed for the staff who served our school district. I felt a passion to encourage them and help them in their daily struggles to bring knowledge and hope to children, sometimes under severely challenging circumstances.

A deep bond emerged among the school board members. We had relatively little turnover and several of us served concurrently for many years. Together, we celebrated numerous successes and experienced some deeply troubling failures, too. Our backgrounds varied widely. We were affluent and nonaffluent, young and old, black and white, male and female.

We learned to understand and appreciate our differences. Because we each had areas of expertise we all valued, our diversity became a great strength. I looked forward to school board meetings. They allowed me to enjoy the company of like-minded people, including the staff, but especially the other board members.

My board experience spurred me to grow in ways unanticipated when taking those first fledgling steps toward getting involved. I developed confidence and skill as a public speaker, learning to survive with a microphone in front of me, a room full of people listening, and a television camera rolling. The importance of humility and the power of apologizing for mistakes became apparent. I learned *how* to apologize by getting plenty of practice in 15 years.

The value of keeping my ego in check and reviewing conscious and unconscious motives hit home again and again. My personality changed under pressure, and sometimes that knowledge made

me uncomfortable. I learned to use an emotional appeal as much or more than a rational one when needing to be persuasive. My capacity for compassion was broadened. The personal growth experience evicted me from the comfort of my middle-class emotional shelter.

Many people can't seem to see beyond the numerous challenges facing our schools today, so they ask that question, "Why would anyone want to be a school board member?" adding, "It's such a thankless job." They don't understand that it's anything but thankless. I heard "thank you" repeatedly and with increasing frequency as the years progressed. Occasionally, expressions of gratitude still come today from people who care about the public schools. Those words, "thank you," go a long way toward explaining why successful, long-term board members keep running for reelection.

In truth, though, not many people do step forward to be on a school board. The job requires someone who cares about the schools and has the confidence to believe he or she can make a difference, someone with the will to lead and the ability to be a part of a team, someone with the desire to meet a challenge. I tell strangers that school board service requires someone unusual, but board members hear me say that it takes someone extraordinary.

3

PERSONAL CHARACTERISTICS OF SUCCESSFUL SCHOOL BOARD MEMBERS

Ask yourself, "Do I enjoy being on the school board?" If your answer is "no," you need to make a change. Either modify the way you handle the job or remove yourself from the board. Life is too short to voluntarily do something you don't enjoy. If you approach the job correctly, being a school board member is an incredibly rewarding experience. If you don't enjoy the job, you're doing a disservice both to yourself and to the schools.

Successful board members feel a sense of satisfaction and enjoyment because they know they're helping students and because they often hear the words "thank you" from people in their school district. Their enjoyment is projected as enthusiasm to everyone they meet.

So why is it these enthusiastic board members hear "thank you" so often, and what can you do to emulate their success? Here are some recommendations.

Don't publicly or privately criticize the schools or school employees. When someone voices a complaint, listen carefully and politely. Let them know you understand the situation and, if their criticism seems legitimate, encourage them to take their concern

to the proper authority. However, under no circumstance should you offer anything critical about the schools back to them.

Any negative comments you make will be magnified and likely spread as gossip. Your condemnation accomplishes nothing positive and damages your reputation for discretion. You need to be known as someone who will listen thoughtfully and respond appropriately without slamming the schools. Successful board members are good listeners, but they're discreet. You absolutely should refuse to participate in gossip.

Respond quickly to problems. Listen carefully when someone brings you a concern, and then suggest some form of immediate action. However, under no circumstances undertake to do an administrator's job!

Often, your best option is to make a referral to the superintendent. A good technique is to tell people, "Call the superintendent and use my name to set up a meeting. Then, call me back if you aren't satisfied after your conversation." Using your name allows people to feel empowered, and inviting them to call you back says you aren't just passing the buck. In fact, you later may want to ask the superintendent privately whether the person called.

Responding *quickly* is critical. A slow or nonexistent response tells the person you really aren't too concerned.

Don't be evasive with people. Be honest. If you don't know the answer to a question or if you don't know how you feel about something, just say, "I don't know." Don't try to stall for time by giving a vague answer, and don't take an uninformed guess. You aren't expected to be an expert on all matters within your school district. People will respect your honesty.

Make the hard decisions even though you find them distasteful. For instance, if a student has violated a policy that calls for expulsion, you must be able to vote to expel the child, no matter how influential the parents. Or a more common example, if a friend is angry with the schools and is in the wrong, you need to be able to tactfully, yet firmly, disagree and explain your views.

Once, a friend from church was upset when our board expelled her young son. She was in tears. The boy's father had died a few years earlier. Basically, he was a good kid, but we couldn't make an exception for one child. A few years later the mother told me how well her son was doing and mentioned her appreciation for the schools' role in helping him. (We didn't just put kids on the street. Specialized learning programs were created for students removed from the normal classroom environment.) Ultimately, our difficult decision had proven beneficial to everyone.

Be humble. Highly successful school board members project an understated, quiet sense of self-confidence. No one enjoys dealing with an oversized ego, and some people are particularly sensitive to signs of egotism in elected officials. So guard against appearing self-important. You have a right to be proud of your service on the school board, but people will be impressed only if serving others is a higher priority than gratifying your own ego.

Learn the art of apologizing. I promise, you're going to make mistakes as a school board member, and some of your mistakes will be made in a very public setting. Sometimes your local news media will tell everyone about your error. When you make a mistake, it's important that you recognize and accept your own humanity rather than becoming defensive or angry when someone points it out.

A sincere, well-worded apology is one of the most powerful forms of communication you can employ, especially when it's done in public. If you make a mistake in public, you should apologize in public. And if your school board makes a mistake, the president should apologize for the board. Good apologies are so powerful because they're so rare.

Here's how to make a good apology:

1. Acknowledge that you made a mistake. If you're embarrassed, say so. People immediately tend to empathize with someone who admits embarrassment.

2. Specifically describe the nature of your error. Don't make any excuses for your action. In fact, a good statement is "I have no excuse for my actions." If you explain your mistake, be careful that your wording doesn't appear to be rationalizing or justifying your actions.

3. Then say, "I want you to know how sorry I am. I apologize." Don't use the word, "if" as in "If I've offended you, then I'm sorry." An "if" statement sends a message that you really don't believe you did anything wrong.

Usually, your apology will be quickly and enthusiastically accepted. Repeat it, if necessary, though. A sincere apology almost always will be perceived as genuine, and as a result, you're likely to build a strong, productive relationship. People know how difficult it can be to openly admit a mistake, and they'll respect your courage and honesty. Errors are great opportunities to build positive relationships. The opposite is true, too. If you refuse to acknowledge a mistake, people will lose respect for you.

Develop and use your sense of humor. People like to laugh and are drawn to others who are appropriately witty. A funny remark at the right time can enliven a dull discussion or sometimes defuse a tense situation. Self-deprecating humor works particularly well because it says you don't take yourself too seriously. Conversely, never use humor that cloaks criticism directed toward someone else.

Dress in a manner that invites respect. The reason a judge wears a robe in court is to foster an atmosphere of respect for his or her position and for the court's proceedings. Similarly, your manner of dress should create respect for your position and for the work of the school board. You should dress at least as well as the superintendent whenever you're together in public. Think about it. How often do you see employees dressing better than their supervisors in any organization? So, men, if your superintendent wears a coat and tie to school board meetings, then put on a coat and tie! A school board meeting isn't a backyard barbeque!

4

THE SUPERINTENDENT: A BOARD MEMBER'S MOST IMPORTANT RELATIONSHIP

If you build mutual respect and trust with your superintendent, you'll have a significant opportunity to help the students in your schools. Conversely, a poor relationship with the superintendent will be a detriment to the schools and overshadow any good you try to do. In a short time, a festering, substandard relationship will infect the entire school district and can do long-term harm.

The quality of your school district directly reflects the competence of your superintendent and school board. No organization can be better than its leadership. If you want an outstanding learning environment for your students, you must have an outstanding superintendent and school board. Here's why.

Leaders attract employees whose competence level is similar to their own. Strong leaders seek highly skilled employees whose expertise complements that of the organization. Weak leaders set lower standards of performance and feel threatened by employees who challenge the status quo. Rarely does a leader hire and *retain* someone who generally is more competent than himself or herself, because a superior employee will be dissatisfied working for an inferior supervisor.

An outstanding employee who works for a less-competent leader will be unhappy. Normally, the employee at some point will seek another job. Eventually, the inferior supervisor will be left with mostly low-performing employees. Superior employees are attracted to and inspired by leaders who exhibit exceptional skills and set high standards.

Over a period of years, your entire school district will be filled with employees who reflect the competence level (either high or low) of your superintendent and school board, so it's critical for students that your board fosters an environment that inspires the superintendent to attract outstanding administrators and principals. Then, these superior front-line leaders will in turn hire and retain outstanding teachers.

Successful school board members help create successful superintendents. How do you as an individual board member help create a successful superintendent? You provide an atmosphere that encourages growth to his or her full potential. Here are three keys:

1. *Provide encouragement.* Superintendents routinely hear more complaints than compliments. It's the nature of the job. Tell the superintendent how much you appreciate him or her. Say it privately, publicly, and often.

   Your public comments tell the superintendent you're serious, and they inspire the community's confidence. Point out specific actions you appreciate. Send him or her a card every year during the holidays and take a few minutes to write about the good things you've seen happening in the schools the past 12 months. Remember how much you appreciate recognition yourself for a difficult job well done.

2. *Avoid criticizing the superintendent to someone else.* You, along with the other board members, are his or her boss. Think about how harmful it would be if your own supervisor were to speak critically about you to other people where you work. Certainly, your superintendent will make mistakes

from time to time. You've hired a human being, and one of the problems with humans is that they all make errors. An important part of your job is to tactfully and constructively make the superintendent aware of mistakes so they aren't repeated. He or she needs your guidance. However, it's essential that this be done between the two of you alone in order to preserve the community's confidence and prevent embarrassment. Furthermore, you're human too, and you may be wrong. Preserving confidence and preventing embarrassment works in both directions when you deal with mistakes privately.

3. *When you want the superintendent to take some specific action, politely phrase your request in the form of a suggestion or recommendation.* Your tactful request is a matter of respect for the superintendent and for your fellow board members who share your formal authority.

Your formal authority as a school board member gives you no power to issue orders for the superintendent to follow. However, your informal authority developed over time will give you significant power to make suggestions and recommendations, which the superintendent will take seriously and usually follow because you've gained his or her trust and confidence.

If possible, first make your request privately, not in a public meeting. A private discussion allows a comfortable conversation to explore your reasoning and the superintendent's reaction. Sometimes, on particularly complicated issues, it's helpful to put your desire in writing. You can carefully phrase your thoughts and enable the superintendent to thoroughly understand your intent. Remember, though, any written communication legally is available to the public. You should consider how your words might someday sound in a newspaper article or on the air.

Now, let's discuss the personal characteristics of outstanding superintendents. Certainly, successful superintendents need the same attributes as successful school board members. Some of these are obvious, such as honesty, knowledge, good grooming, and self-confidence. Most superintendents possess these qualities. However, other equally important but less common characteristics include:

- Commitment to the community
- Humility; no ego on display
- A passion for education
- Patience with people

These four fairly rare characteristics allow a superintendent to build support with the schools' staff, parents, and the community as a whole. They make the difference between an average superintendent and an excellent one. One more quality remains, though, that sets outstanding superintendents apart from all the others. *Truly outstanding superintendents consistently make good decisions.* A superintendent who can process large amounts of complicated information and routinely arrive at optimal conclusions will limit his or her mistakes and keep the board members, staff, and community happy. This ability may be innate but also can be enhanced through practice.

Your job as a school board member is to help your superintendent develop all these qualities to his or her full potential. In part, you facilitate the growth process through private coaching. More importantly, though, you must exhibit these traits yourself. You should set an example, not just for the superintendent but also for the entire school district. Remember, the staff will reflect their leadership, and you're at the top.

It's critical that you fully trust the superintendent's decision-making ability. Your superintendent is *the* education expert in your

school district, and you must be able to rely on him or her to solve problems. When a citizen brings you a concern, your first thought should be to make a referral to the superintendent. Tell the individual to call the superintendent and that you have complete confidence the problem will be handled fairly. Also, it's helpful to suggest the person use your name when making the call. Your personal referral empowers the caller and tells the superintendent you trust him or her to make a good decision.

Effective superintendents easily handle complaints and problems, but to be effective, board members must allow superintendents to do their jobs.

If you don't trust your superintendent to make good decisions, you have a serious problem, and you absolutely must take action. If you grumble and complain about the superintendent rather than work to resolve issues of conflict, you'll be failing miserably as a school board member. Here's what you do.

First, examine your values and judgment. What are your board's core values? What is your shared vision for the school district? It's quite possible you're wrong in your judgment of the superintendent, especially if others on your board don't seem to share your opinion.

Regardless, whether you're right or wrong, you owe your superintendent the courtesy of privately discussing issues of conflict. You must address your concerns with him or her *every time* one arises. Most issues can be resolved when people honestly and tactfully discuss their differences. Similarly, most problems will worsen if they're ignored. It may be hard to look someone in the eyes and calmly, politely point out a mistake, but you aren't doing your job if you see problems and say nothing.

Most superintendents will receive constructive criticism happily in private conversations, and they'll grow as a result of your direct suggestions for improvement. The worst thing you can do is talk about your concern to others. Remember, your comments will be magnified and spread as gossip. This hurts not only the school district and the superintendent but also your reputation. You need to

be known as a leader who quietly solves problems, not as a gossip who talks about them.

Occasionally, problems can't be remedied in spite of a board's best efforts to resolve differences with the superintendent. If you find yourself alone in concluding that issues are impossible to resolve, then you're dead wrong. However, if a significant portion of your board and community are irreversibly dissatisfied with the superintendent's leadership, then he or she also must be unhappy in that situation.

Rarely do people desire to continue in positions where they're unwanted. Your board needs to be forthright when you, as a group, conduct the superintendent's annual review. At the same time, you also need to act with kindness and compassion to find a solution that helps everyone.

Certainly, though, my definition of "kindness and compassion" absolutely never includes buying out a superintendent's contract. A buyout sends a terrible message of waste to the public. Voters frequently hear school officials say they need more funding, but a contract payoff implies money is available to throw away to someone who is leaving. Also, think about the impression an expensive settlement makes on our underpaid teachers! Without a doubt, if you authorize a buyout, citizens and employees will decide the board is incompetent. They'll rightly conclude that you mishandled either the hiring or the firing process, or both.

If the superintendent has committed an illegal or immoral act, you have just cause for outright termination with no financial payoff. However, if termination for cause isn't merited, you patiently should help him or her make a change to a new professional situation. Taking time to allow for a quiet, normal transition is much less damaging to your schools and community than a public bloodletting. A rancorous exit by a superintendent will destroy your citizens' faith in the schools and in the board's leadership.

Regardless of the circumstances or reasons for a vacancy, hiring a new superintendent is the most important decision a board

member normally ever will face—so proceed with great caution. Here are some guidelines on how to make a good choice. *If possible, hire someone who already is a part of your school community.* No substitute exists for a superintendent's local commitment. It's relatively easy to help an eager yet humble new superintendent overcome his or her known weaknesses, but it's difficult to identify the character deficiencies in an unfamiliar job candidate. Further, even a brilliant and honorable outsider is a fatally flawed candidate if he or she leaves your school district a year or two after being hired.

If you're certain that a viable local candidate is unavailable, give serious consideration to using a professional search firm. A paid recruiter can provide the largest possible selection of candidates and lend expertise in the selection process. Be mindful, though: The search firm's primary goal is to earn a fee. Finding an outstanding leader for your schools is a secondary consideration for them.

A new superintendent needs the board's unanimous support. Naming a new leader for the schools with a split vote almost inevitably will create distrust on the board and dissension in the community. On such an important decision, the board members in the minority won't soon forget that their concerns were disregarded by the majority. They'll be tempted to prove they were right by continuing to find flaws after the new superintendent is hired. At best, relationships among the school district's leaders will be strained. At worst, the new superintendent ultimately might fail to be successful through no real fault of his or her own.

For example, if a seven-member board makes a hiring decision by a vote of four to three, the minority needs to find only one ally in a future election in order to form a new majority. Then, they'll be able to terminate the superintendent's contract at will. Even a five-to-two vote easily can be rescinded by a change in just two board positions. The high risk of alienating board members and creating disharmony in the schools makes hiring by a split vote unwise.

Don't be too impressed by a candidate's physical appearance or by how confidently he or she speaks in public. Attractive, confident candidates frequently are adept at presenting false images. Be particularly cautious if the candidate has worked in several school districts over the years. Often, the best superintendent is a quiet, humble person with a sincere desire to help students, although these qualities may not be readily apparent or particularly impressive in job interviews.

When you narrow the field to just a few candidates, be sure to talk individually and privately with people in the candidates' previous school districts. Find these people yourself. Don't rely on a candidate's list of references! You'll need to expend some extra effort, but your work will pay off. A comprehensive list of PTA officers is a good place to look for help. You'll need honest, unbiased recommendations from people who know the candidate's reputation firsthand. It's a red flag if someone voluntarily mentions "a small problem." Pay attention! The problem may not be all that minor, and you can bet it will reappear in the candidate's next position.

You normally don't hear many, or any, unsolicited negative comments about truly outstanding leaders. Still, it's critical to learn about a candidate's shortcomings. If you receive only glowing comments, ask specifically about deficiencies. Say, "I appreciate hearing about this person's strengths, but we all have weaker areas, too. How would you describe his or her weaknesses?"

Our board tactfully removed an abrasive, divisive superintendent in a compassionate manner, which some members of our community felt he didn't deserve. We resisted a union-led effort to have him dismissed by vote of the board. He continued in his position while quietly looking for other employment. In fact, we helped him find another job. He was articulate and attractive and quickly convinced another school district to hire him. Our failed superintendent was hired by the first board who named him a finalist among their candidates.

We found his replacement within the ranks of our own school district. The new superintendent had been overlooked as a candidate by previous boards. He was soft-spoken, low key, and humble. He went on to lead our district for 15 years and became known in Arkansas education circles as one of the top superintendents in the state.

You'll find our tale of two superintendents in chapters 11 through 13. The steps we took weren't easy, but they were effective. Our community rallied behind the board. The change in superintendents, particularly the manner in which we handled the task, triggered an immediate turnaround in our district's history of failed tax increase initiatives.

One final, significant point: *If you have an outstanding superintendent, pay him or her an outstanding salary.* Being a superintendent is one of the hardest and most important jobs in America today. If your superintendent is truly superior, your community won't complain about his or her earnings. An excellent superintendent deserves to be recognized and rewarded.

5

BONDING WITH FELLOW BOARD MEMBERS

Successful school boards are composed of members who trust one another and know they're all working toward the same goal. These boards exude a spirit of cooperation, mutual admiration, and pride. If you've been on a championship-winning athletic team, you know the feeling. It's gratifying to know you've contributed to something bigger than yourself, something unquestionably successful. Here's how you can help create that same atmosphere on your school board.

Treat all members with equal respect. Don't let friendships be perceived as cliques. Be friendly and warm to everyone, especially to new members, those who may be a little shy, and others whom you don't know especially well. As always, your actions speak louder than your words.

Particularly in a racially or culturally diverse school board, your actions should demonstrate clearly your concern for all children in the school district. Whether you're among an ethnic majority or minority, your words will be perceived as empty platitudes if they aren't congruent with your behavior. In fact, too much talk may create suspicion of ulterior motives. Over time, focusing on the

well-being of all students, not the machinations of adults, will establish a basis for trust.

Initially, our school board included just one newly elected African American. (See the Historical Note and Timeline at the end of this book for a recap of the major desegregation events in the Little Rock and North Little Rock school districts.) We appointed a second black member in my first year. Thereafter, for 14 years, we always had at least two and often three African Americans on our seven-member board. We never once split our votes along racial lines on an issue of substance. (Actually, I can't recall racially splitting our votes on any minor issues either, although it's possible I've forgotten some insignificant proposals.)

Our board truly believed and practiced the credo "The students come first," and we meant *all* students. We came to trust one another and to be respectful of the full board's concerns. A key factor was that our superintendent inspired the confidence of the entire community. His priorities included fairness and equal opportunities for children across the breadth of our diverse student population. The schools' staff followed his lead.

Let the other board members know from time to time how much you enjoy serving with them. Send holiday cards with a personal note, honestly telling individual members of your admiration for their specific contributions to the board's success. Also, a small gift exchange a few minutes before your December meeting is a fun way for everyone to express mutual appreciation.

If this sounds trivial, you might be surprised. I wasn't the least bit enthusiastic when the women on our board began giving Christmas gifts. However, the next year the men reciprocated (out of guilt, I suppose), and the practice became an annual tradition. Eventually, I realized we had established an important ritual symbolizing our regard for one another. I actually began looking forward to our brief little Christmas party every year.

If you disagree with another member, address the issue, not the person. Allow the other member to defend his or her position at

length without interruption. Make eye contact to show you're listening. When you reply, don't begin a sentence with the word "you," as in "You are mistaken." Instead, talk about your own perception of the issue. Also, don't feel you must have the last word.

If you're on the losing side of a split vote, don't take the loss personally and don't carry a grudge. If the debate is spirited, it's particularly important to demonstrate your goodwill. Be friendly. This is a good time to display your sense of humor. (Sometimes when I see the futility of a losing argument, I'll throw up my hands in exaggerated defeat and declare "I surrender!" This usually gets a laugh and breaks any tension.) It's important for you to take your loss well, because you'll want the other board members to do the same at some point in the future when they're on the losing side.

Lose the little battles now so you can win the big ones later. Learn the art of compromising. If you disagree with another board member's motion, but the issue isn't particularly important to you, consider supporting the proposal. Your flexibility demonstrates your respect for his or her opinions and confirms for everyone that you're a team player. The payoff comes when the other board members reciprocate on proposals that are meaningful to you.

Except on rare occasions, don't be the only "no" voter. If you frequently find yourself alone in voting "no," you need to examine your motives. What are you accomplishing? You're telling the community that you're not a team player, and you're hurting your rapport with the other board members. You're also damaging your informal authority to get things done.

Pass around the presidency. All members who stay on the board a few years should get a chance to be president. Rotating the leadership position makes clear that everyone is equally important and respected. Help a struggling chairperson learn to do the job. Lend your assistance, and you'll be repaid with his or her support at some future point when you need it most.

If an effective fellow board member is up for reelection, lend some help to his or her campaign. The return of a good board

member will help the schools, and you'll have demonstrated your support in a real way. Be up front about your assistance, though. Don't be coy or try to hide your involvement.

Actively encourage bright, energetic citizens to run for open school board seats, and then help them get elected. You'll have earned the new member's trust, and he or she will look to you as a mentor. More importantly, though, you'll have done the schools a favor. If the candidate you backed (either incumbent or first-timer) happens to lose, be especially warm in welcoming the victor to the board. Don't try to downplay your past allegiance. Instead, show the new member genuine support through your words and actions. Demonstrate clearly that you're giving the winning candidate every opportunity to prove the voters correctly selected the best person for the job.

6

PRODUCTIVE RELATIONSHIPS WITH SCHOOL DISTRICT EMPLOYEES

A successful school board member walks a tightrope in his or her relationship with the employees of the school district. You must show employees your interest in listening to them while simultaneously making clear your support for the authority of the superintendent and other supervisors. Here are some suggestions for maintaining this balance.

Be active in the schools. You need to know firsthand about the atmosphere and the personalities in the district. Basic business management textbooks often recommend MBWA, or Management by Walking Around. You must be able to form your own opinions about the strengths and weaknesses of the district, but you can't be objective unless you have direct contact with the employees.

If you have a child or grandchild in school, some contact comes naturally and easily. If you don't, you'll need to work a little harder. You can volunteer in the schools and schedule visits through the principals. Either way, you want employees to feel comfortable telling you what works well and what doesn't. So be friendly and approachable. A board member can learn a lot by listening, and an employee can be helped just by being heard.

An employee with a complaint should be encouraged to use the chain of command. However, first listen carefully to the employee for as long as he or she wants to talk. Then, play back to the employee what you've heard. Say something like "I understand you feel . . . (a brief summary of the complaint). Is this correct?" After you're sure the employee feels heard, say "Here's what I'd like you to do. Go see your supervisor. After that, see his or her supervisors all the way up to the superintendent if necessary. The superintendent is easy to talk with and will listen to you. Once you've done that, call me back if you're still unsatisfied."

If the employee uses the chain of command, the problem normally is solved without coming back to you. Frequently, though, the employee will reply, "Oh, I didn't expect you to do anything. I just wanted you to know what was happening." In that case, simply thank the employee for the information, and say, "Call me again if I can ever help you in the future."

Don't be a pest in the schools. Sometimes a board member will become so emotionally committed to serving that he or she will become too involved in daily school activities. Employees will perceive these actions as meddling, and fellow board members likely will resent the offending member's overinvolvement. If you're guilty of being a pest, people probably won't confront you directly, but you'll lose their respect. So, be sensitive to signals from school employees and board members implying that you need to back off. Little "jokes" about the frequency of your presence may be intended as hints, not humor.

Don't ever give school employees orders, and don't offer any unsolicited suggestions or recommendations to anyone other than the superintendent. Employees will take offense to your presumptuousness, and you'll damage your relationship with the superintendent. Meddling in this manner is one of the worst errors a board member can commit. You absolutely will destroy your ability to affect the schools in a positive way because undoubtedly people will

lose respect for you. It's ironic, isn't it? By giving an instruction you hope will cause a positive change, you're actually destroying any possibility of significantly helping the schools.

The payoff for actively listening to employees and conducting yourself with professionalism is that you'll be trusted. Lack of trust is the primary cause of internal strife and is a driving force in the formation of labor unions. It's reasonable and necessary for employees to look to a union for support if they don't expect their schools' leaders to act in the staff's best interests.

Our school district was able to discontinue collective bargaining primarily because most teachers believed that the superintendent and school board would be honest and fair in our dealings with them. The staff especially trusted the superintendent, and they knew he had the respect of the board, too. Teachers also had learned to trust the individual school board members through our years of involvement in the community. Many teachers knew us personally and told their coworkers about our values.

As a result, most of the staff realized that a union wasn't necessary to protect their interests. Our school district became the only one of three in the greater Little Rock area that didn't participate in collective bargaining. We also were the only district where the superintendent and most board members had long, close ties to the teachers.

The important events that enabled our board's dissolution of union contracts and ultimate resolution of continuing conflicts took place over seven years. You'll find the details in chapters 15 through 17. For me personally, though, my thoughts regarding the teachers' union evolved during a much longer period. I originally had been elected with the union's enthusiastic endorsement and had been loudly cheered in board meetings. Actually, all seven of our board members initially had been elected with the union's endorsement.

Eventually, we unanimously voted to drop union recognition, knowing we would face contested reelection campaigns and a

strike attempt. We came to see the never-ending power struggle of union negotiations as detrimental to the education process in our school district. In the end, the large majority of our teachers agreed with us. The evolution of my personal thinking is a thread that runs throughout the second part of this book.

7

CREATING A POSITIVE
RELATIONSHIP WITH THE PUBLIC

The quality of a school district's relationship with the general public results from the continuing collective experience of people all over your community. Most important are the many small interactions happening in the schools every day. It's critical that your staff members are enthusiastic and effective in their jobs. A well-functioning school board is indirectly responsible for these day-to-day activities.

The board must create an atmosphere that fosters high-level performance throughout the schools. However, a successful board can enhance the community's perception directly through well-thought-out programs and procedures that encourage public participation in all phases of school activity. Here are some suggestions to help create a positive relationship with your community.

Welcome people to your board meetings and encourage input from visitors. Anyone who takes the time to attend a school board meeting should be treated as an honored guest. Members of the public are seldom motivated to show up, and when they do, their first impression often is negative. So welcome your visitors warmly. Their ability to address your board should be a painless process for the visitors, if not for the board members also.

It's sometimes hard to listen attentively and patiently to a citizen's unusual or unexpected concern. It's important to remain polite, though. One evening, an irate speaker concluded his remarks to our board by slamming the lectern to the floor. As he departed, our angry guest heard the president simply reply, "Thank you for your input." Nothing more needed to be said.

Hold open-forum meetings in the schools when you're faced with controversial issues such as school closings, budget reductions, and so forth. Send notices home with the students. Ask parents to come hear and respond to the details of the situation. These meetings work best if administrators present the facts, the public replies, and school board members just listen.

Involve parents and other citizens on special committees to consider potentially controversial decisions. Such issues might include dress code changes, hiring coaches, discipline policies, and the like. Give the participants meaningful work, not busywork. Then listen respectfully to their recommendations.

Use study circles to build community involvement and support. The study circle concept is a type of deliberative dialog that involves inviting large numbers of citizens to look at a broad range of school issues in small groups. People volunteer to participate in areas that interest them. Each group does an exhaustive study of its topic over a period of several weeks and then presents recommendations to the school board for changes in district policies or procedures.

Study circles serve the dual purposes of informing the public and seeking input. Even more importantly, they build understanding and trust among all involved. Our district was a pioneer of the concept in Arkansas. We found the extensive community involvement particularly helpful in advance of openly campaigning for a tax increase.

The Arkansas School Boards Association has an excellent program to help school districts get started with study circles. The organization's director of the Arkansas Study Circles Project will be happy to speak with you. The National School Boards Association

also can provide more information. A primary reference is the Study Circles Resource Center in Pomfret, Connecticut. Website addresses for these groups are furnished in chapter 9.

Create a good working relationship with your local news media. A part of the media's job is to be watchdogs for the community. If they bark a little too loudly in criticizing you, try not to take their noise personally. No one will read articles or listen to television and radio reports about the school board as closely as you do. If a reporter asks for a document, supply it quickly. All school documents, except personnel files, normally are available to the public by law. And when the newspaper prints a particularly good feature or a television station airs a complimentary report on one of the school's programs, be sure to send a thank-you note.

I've been fortunate to spend most of my professional career working for a large statewide newspaper. I've had an opportunity to see firsthand the integrity and dedication of journalists. My job doesn't directly involve the newsroom, but I've gotten to know reporters and editors personally. I feel close to some of them and also admire many others.

Newspaper professionals observe an invisible wall that separates the news-gathering functions from the business areas. This division is important to ensure that the newspaper's profit-making goals and management's biases don't affect the reporting process. So, as a school board member, our reporters treated me the same way they would any other public figure.

When our board made mistakes or when we had problems in our schools, the reporters wrote about them. If I said something particularly dumb, they recorded my words verbatim. I suppose, though, tolerance came a little easier for me than for most board members. When I felt under attack, I knew with certainty that an article wasn't intended to be personal. Even so, I sometimes had to grit my teeth to remain cordial.

On vacation out west a few years ago, I read about a school board member who dumped a truck load of horse manure at the

front door of the local newspaper. I admired his creativity of expression (although I don't recommend his action) and identified with the emotion he must have felt.

Initially, our school board had a superb working relationship with the local hometown weekly newspaper, the *North Little Rock Times*. Midway through my board tenure, though, the paper was sold to a husband-and-wife team from the East Coast. They came south to realize the dream of owning their own business. The husband ran the financial side of the publication, and the wife, who was a former reporter at the *Washington Post*, became the editor. They had no understanding of our school board's history.

The new editor improved the paper remarkably, winning several national awards. She was on a quest for excellence in remaking the newspaper. In the process, she became convinced that our board was doing backroom deals in violation of the state's Freedom of Information Law. The teachers' union leaders had her ear. Her editorial columns frequently focused on the school board, and the news articles seemed slanted against us. Her first name was Kitty, but she was more tigress than kitten.

One of our wittier board members privately began referring to the editor as Kitty Litter and calling the newspaper the "cat box liner." I still smile when I recall the twinkle in the board member's eyes the day she first came up with the expressions.

Ms. Litter, the editor, at one point thought she had caught our board in a major misdeed. A front-page story criticized us for excessive travel expenditures. A headline declared that the board members had recorded $40,000 in personal expenses the past year. She didn't know, though, that her reporter had the amount wrong. We had spent only $4,000 for several members to attend the National School Boards Association's annual conference and training sessions in San Francisco. Her glaringly incorrect article was published on a day when we had scheduled one of our regular monthly meetings. The school district's financial staff called her immediately, so she had all day to think about her error before our board session that night.

The editor showed up unannounced at the school board meeting and asked to speak to us. She apologized profusely for perhaps ten minutes, emphasizing her contrition at least twice with the term *mea culpa*. I always had appreciated the editor's talent and admired her intelligence, but my esteem rose quite a bit higher that evening. She didn't let up on the board afterward, although she seemed more approachable, perhaps a little more human.

The two of us developed a warm relationship away from school board matters. Occasionally, I would send her a note about an item in her newspaper, sometimes a compliment, sometimes a tactful complaint. I always stated that my words were not for publication, and she honored my request.

Learning to thrive in spite of public criticism is one of the great lessons of school board service.

Consider televising your school board meetings if you have access through a local cable television provider. Your cable company might supply the equipment at little or no cost, particularly if it's nearly time for your community to renew the company's contract. Only a small group of people actually will watch your meetings, but a large group will appreciate knowing they have the opportunity.

Our school district began televising board meetings as a direct result of input received from a study circles group. We contacted the city's cable television contractor immediately upon hearing the group's suggestion. Our meetings were being televised by the time the idea was included among the group's final recommendations. The cable company provided the initial equipment free of charge. We eventually expanded our high school's radio and television education curriculum to include student filming and production of the broadcasts.

Encourage unhappy citizens to take their concerns directly to the superintendent. Hearing that they can go straight to the top delivers a message that the public is taken seriously. See the specific referral technique recommended in chapters 3 and 4.

8

EFFECTIVE SCHOOL BOARD MEETINGS

Successful school boards follow a simple set of parliamentary rules that allow for fair and efficient meetings. You need not worry about most of the complicated procedures explained in great detail by *Robert's Rules of Order.* Your school board can use any legally defensible guidelines you choose to set in place. Here are some suggestions that will help promote effective meetings.

Limit your use of the more complex parliamentary motions. Normally, you only need:

- main motions
- amendments to main motions
- motions to table (temporarily delay consideration)
- motions to postpone to a specific date

You risk confusion by using procedural techniques more complicated than these. Your meetings can be reduced to a nightmare of misunderstanding where form becomes more important than education substance. Keep your motions simple so you can focus on school issues rather than parliamentary rules.

Use the "unanimous consent" procedure. This concept eliminates meaningless votes and wasted time in meetings. For instance, if a member asks for a brief recess, the president can say, "With the board's consent, we will take a ten-minute recess." Then after a brief pause, he or she can follow with, "Hearing no objections, we are recessed and will reconvene in ten minutes." The president makes the assumption that the board has given its unanimous consent when no one objects. If a member does object, the board still has the option to consider the question by the usual procedure of a motion and majority vote.

Apply your rules consistently to be certain all board members are treated fairly and equally. For example, the president shouldn't allow board members sometimes to talk freely without formal recognition, then at other times insist that members wait to be recognized before speaking.

Allow each board member to feel fully heard on any topic. The board has the right to set a time limit on the discussion of any item (by majority vote), but this may damage the spirit of trust and cooperation among your members as well as alienate the public. If you choose to set a time limit, the restriction should be established before discussion commences on a particular item.

Commonly used parliamentary procedure requires a two-thirds majority vote to prematurely terminate debate. This is one of the most misunderstood basic rules in formal proceedings. Sometimes, in organized committees and groups of all types, a member who's eager to vote and get on to other business will shout, "I call the question." Often a sympathetic chairperson will then proceed to voting by stating, "The question has been called." This tactic has no place in a school board meeting.

Abruptly ending a period of discussion is simply unfair to members taking a minority position on an issue. The president of the board should remind everyone that a majority of at least two-thirds is required to close comments. For example, on a seven-person school board at least five members must agree to conclude discussion on any

item. In actual practice, though, terminating debate by a parliamentary motion rarely makes sense. The voters in the minority will long remember their fellow members' unwillingness to listen. *You must maintain a quorum to conduct business.* If a board member leaves the meeting after it begins, causing you to fall below the number of members required for a quorum, you can continue discussions, but you can't take a legally binding vote.

Put your board meeting rules into your written policies. Establish these written regulations and procedures when things are calm among members. If you don't have a current policy manual, see your state school boards' association for help, quickly!

Use titles when addressing people in your meetings. This creates an atmosphere of respect and equal treatment for everyone. For instance, refer to "Mr. Jones," not "Jim." This allows you to treat an irate parent or employee with the same indication of dignity and friendliness you treat the superintendent or a board member.

Arrange your meeting in a "C" formation with the president and superintendent sitting at the center of the "C" and the remaining board members at either side. Members are able to see each other when speaking, and the audience has a good view of the proceedings. Also, locating the president at the center, with the superintendent nearby, allows for better control of the meeting.

Use a prepublished, written agenda of business. This helps you prepare in advance, stay on track during meetings, and defer introduction of items the board is not prepared to address.

Start your meetings on time. Your promptness sets a tone of efficiency and is a matter of respect for the people who have come to hear you.

Allow time for public comments just before officially calling to order the routine monthly meetings of the school board. This practice serves several purposes:

- The board's interest in listening to the public is clearly demonstrated. Anyone is able to speak on any topic.

- Visitors are allowed to make their presentations immediately. They aren't required to wait to be heard.
- The board can explain that the public's time to speak is before the meeting, not throughout the board's discussion of items on the agenda.
- A speaker can be told that his or her concern will be reviewed and can be considered privately or in a future meeting. The item can't be considered immediately because the board is not yet officially convened for business. Moreover, the board normally works from a prepublished, written agenda.

It's certainly workable if your board wants to take public comments after the meeting is officially opened. However, the visitors' presentations should come early, and the speakers must understand that their time ends when items on the written agenda begin.

Members of the public shouldn't interrupt the board's discussion. If you choose to allow public comments on a particular item during the business portion of the meeting, ask for those comments prior to the board's debate.

Use board meetings as a time of recognition for outstanding students, employees, and school supporters. Special recognitions at the beginning of the meeting set an uplifting tone for the entire proceedings. Also, scheduling the recognitions early allows visitors to see the presentations and then leave before the business part of the meeting begins.

Don't keep visitors waiting a long time to hear you consider a particular item on your agenda. For instance, if a large group has shown up to hear about a new athletic program, amend your agenda to put this item ahead of other nonrelevant business. Making visitors unnecessarily sit through long discussions of other topics sends a message that the board is insensitive and feels self-important. Furthermore, if the crowd is angry, you'll make the situation worse by causing them to wait.

In one of the first meetings I attended as a visitor (and potential candidate), our community's school board required a gathering of about ten angry parents to wait until the end of the formal agenda to be heard. The parents were made to sit through long academic discussions, financial reports, and so on. When they finally did get to voice their concerns, they were even more upset than before, and then the board brusquely dismissed their complaint. I thought, "Why couldn't the board have heard these people earlier?" From an uninformed outsider's perspective, making unhappy visitors wait to speak seemed merely an exercise in self-importance and terrible public relations

Some time after joining the board, I came to realize that first-time candidates often say that the group they hope to join is not sufficiently open to the public. In fairness, that complaint was made about the boards on which I served, too. However, the treatment those visitors received that night lingered with me. Not too many years later, a place was created in our program for people to speak before we began the formal agenda. Visitors never again were forced to wait for a prolonged period to address the board.

Do your homework. Individual members should be fully informed before a meeting starts. Review your agenda in advance and where appropriate ask questions before a meeting begins. Thorough preparation saves everyone's time and allows you to get answers much more quickly and comfortably.

If possible, avoid asking potentially embarrassing questions in board meetings. Again, ask your questions of the staff before a meeting or maybe afterwards. Administrators and teachers will appreciate your thoughtfulness, and the school district will present a better image to the public.

Don't needlessly drag out meetings by rambling off track in your discussions. Also, don't feel that you must comment on every, or even most, issues. If you're talkative by nature, be sensitive to others' time and energy levels. A loquacious board member may be perceived as too often a long-winded pontificator on minor issues.

It's every member's job to police his or her own words, but it's particularly the president's job to keep all members speaking to the point under consideration.

Here are some other ways the president can promote successful meetings:

- *The president sets the tone for the entire board, both in and out of meetings.* In this role, it's critical that you be impartial, fair, and conciliatory.
- *Be sure to equally include and recognize all members during board discussions.* Be as neutral as possible when an issue is being debated so no member perceives any favoritism on your part.
- *Go slowly when first taking office as president.* Later, you can move faster after you've gained the members' trust. Trying to do too much too soon creates the impression of a hyperactive ego.
- *Talk very little in meetings.* As president, it's tempting to speak often because you can claim the floor whenever you like. A board chairperson who submits to this temptation may be perceived as domineering. An effective president actually uses few words to facilitate the equal participation of all members.
- *Refer problems to the board.* Some presidents mistakenly assume they should provide answers to all procedural questions. A better course is to ask the other members, "Board, how would you like to proceed?" Deferring to the board helps create an atmosphere of cooperation and sets an example for the other members to follow.
- *Invite differences of opinion.* If as president you sense that a board member holds a minority opinion yet for some reason is reluctant to talk, you tactfully should encourage him or her to speak out. Also, in a spirited board debate, it's a good idea to alternate your recognition of members so that all sides are evenly presented.

9

CONTINUAL GROWTH AND LEARNING

You gradually become a better school board member over time as you gain experience on the job. This is a natural process and can't be replaced. However, the most successful school board members grow faster and learn more than other members who are less focused on personal progress. Here's how.

Your state school boards' association offers a wide variety of services, seminars, and materials for school board members. Learn about these and take advantage of them. Also, your state association meetings provide an excellent way for you to meet and learn from other school board members. You'll help the other members by sharing your experiences, and you'll hear about their success in overcoming problems similar to yours.

The National School Boards Association (NSBA) offers exceptional growth opportunities at an annual conference for school board members from all over the country. A separate NSBA technology and learning event also is held annually. These meetings provide a wealth of information. The editorially independent *American School Board Journal* is an NSBA-affiliated monthly magazine with informative features you'll find useful. Be sure to check out the

NSBA website, too. In particular, see the "News and Issues" section. Their Internet address is included in the list below.

Read your daily newspaper and news magazines to keep up with current trends in education and proposed changes in law. Intentionally look for articles related to education. You'll be amazed at how many you find, as well as how much you learn. Your knowledge of school issues will enhance your credibility among citizens, staff, and other board members.

Pay attention if you happen to hear a television or radio report on an education issue. If it's on the air, it's likely to be important, particularly if it's on a local channel. You won't be alone in hearing the report. You'll want to be able to intelligently discuss the issue.

In recent years, the Internet has dramatically changed our ability to gather information in all phases of our lives. Your school board service is no exception. Have a question about an education issue? The answer literally is at your fingertips. The research that once required several hours at a library can now be done at home in just a few minutes. If you aren't already proficient in searching the Internet, set an immediate goal to learn. You need ready access to a computer and the ability to skillfully use it. Public libraries commonly offer Internet availability at no charge.

Explore the following websites:

- www.nsba.org (National School Boards Association)
- www.asbj.com (*American School Board Journal*)
- www.studycircles.org (Study Circles Resource Center)
- www.centerforpubliceducation.org (Center for Public Education)
- www.americansforthearts.org (Americans for the Arts)
- www.arsba.org (Arkansas School Boards Association; see Study Circles listed in the Programs section)

And, of course, become familiar with your state school boards' association site.

Routinely review and evaluate your own performance. Just as you should intentionally and consciously build positive relationships with others, you similarly should work to enhance your own performance. Be intentionally introspective. Consciously think about how you can improve the way you function as a board member.

When a meeting doesn't go smoothly, or an interaction with a parent seems awkward, you should actively consider how the situation could have been better handled. Set a personal goal never to make the same mistake twice. Sometimes we learn from mistakes without consciously thinking about them, but we learn much more quickly by intentionally considering and adjusting our future behavior.

⑩

THE TOP TEN SECRETS
OF A SUCCESSFUL
SCHOOL BOARD MEMBER

If you count the recommendations, this book suggests something well over 100 specific ideas to help school board members function at a high level. Which of these are the most important "secrets of success"? I have no definite answer, but here, with a nod to David Letterman's Top Ten List, are my personal favorites:

Successful school board members:

10. *Create successful superintendents through encouragement, support, and private suggestions.* Any employee needs a supervisor's support and advice in order to grow to his or her full potential. Your superintendent is no exception. However, the unique nature of the board relationship requires that you, as one member, present your ideas for improvement as tactful suggestions, not instructions. Once a bond of mutual trust has been created, your superintendent will respond with enthusiasm to privately offered constructive criticism.

9. *Actively listen and respond quickly to citizens and employees without gossiping or criticizing the schools.* A successful

school board member acquires a reputation as a patient yet active listener, not as a gossip. Your negative remarks will be spread throughout the community and can only do harm to the schools. However, your rapid referral of a problem to the property authority will demonstrate your concern and inspire confidence in the staff.

8. *Intentionally and consciously work to improve relationships and job performance.* Growth will occur naturally in relation to length of board service. More rapid progress, though, occurs when you conscientiously review your own activity. When a conflict occurs or an interaction doesn't go smoothly, spend some time considering how a better outcome might have resulted. Never make the same mistake twice.

7. *Trust the superintendent and often refer people to him or her, saying "Use my name when you call."* Using your name empowers a citizen with a complaint. Referring the person to the superintendent empowers the leader of your schools to solve the problem. You may want to quietly follow up with the superintendent a few days after making the referral. Keep your conversations private though.

6. *Strive never to ask a potentially embarrassing question in public.* Do your homework by asking most of your questions of the schools' staff before a board meeting begins. The employees will recognize your thoughtfulness, and everyone at the board session will appreciate a shorter meeting.

5. *Encourage and support fellow board members through gestures of genuine appreciation and admiration.* Put some effort into acknowledging the other board members' individual contributions to your group's overall success. Look for their strengths that complement your weaknesses. Occasionally, send them handwritten notes expressing your genuine appreciation.

4. *Always dress at least as well as the superintendent whenever together in public.* How often do you see an employee dress

better than a supervisor in any organization? Your attire will foster respect or disrespect for both you and your position on the board. A judge doesn't wear a robe to court because it's comfortable.

3. *Understand and employ the power of apologizing.* An error is a great opportunity to build trust. Sounds counterintuitive, doesn't it? Yet a humble, appropriate apology creates respect. Likewise, a refusal to acknowledge a mistake will create disrespect and reinforce the offended party's negative opinion.

2. *Impress people by serving, not with a title.* No one enjoys the company of an egotist. And, after all, a board position isn't the U.S. Senate. Your friends won't be impressed with the title "school board member," although they'll recognize with appreciation your quiet dedication to the community's children.

And, the number one secret is:

1. *Successful school board members frequently hear the words "thank you."* The preceding nine secrets were recommendations for traveling a path to school board success. This last secret, though, is a road sign that says you're on the correct route. People often voice their appreciation when feeling truly grateful. Parents and staff members will acknowledge their gratitude with various expressions of thankfulness if they're pleased with the schools or with a particular occurrence that helps an individual child. You're headed in the right direction when you hear these words. Conversely, when expressions of gratitude come too infrequently, it's time to change course. Begin by examining your relationships.

PART II

LEARNING THE SECRETS OF SCHOOL BOARD SUCCESS THE HARD WAY

11

MY FIRST POLITICAL CAMPAIGN, MY FIRST POLITICAL DEFEAT (1986)

I truly didn't know what I was doing when considering my initial school board race in 1986. A first step was to attend a board meeting to see what it was like. Entering the boardroom and taking a seat with about 30 other people, my hope was to blend in with the crowd. I knew no one there, but everyone else figured me out immediately.

The leadership of the teachers' union sat on one side of the room; the administrators, whose attendance was required, sat on the other side. They all realized that members of the public never showed up for a meeting without an obvious reason, so it was an easy assumption, with elections coming up, that a candidate was surfacing. The meeting lasted several hours and left me with a desire to learn more about the inner workings of the schools.

A few days later, I called for an appointment and met the superintendent for the first time. An attractive, well-spoken man in his early 40s, he had been hired from out of state two years earlier in spite of never having held the top job in any school district. He had replaced a longtime superintendent who had retired after several rough years of disagreements with board members and some parts

of the community. The board had brought in the new superintendent with a shared enthusiasm for major reforms, and although many of his changes were positive, he was in the process of alienating even more people than his predecessor.

At the time, I knew little of this information due to my lack of involvement in the schools. My immediate impression was that the superintendent was talkative, bright, and self-confident to the point of being somewhat arrogant. He had ready answers to all my questions, informing me also of how hard he worked and how disappointed he was in the community. He spoke at length about his boat and sideline fishing business at his last location.

Superficially polite, the tone of the superintendent's conversation carried a note of condescension. A sign on his desk faced visitors and read, "Lead, Follow, or Get Out of the Way." The desk plaque, along with his demeanor, communicated that he was the leader, and the rest of us were left with the other two choices.

I began attending more school board meetings and talking to acquaintances about entering the next election. The board sessions encouraged me and so did my friends. One neighbor had been a school board member some years earlier. The atmosphere at the meetings seemed to reflect the superintendent's personality. My astonishment at the treatment received by a group of parents one evening is described in chapter 8. The board's callousness left a lasting impression on me and probably on some of the parents, too.

So I made a decision to enter the race, with the encouragement of a few friends and neighbors and with the growing idea that room for improvement existed at the top of the school district. I didn't recognize the incumbent candidate's strength until beginning to make calls in search of support. My opponent had lived in our community all her life and had been active in the schools for years. She also was running a family-owned business her parents had operated before her. A neighborhood acquaintance informed me that she would vote for my opponent because of their work together in the PTA. Her remarks were sobering.

My neighbor, the former school board member, agreed to be my campaign manager. His advice and support were invaluable because of his experience and contacts with others who had been active in the schools during the tenure of the previous superintendent. Many of our community's longtime public school supporters felt alienated by the present board and superintendent. This older group provided considerable support in mailing personalized postcards. My sons also helped me by knocking on doors and putting out handbills.

But, largely, I ran a one-man campaign, asking for contributions and spending some of my own money, erecting yard signs, placing ads in the weekly newspaper, and attending candidate forums. I introduced myself all over the school district and interviewed with the teachers' union committee that selected candidates for endorsement.

My meeting with the union committee was my first contact with anyone affiliated with the National Education Association (NEA) since the end of my brief teaching career 16 years earlier. During those years, the NEA had shifted from functioning as a professional association to operating as a true union. Going into the interview, I wasn't aware how much the character of the organization had changed, but I would have had no misgivings about meeting with the union in any case. On the contrary, I would have been just as hopeful of getting their support because I had been a teacher and all my life I had identified with working people.

My mother, father, and older siblings had joined the migration of Okies to California during the Depression. Mom eventually was allowed to be an uncertified schoolteacher after the state was overloaded with too many children for far too few accredited teachers. She loved the work. Her teaching career was the highlight of her life. My father was a blue-collar worker, holding various jobs in his younger years, but settling into janitorial positions by the time I was born late in his life. For a while, he was a custodian in the same small-town school district where Mom taught and I attended class.

As a young child, I rode atop the power buffer while Dad polished the tile floors in the administrative offices.

Brought up to work hard, I'd been a church janitor myself as a teenager and had picked cantaloupe in the fields of California one summer. An older brother had been a welder in the ironworkers' union before he died in a job-site construction accident in his mid-30s. Although never active in a union myself, I always had supported the rights of organized labor and still do today. It's a no-brainer for me, something I feel on a gut level.

In my view, the session with the union committee had seemed to go well. My answers to their questions were reasonable. As a board member, fair dealings and thoughtfulness would be priorities for me. The interests of children would come first in all matters. Some of the committee members seemed skeptical about my ideas on specific points, but everyone was polite and friendly.

The interview concluded after about two hours and left me hopeful of receiving the union's endorsement. Several years later, one of the committee members confided that I never had any chance at all. The committee members had seen me as a typical white, middle-aged male working in management who would always take the side of the superintendent in any conflict.

My empathy and compassion for working people went utterly unrecognized. I had approached the interview process from a rational viewpoint, explaining my abilities and ideas in a well-reasoned way. No emotional connection had been made with the committee members. They endorsed my opponent, even though she had been on the board that hired the superintendent and had every reason to continue supporting him.

On paper, my formal education and experience in teaching and management made me a better candidate. My opponent had limited formal education, and her professional experience came from a family business. However, she had lifelong friends who were aware of her values and character. They knew she had been active

in the schools and that she cared about the community. Many of them didn't know me at all.

I ended up with the endorsement, either explicit or implicit, of both Little Rock daily newspapers and the weekly *North Little Rock Times*. She ended up with more votes.

The election was held citywide in our town of about 60,000 people. I had tried to cover the entire community, making contacts and putting out signs. All 25 of the schools had been visited and meetings had been held with principals and as many teachers as possible. I took off work on election day and spent time at various polling sites greeting voters and handing out cards.

Since Arkansas school elections are held on entirely separate dates from all other political races or referendums, the voter turnout tends to be fairly light. However, this particular year the school district had a tax increase on the ballot, resulting in more publicity and voters than usual. Many citizens were coming to vote who didn't know either me or my opponent. As the day passed, though, people leaving the polls kept relaying that they had voted for me. My optimism grew.

After the voting sites closed, my family and I dined at a local restaurant and were greeted by a principal from another school district and his wife, a teacher in our district, who also said they had voted for me. I had never before met either of them. Of course, my family was supportive, and we were in a festive mood. My optimism reached a peak. Victory seemed certain.

We hadn't planned any sort of postcampaign party, probably due to a subconscious fear of losing. In any case, not many people would have attended. A mostly one-man campaign doesn't generate large crowds. So, that night I was sitting alone in our darkened den watching television when the election results were posted. I was stunned by my loss, having never previously failed at anything I had worked so hard to accomplish. The darkness seemed to engulf me.

In a few minutes, my son Logan, who was about 12 at the time, came in to inquire about the election. On learning the results, he replied with surprise, "You lost?" He couldn't believe it, and neither could I.

Defeat came by a somewhat respectable 439-vote margin. My opponent received 2,519 votes to my 2,080, or 55 percent to 45 percent. A complete landslide had been averted, but I felt terrible about losing.

The tax increase question was defeated, too, by a vote of 2,672 in favor to 2,817 opposed. (A tax proposal had been defeated by 2,665 to 3,592 the year before.) I was depressed for several days and at first embarrassed to face my friends, neighbors, and coworkers. However, their supportive comments were appreciated, and the worst was over with the last of the condolences. I didn't know at the time whether I would ever run again.

⑫

REDEMPTION!
A WINNING SCHOOL BOARD
CAMPAIGN (1988)

My phone started ringing not long after losing my first school board race. The public schools have a nearly desperate need for volunteers of all kinds. Through my campaign, people connected with education had figured out I was interested in their cause, and they began calling to get me involved.

Before long, I was active in numerous school-related groups, including the Band Parents, the Association for Gifted and Talented Education, the PTA, a Dropout Task Force, a Key Communicator Committee, and a Committee of 100 Citizens that provided feedback concerning a districtwide reorganization plan. My contact with school employees and supporters grew significantly, as did my understanding of the issues and problems.

I actually never quit thinking about another school board race after that first loss. As my involvement with the schools increased, so did my interest in running again. I could see even more clearly how the schools' leadership might benefit from my organizational abilities and my background in management. Also, the people I met through the various school groups often would encourage me

to give it another try. Yet I was reluctant to work that hard again in what might prove to be another losing effort.

Not too long after my defeat, I was at a business-related dinner meeting when I happened to be seated next to a woman who was an aide to Dale Bumpers, the former Arkansas governor and then the current U.S. senator. After mentioning my losing school board race, I understood her to reply, "You know, Senator Bumpers lost his first school board race, too." The knowledge immediately inspired me.

After that evening, a second school board race became just a matter of time. I didn't want to live the rest of my life knowing I had lost and given up after the first attempt. (I learned much later that Senator Bumpers hadn't been defeated in a school board campaign. He had lost an election for the state house of representatives as a young man, although he did serve for a while on a small-town school board.)

When election time came around again in 1987, I considered the school board races but decided to pass. Two positions were on the ballot, both held by board members who were well connected and running for reelection. One was opposed by a former teacher who was active in the schools and backed by the teachers' union. The other incumbent candidate was a lifelong resident of the community who was well liked and heavily involved in various civic groups.

I assumed he also would get the teachers' union support, given the choice between him and me. I might have been able to beat him because even more discord existed in the schools than the previous year, but I was uncertain. So I opted not to run and felt good about my decision. I didn't know when another opportunity might arise, but I decided to wait for an opening on the board with no incumbent in the race. I was beginning to learn a little patience, a quality that has never come easily for me.

In 1988, two different board positions were open, and this time both incumbents chose not to run again. The disharmony in the

schools had worsened, and these two board members wanted out. The lawsuits concerning possible consolidation of the three school districts in our county had been costly for the schools, demoralizing for the community, and time consuming for the board members. In the 1987 election, the school district hadn't bothered to try again for a tax increase, and the former teacher with the union's backing had defeated her incumbent opponent. Also, by 1988 the superintendent was more unpopular than ever in the community and with his staff.

Learning both positions would be open with no returning board member in the race, I quickly concluded this would be my year to either win a school board seat or give up the idea forever. My opponent was undetermined at the time, but he or she wouldn't have the advantage of incumbency.

Eventually, I ended up with two other people in the race. One was a woman with two master's degrees in education who had retired as director of a community center after finishing a career in education. She had been among the first black teachers employed in our school district after desegregation in the early 1970s. My second opponent was a white woman who worked in real estate and was active with her children in the schools.

When the slate was settled, my chance of winning the election seemed favorable, but I remembered, too, my optimism in my first race. Arkansas law requires a majority, not just a plurality, to win a school board election. If one candidate doesn't receive more than 50 percent of all votes, a runoff election must be conducted.

I wasn't sure at all that I could even win; however, I wanted to avoid a runoff if at all possible. It's hard enough to work full time, keep family commitments, and run one race, so I really wanted to escape the added burden of conducting a runoff campaign, too. My motivation was intense to work harder and smarter than in my first race.

I employed the same campaign strategies as in my first election but doubled my efforts. I raised and spent twice the money (about

$1,500), mailed twice as many postcards, put out twice as many signs, and knocked on twice as many doors. Since no incumbent was in the race, I had a slight name recognition advantage.

At least a few people remembered my campaign from two years earlier, and others knew me through my activity in the schools and community during the intervening years. One thing I did differently, though, was involving more people, which came easily and naturally. More people knew me, and the community's unrest concerning the schools was greater than two years earlier.

Moreover, this time I was successful in getting the teachers' union endorsement. The union leaders had been hearing good reports about me, including a recommendation from the new board member who had beaten an incumbent one year earlier. She and I met for the first time during the campaign, liked each other immediately, and became friends. She had spent a frustrating first year on the board suggesting changes and often finding all five of the other members disagreeing with her. She was looking for help and saw me as a potential ally.

Her support no doubt played an important role in my receiving the union's endorsement, although I had learned my lesson about relating emotionally, not just rationally. I told the endorsement committee the same truths they had heard from me two years earlier, though this time emphasizing the parts of my message that were important to them.

In particular, they felt they were being treated in a cavalier manner by the superintendent, and my own experience with him made empathy easy. I agreed that he would need to improve his methods of dealing with people, or he would need to leave the district. I made a commitment to work cooperatively with them, promising to do my best to change the combative atmosphere that pervaded their relationship with the board and the superintendent.

Clearly, the union wanted to see the superintendent leave, and they would be looking to me for help. I had no misgivings about that goal, because I agreed he was damaging the schools. He also

seemed to be hurting himself. In addition to professional problems, he was rumored to be having family difficulties and stress-related health issues. I hoped that once elected, I could help him alter his approach to his job and to people in general. Realistically, though, without a significant change, he would need to leave the district for his own good, as well as for that of the schools.

The union also endorsed one of my opponents, the African American woman who was a former educator. Naturally I was disappointed not to receive their sole endorsement, but my opponent's qualifications were beyond reproach. I had met her early in the campaign and knew she was sincerely dedicated to helping children. She was a life-long resident of the community and well qualified to serve.

However, our citywide electoral process made difficult the election of an African American to the school board. Other highly qualified candidates from the black community had been defeated in fairly recent years because people tended to vote along racial lines. The minority population was growing, but the white majority still dominated the polls.

In the community's history, only one black person had ever been elected to our school board. In fact, the all-white board nearly always came from one relatively small residential area where the city's money and political power tended to cluster. The lone black board member had been elected 20 years earlier in 1968, when the city fathers had quietly decided that the board "needed" a minority member during the early years of desegregation. His son had been turned away as one of six African American students who had attempted to integrate the high school in 1957.

Our district's initial black school board member came to office with the enthusiastic support of the community's thought leaders. However, when running for reelection in 1972, he was defeated by a white man who campaigned against busing students to help desegregate the schools. (See the Historical Note at the end of this text.) In the years after 1972, several other African Americans had tried for a seat on the board, but all were defeated.

By 1988, activist attorneys representing the NAACP had experienced considerable success in creating representative voting districts in communities across Arkansas. Their legal action had forced school boards to terminate citywide voting and set up neighborhood electoral zones. The city leaders knew that a lawsuit was in the offing for our community, too. In an apparent effort to assuage the African American community, they decided the time was right for the board to have another minority member. They didn't pick my well-qualified opponent, though.

They chose a relatively young minister who was fairly new to our town. He was highly verbal, comfortable in front of a crowd, and he loved politics. A local politico (later discredited for dealings of questionable ethics) hosted a *post*-election $25-a-plate dinner fundraiser for him that attracted a crowd estimated at 300 to 400. He entered the race for the other open school board position against a young white woman who had few connections and little money.

I campaigned on a seven-point platform that included the concept of electing school board members by geographic area rather than through citywide elections. Most people realized this was an idea whose time had come. In my first year on the board, we made the switch to zone representation. The 1988 elections were to become the last that elected school board members by citywide voting.

One of the great experiences of being a candidate was the opportunity to meet interesting people I otherwise would have never gotten to know. In my campaigns, I tried to make contact with everyone in our town who ever had been active in the community to any significant degree. In particular, I make a point to meet past officeholders, get their advice, and ask for support. I introduced myself to all the former school board members and really enjoyed hearing their stories.

One such person was the former mayor for whom our town's public library is named. He was in his 70s at the time I met him. He had lived his entire life in our community and married his childhood sweetheart, whom he had met in elementary school. He

began his political career by serving nine years on the school board in the 1950s. I had heard of him for years but didn't know him except by reputation.

Our city's former leader had been a combative, outspoken officeholder, and the newspapers painted him as a typical small-town politician. He was anything other than typical, though. He never lost an election, and he remains to this day the brightest, most genuine politician I've ever met. He regaled me with entertaining tales of growing up in North Little Rock and fighting political battles of which I was only vaguely aware.

When we met for lunch, he would pick a table in a far corner of the restaurant and sit facing the wall so his back was to everyone except me, hoping to limit the number of interruptions from his friends who wanted to say hello. I was among a long list of his admirers from all areas of our town. (I later learned my black school board opponent gave him credit for making possible her first teaching job in the white community.)

So the former mayor and I became friends. He, too, was unhappy with the superintendent, and he passed along his support of my candidacy to his many acquaintances, including those in the teachers' union.

As election day drew near, I felt hopeful but hadn't forgotten my unjustified optimism in the first election. And the former mayor had given me some advice: "Always campaign like you know you're ten votes behind." I wasn't taking any chances at this point and wasn't at all certain that a runoff could be avoided. I had two new ideas for election day.

First, I decided to put out updated yard signs along the major thoroughfares the night before the election. The signs read "Vote TODAY for Lynn Hamilton" as reminders for commuters that today was indeed an election day for the schools.

Secondly, I had recruited about 25 friends to hold posters and hand out literature near each polling site between 5:00 p.m. and 7:00 p.m. when traffic would be heaviest. The school district had

decided to make another attempt to pass a property tax increase, and I knew that once again many voters would be showing up to cast a ballot on the tax question with absolutely no knowledge about any of the board candidates.

My plan to put out new signs on election eve was going well until about 11:00 p.m. when alone, exhausted, and anxious about the next day's outcome, I smashed the index finger on my right hand while driving a wooden stake into the ground with a sledge hammer. My finger was caught between the hammer's shaft and the top of the stake, tearing an inch-long gash across the middle knuckle. Blood gushed down the yard sign and onto the ground.

After getting beyond my initial fright, I realized the finger could still be wiggled, so the bone probably wasn't broken. I decided not to go to the emergency room in order to get some sleep. The next day, I took off from work as I had two years earlier and spent my time greeting voters and shaking hands, this time with my index finger wrapped in a large bandage. All day long I kept thinking about taking time to let a doctor check my wound, but I couldn't tear myself away from the polls.

I never did get to the doctor—so today a jagged scar crosses my knuckle, and the finger bends to the right a little unnaturally at the joint. When I occasionally happen to notice the scar, it's not an unpleasant reminder of that winning school board campaign. I think of it as my own personal red badge of courage. I kept the blood-stained yard sign, too. I'm thankful, though, that my first campaign didn't leave me with any scars, at least not visible ones, to remind me of the election I lost.

My wife and I held an election night party for the poll workers, other supporters, and their children. We had a great time, and when the party got going I felt joyous with relief that the campaign had ended. Win or lose, I was glad the time had come to relax for a while. I cringed at the thought of gearing up again the next day for a runoff election, though. I dreaded the possibility of continuing the campaign more than I feared losing.

Soon after the polls closed, we began getting results from a few voting sites. I was clearly in the lead, although an outright majority was doubtful. My vote count was hovering around 50 percent of the total. Finally, perhaps an hour and a half after the polls closed, we heard from the election commission that I had received 3,319 votes to my two opponents' combined 3,191. I had 51 percent of the total, avoiding a runoff by 127 votes. The room erupted in cheers and congratulations. I felt a staggering sense of appreciation for the people who had helped me and also an immediate sense of vindication for the loss in my first election.

I know how the coach of a winning Super Bowl team feels immediately after the game. No Super Bowl coach (and no winning political candidate, regardless how important the office) could feel more exhilarated than I did that night.

In the other school board race, the African American minister, with the city leaders' support, won his seat by a dominating margin of 3,505 to 2,579. Unfortunately, though not surprisingly, the school district again lost its bid for a tax increase, even after paying a large fee to a political consultant whose strategy had included costly mass mailings. The vote this time was 3,235 in favor and 3,812 against. This was the district's third consecutive failed tax proposal in the past four years. The margin of defeat had widened to 54.1 percent in this election, up from 51.3 percent two years earlier.

A few days later, the superintendent said he thought the loss was caused by the various board candidates attracting so many people to the polls (without mentioning the expensive, impersonal mass mailings he commissioned). The turnout was indeed our city's largest ever for a school election, but the superintendent seemed oblivious to the real problems causing the public to repeatedly reject his funding proposals.

⓭

A TUMULTUOUS FIRST YEAR
(1988–1989)

Autumn 1988 was an exciting time for our family. Our oldest son, Logan, was entering high school, a tenth grader. Our younger son, Wyatt, was starting junior high as a seventh grader, and our daughter, Kelly, was just beginning elementary school in kindergarten. Our kids were entering all three major stages of public education at the same time I was getting started on the school board. The significance of all these new beginnings wasn't lost on me. I was proud of my children and happy to play a role in an institution so important in their lives.

However, my school board career began with an acrimonious first meeting, clearly portending a tumultuous year ahead. Anyone who knows much about school boards easily might have predicted what was about to happen. Our six-person board consisted of three veteran members, who routinely supported the superintendent, and three newcomers, who were elected in large part by voters who were upset with the current administration.

The three long-term members were the woman who had defeated me two years earlier, a man who was a lifelong friend of hers, and a second woman who had been active in the PTA for

many years and was also a good friend of my former opponent. The three newcomers included myself, my election campaign ally who had had just finished her first board year fighting losing battles, and the politically motivated minister.

As only the second minority ever to serve on our board, the minister's broad-based support came from the city's political establishment and the teachers' union as well as the African American community. Our politically motivated newcomer was something of a wild card, though, because no one knew with certainty where he stood on several key issues, including his position regarding the superintendent.

Arkansas law requires school boards to select new officers as the first order of business in the initial meeting after an election. The presidency is the only meaningful office, because the president sets the tone for meetings by introducing agenda items and hopefully keeping everyone on track. However, the board's leader has no more authority than the other members, making the job mostly ceremonial in nature.

For this reason, I've always thought the position should be passed around, allowing all members a chance to hold the title at least once during their board careers. The damage that any single, weak member can do as president is miniscule compared to the harm that might stem from ill feelings created by denying someone the role. Board members who support and help a struggling president grow into the job have the opportunity to create an ally. Those who exclude a fellow member from the minor limelight of leading meetings may entrench someone as an obstacle to progress on much more important issues.

In my first meeting, our board faced choosing between two women who wanted the presidency. One was the veteran member who was a friend of my former opponent, and the other was the board sophomore who had supported me in my just-completed campaign. The animosity between them probably ran deeper than I realized at the time.

Nominations for board president opened with the three veteran members nominating and voting for the board veteran; the three newest members voted against her. Until the role was called, I had no idea how our wild-card member would cast his ballot. I was mildly surprised when our votes split three to three. A majority vote of four "ayes" was required to pass any motion, including the election of officers, so the nomination went down to defeat.

Next, I nominated my ally, explaining that she was the only member with previous board experience who never before had served as president. Given our predicament of having to choose between two nominees, I reasoned that the position should go to someone who hadn't yet done the job. The three senior board members were having none of that, though, so we deadlocked three to three for a second time.

Then, once again, the veterans nominated their friend, and we rejected the nomination. I followed by renominating my ally, and again she was rejected. We rocked back and forth in this manner several times, getting nowhere. Finally, the veteran who wanted to be president grabbed her microphone. With a scowl on her face and anger in her voice, she nominated our wild card, the politically motivated board rookie.

So, in short order, our board had a leader who not only hadn't yet served an hour as a member, but also was the first African American president in the history of our school district. I wish I'd had the good sense to have voted for him, but still caught up in the emotion of my just-ended campaign, I cast the only negative vote. I wasn't thinking clearly enough to realize his selection was an appropriate compromise, and I hadn't yet learned that it's rarely a good idea to be the lone dissenter on any proposition.

Our newly elected president did an excellent job leading the remainder of that first meeting and all those that followed. He obviously was familiar with parliamentary procedure and appeared quite comfortable in the role, much more than I would have been. He did so well, in fact, a year later we elected him to a second term

as president. The next editions of the newspapers (two metropolitan dailies and a hometown weekly) duly reported the unusual nature of that first meeting, but they didn't capture the reaction in the board room that night. The school administrators seemed uneasy and taken aback. They appeared uncomfortable and, in some cases, embarrassed for us.

Obviously, the old order on the board was changing, but no one knew in what direction. Rarely are controversy and conflict perceived as good in an institution that depends on the public's support for its existence. The union leaders were buoyant, though, because they saw the possibility of change and also because the board and administration were in disarray. A bickering board and weakened administration nearly always give a union added leverage in collective bargaining.

Not too long after my first board meeting, I met with the superintendent in his office to tell him my perception of his administration. Our conversation lasted perhaps 30 minutes. I'd tried tactfully, yet directly, to let him know that, to an outsider, he and some of his staff members seemed somewhat aloof and arrogant. He responded defensively, perhaps even uncaringly. In so many words, he replied that some people inevitably would be upset by change, regardless how beneficial.

So, I gave him an example. I told him my reaction to the sign on his desk in front of me, the one that read "Lead, Follow, or Get Out of the Way." He looked a little startled when I explained that, due to his demeanor, some people who came to see the leader of the school district might perceive the sign facing them to mean they were expected to follow or get out of his way. He said he had never thought of that interpretation, that he was trying to encourage people to be leaders, that the sign reflected his personal philosophy, and so on.

I'm certain he was telling me the truth. He consciously never had considered the unspoken message behind his sign. It was his unconscious behavior, though, that was getting him in trouble. To

his credit, he removed the sign from his desk after I left his office, and I never saw it again. He couldn't change his personality as easily, though.

Looking back, our superintendent possibly never intended to stay long in our community. He may have seen the job as an opportunity to build a resume. I doubt he cared much what we thought of him. His motive may have been to remake the school district in an image that would be perceived as progressive by a search committee from a larger city. He certainly didn't want to alienate his board members, though, because he needed our support and positive votes when time came to renew his contract.

I must have made him angry and frustrated at times, and although he was never warm, he was always professional in his dealings with me. In truth, our school district made progress under his leadership. In the business world, sometimes a "hired gun" purposely is brought into a corporation to make unpopular but necessary decisions, knowing up front his ill-favored tenure will be short-lived. Our "Lead, Follow, or Get Out of the Way" superintendent may have seen himself in that light.

During the first year or so of his employment, our superintendent replaced nearly every key administrator in the district. Many of the former staff members were longtime district employees who simply retired when pressed by the requirements of a new, unpleasant superintendent. Most of their replacements were hired from outside the school district, and many presented personalities similar to the superintendent's. They appeared to be bright and confident while short on patience and modesty. Any organization tends to reflect its leadership.

Many years later, I had opportunities to visit with several of the education professionals hired by the superintendent long after they had left our school district. I was impressed with their soft-spoken humility about the days of their employment with us. Probably at the time I first got to know them, they were responding to the pressure they felt in their jobs, and they also must have been

reacting to the conflict on the board. I'm sure they perceived some significant negative characteristics in my performance, too. Their presence in our district brought in new ways of thinking and innovations that otherwise might never have occurred.

Among the superintendent's new appointees was an attractive, intelligent young woman who was named the district's public relations coordinator, a newly created position. She had no previous school experience and was paid a salary far exceeding that of most teachers. She was a lightning rod for criticism by the union leadership.

A few years into her employment, she was indicted and tried for the murder of her flamboyant, older husband. She was acquitted of the charge, though only after a locally sensational trial that included detailed, sordid descriptions of sexual and drug-related incidents. She resigned from her job as the trial began. When she left (some time after the superintendent's departure), we filled her position with a lower paid, highly competent, and less controversial replacement. Our original public relations director turned out to be a poor choice, but the position became an important one for our district.

Our unpopular superintendent made at least one unquestionably good appointment. He promoted a loyal, longtime district employee from a high school principal's slot to assistant superintendent of personnel. The new assistant superintendent eventually would lead our district.

I first met him when he was still a high school principal during my initial school board campaign. I liked him immediately. He was low key, humble, a good listener, and obviously highly competent. He had a great reputation, too. He and his wife both had worked in our school district for about 25 years. They lived in our community and sent their two children through our schools. He had started his career as a social studies teacher, moving up gradually—probably too gradually given his competence.

The superintendent likely recognized these facts and may have planned from the beginning for the new assistant superintendent to

become his eventual replacement. If so, our community certainly owes him a debt of gratitude for recognizing his successor's talent.

Also in defense of the superintendent, he came aboard just before one of the most trying times in our schools' history. Soon after he was hired, the neighboring Little Rock School District filed suit to consolidate the three separate county systems into one large, metropolitan district. A local federal judge initially ruled favorably, triggering a spate of countersuits, extended courtroom distractions, dramatically escalating legal expenses, and a loud outcry from the white community.

Eventually, a federal appeals court overruled the local judge's order for consolidation but did support a finding of racial disparities in all three school districts. So, even though our schools' autonomy was preserved, the three districts remained under supervision of the federal court system. The ensuing legal proceedings continued as a constant distraction from day-to-day school activities for years into the future.

Our superintendent was struggling through these complex legal issues at the same time he was hiring a new staff and reorganizing the entire school district from the top down. He was straining under a workload that would have caused anyone to suffer. The combination of the workload, the white public's unhappiness with the federal court system, and his own arrogance made for the superintendent's eventual undoing, in spite of the creativity of his ideas and the effectiveness of some of his initiatives.

In just a few months, the superintendent's contract would be coming up for renewal. Our board's contentious first meeting led to speculation that once again we would evenly split our votes three against three. Four votes were required to extend the contract. The issue essentially was a referendum on the superintendent's performance, not a direct question of immediate dismissal from the position.

His contract called for a three-year term and required the board to vote annually on a one-year extension. Thus, the agreement's

length could be reset repeatedly to a total of three years. If the board failed to renew the contract and in a separate decision chose to dismiss the superintendent outright, we would need to pay him for the remaining two years. He would walk away with slightly over $200,000, an enormous sum for an Arkansas school district in 1988.

Soon after being elected, I had decided privately that I would vote to extend the agreement. Nonrenewal would put us in a no-win situation. If the board told the public that we had no confidence in the superintendent, how could we expect our community to the support the schools during the two lame-duck years? We would be forced to pay him off. Yet how could citizens and staff possibly think us wise to give $200,000 to someone who had failed in his job (especially since the school district routinely had pleaded for tax increases in recent years)?

With either option, the board would be perceived as incompetent. I decided to work cooperatively with the superintendent, confidentially telling him when I was concerned with his performance and hoping he eventually would improve or find another job. I told no one of my decision, though, keeping my options open and delaying the need to explain myself to the union leaders who were pressing me to vote against renewing the contract.

The months immediately following my first board meeting were a honeymoon period. The union leaders liked me because I wasn't shy about voicing my opinion or voting against an occasional ill-conceived proposal from the superintendent. And the local weekly newspaper often gave me favorable coverage because I encouraged an atmosphere of openness in board meetings. I made a point, too, of trying to find middle ground when problems or concerns were brought to the board by individual teachers or citizens.

Some of our meetings ran long into the night with presentations concerning the federal court proceedings, policy questions, collective bargaining issues, and planned restructuring of the schools. Often our discussions would be argumentative, resulting in split

votes. Through the conflict, the importance of level-headed thinking and professionalism became clear. When I disagreed with someone, I learned to do so tactfully and respectfully.

Sometime after the first of the calendar year, probably in January or February, the superintendent's contract came to the board for our vote. The lobbying against renewal intensified as the time approached. I don't recall any advocates *for* renewal.

A week or so before our decision, I was at home on a Saturday evening when I heard the doorbell ring. I went to the door and there stood our superintendent. He hadn't called ahead, and he previously never had been to my house. I tried to contain my surprise as I invited him inside. He got right to the point: he only had the votes of the three veteran board members. He had learned our wild-card board member was not going to support him. The superintendent had a deal for me. He said if I would vote for the contract extension, he would do his best to find another job and leave our district quickly. He needed my vote to leave, though, because he rightly felt that nonrenewal would make other school districts reluctant to hire him.

I replied immediately that he had my support. I wished him well in his job search and told him he was making a decision that was good for everyone involved. I felt compassion for him. I didn't ask, nor have I understood, why he approached me at the time and place he chose. Perhaps his problems were closing in that Saturday night, and he felt compelled to resolve them right away.

Not long after my conversation with the superintendent, the president of the teachers' union asked me to meet with the union leadership one evening just days before the upcoming contract renewal vote. The meeting was to be held in a conference room in our town's library, the one named for my friend and supporter, the former mayor.

Apparently, the union leaders also had figured out that the superintendent didn't have the votes to renew his contract without my support. I arrived at the meeting and saw 20 to 30 union lead-

ers, teachers, and community members, many of whom were my acquaintances, including the former mayor. The union president began the meeting, and her message was clear: They wanted me to vote against renewal. They were unhappy with the superintendent and wanted him to hear that message loud and clear.

When the union leaders finished talking, I used my first words to inform the group that I would be voting in favor of renewal. Relating my conversation with the superintendent, I told them he would be trying hard to leave the district soon and that renewing his contract would expedite his exit. If the board failed to renew his contract, we would be stuck with a lame-duck superintendent for two years because no one would hire him. Further, we would be morally wrong to buy him out at a cost of $200,000. Either scenario resulting from nonrenewal, whether lame-duck or buyout, would be detrimental to the schools and to the community.

The union leaders didn't agree with me. They argued that they didn't trust the superintendent to keep his word. He might not leave. So, I countered that I thought he would be gone within a year and that, if not, I certainly had the option to vote against renewing his contract the next year. That wasn't good enough for them, though. They wanted nonrenewal now. The argument that the community and schools would be harmed made no impact on them.

Perhaps the union leaders didn't particularly mind the negative effect on the community or schools. Several, maybe most, of them lived outside our town in one of the other school districts. Their primary goal may have been to embarrass the superintendent and damage his ability to lead, thereby helping the union in their power struggle with the board and administration. Perhaps they also intended revenge for perceived past mistreatment. The union leaders remained polite throughout the discussion, but they were intractable.

Finally, the normally talkative former mayor, who had said little or nothing during most of the meeting, rose to speak. He smiled

and said, "I can tell Lynn is someone who isn't going to change his mind once he decides what he wants to do. It's probably time for us to end this meeting." Shortly after the former mayor sat down, the meeting did end.

He hadn't openly chosen a side in the contract renewal debate, but he didn't need to. I could tell by his warmth that he agreed with me, and probably everyone else could, too. Perhaps the union leaders weren't worried that the city and schools might be hurt, but the potential harm certainly mattered to the elderly former mayor who had devoted his entire life to our community.

At our next board meeting, we did in fact renew the superintendent's contract. The official vote was four in favor and two opposed, but only after our president tried to find a middle ground of sorts. He walked out of the meeting saying he needed to abstain, giving no justification, leaving the vote to be conducted by the vice president, and returning only after the decision was made. He was told Arkansas law made no provision for a groundless abstention and that his action would be counted as a negative vote.

I never did learn the reason behind his action; however, I always assumed he was feeling pressure from his political supporters on both sides of the issue. Crossing my mind, too, was the idea that his abstention possibly could have come at the superintendent's suggestion. A contract renewal vote with only one board member openly voicing an objection wouldn't sound too bad to another school district's search committee.

Nothing came easily for our board that year. Although our president's attempted abstention made for even more chaos than usual that evening, the net effect turned out to be beneficial because the action ultimately helped the superintendent leave. I felt good about my role, too. My stance had been somewhat difficult but well reasoned. One of my original motives for becoming a board member was to help make decisions that would have a positive effect on our school system. For the first time, my presence had made an impact on a significant issue.

We had a large crowd the night the decision was made, and because of our past split votes, many people expected to see the board deny renewal of the superintendent's contract. It was a gallows crowd, the kind in the old west that came to see justice served at the end of a rope. Many were disappointed when we failed to carry out the execution.

On my way out of the board room after the meeting, several union members openly expressed their displeasure to me. I quietly and quickly explained that my action had been taken in their behalf. My vote was the best way for us to realize their expressed goal of seeing the superintendent leave. I assured them of my conviction that he would depart sooner rather than later, and privately I prayed I was right.

Winter turned into spring, and with the contract renewal question behind us, the contentiousness decreased somewhat among the board members. However, the union's animosity and the federal courtroom conflicts over consolidation remained unabated. One of the litigants in the legal battles was an outspoken African American activist attorney. He began pressing our district to implement election of board members by geographic zone.

The black community wanted more representation on the board. A significant presence would never be possible with citywide voting. The attorney's threats of litigation made clear that our board would either need to comply voluntarily or fight a legal battle that our counsel advised we couldn't win. Even the most reluctant of our board members agreed we didn't need the distraction and financial waste of defending a lost cause. Moreover, needlessly antagonizing one of the primary parties in the overriding federal court case was senseless.

So we voted unanimously to create geographic zones of representation for a seven-member board. Implementation would come late in the school year after an extensive demographic study by an outside consulting firm.

Everyone saw the need for an odd number of board members. Four votes still would still be required to approve any board action,

but never again would a question fail and long arguments ensue due to an evenly divided board.

We assigned all six current members to separate geographic zones, even the ones who didn't actually reside in their assigned area. We also staggered the years of election, so annually only two or three board seats would be on the ballot. In this way, the woman who had defeated me in my first election would remain on the board one year longer than her elected three-year term, thus allowing her to continue as a member during her son's senior year in high school.

After completion of the demographic study, we appointed the retired African American teacher, who had been one of my opponents in the last citywide election, to fill the newly created seventh board seat. This proved to be one of our shrewdest moves. She was recommended by the activist attorney and admired by the entire African American community. She became a highly effective board member whose opinions were taken seriously.

Over the years, our appointed member was reelected numerous times and helped our district become recognized for its desire and ability to provide a quality education for all students. She was a true friend of children and of the schools. She made clear to the activist attorney, and to everyone else, that we were endeavoring to act in the best interests of the entire community. Partially through her efforts, our district generally came to be known as the best functioning of the three involved in the extended legal proceedings. Eventually, the focus of the activist attorney and the federal judge shifted away from us.

This second former opponent of mine went on to serve with me during 14 of the 15 years I was on the board. She stayed a total of 16 years, one of the longest tenures in the North Little Rock School Board's history. Her longevity contributed significantly to our success.

In the spring, the superintendent did indeed find another school district that wanted him, one in his home state. Our community co-

operated enthusiastically in helping make a positive presentation to that district's search committee. They visited our town and met with various designated ambassadors who answered their questions. I'm unsure what was said to them, but I seriously doubt they learned how close the board had come to not renewing the superintendent's contract or how much animosity was felt toward him.

I purposely avoided participating in the hosting process. I don't know exactly what I would have said if one of the visitors had asked me for an opinion, but I would have tried to give the superintendent the benefit of the doubt. I would have been tactful. I wouldn't have volunteered anything too negative. In any case, with our help the superintendent was successful in getting his new job in the first district where he became a finalist for the position.

We learned just a few years later that things hadn't turned out too well for him in the new location. He also had left that position. I don't know much about what happened in his new district, but unless he made a radical personality change, he experienced problems similar to the ones he had with us.

The new district's selection process of formal group interviews had been much the same as our own board's procedure several years earlier. It was the method nearly every school district uses when looking for a new superintendent in a geographically remote location. Group interviews favor smart, physically attractive, eloquent candidates who have a talent for performing in front of an audience.

Those traits are fine as far as they go, but if a district doesn't dig further, the process doesn't reveal the candidates' character or values. The candidates who are the slickest, the smoothest talkers— in short, the most superficial—are the ones most likely to be employed. Watching the other school district hire our superintendent caused me to shudder.

I made a mental note that if ever I were in the position to help hire a new superintendent who wasn't known to our community, I wouldn't trust anything anyone told me in a group meeting. I

would dig into the candidate's background by privately asking probing questions of PTA members and teachers in school districts where he or she previously worked. And the smoother the candidate talked, the more I would dig.

Fortunately, I was never put in that position. In June, our usually contentious six-member board (our new seventh member had yet to be appointed) unanimously named our assistant superintendent of personnel as our new superintendent. We made the appointment without conflict and without a search process, such was the obviousness of our choice.

I've wondered why the school board hadn't made the same choice several years earlier. Our new superintendent was available back then, too. At that time, he would have had to be promoted from his principal's position over several old-timers in the district, although certainly he would have been a good selection.

In any case, the new superintendent went on to serve the next 14 years that I was on the board, and one additional year after I left, for a total of 15 years in the position. He spent 43 years altogether as an employee of our school district. Before I left the board, he already had quietly hand-picked a young educator, the superintendent of a smaller district, who eventually would become his successor.

I'm getting ahead of myself, though. Our new superintendent had a calming effect on our board, but he couldn't work overnight miracles. Shortly after he took office our board saw its most disagreeable conflict. A continuing series of sometimes heated arguments had been carried on by the two women who had wanted the board presidency earlier in the year.

The boiling point was reached one night late in a meeting when, with faces flushed in anger, they both raised their voices and questioned the other's intelligence in a particularly vitriolic exchange. They stopped short of profanity, although for a time I didn't know if they would. The new superintendent didn't say a word, but he didn't need to. The look on his face, particularly the size of his

eyes, displayed a combination of astonishment, anger, and dismay. He must have been wondering how many such displays he would be forced to endure. Looking back, that night seems to have been a turning point for our board. I think the new superintendent consciously set a goal of helping individual board members resolve personal conflicts privately. Certainly, he set an example of humility, flexibility, and cooperation for the board to emulate. Perhaps, too, the entire board was embarrassed by the incident, especially the two participants. Maybe we all individually resolved to conduct our meetings with more decorum. I'm uncertain. Nothing was discussed about making a change, but change we did. Our relationships slowly began to improve.

By mid-1989, we had a popular superintendent, an inclusive board electoral process, and a new highly competent member appointed as the seventh person on our board. This was an exciting time of renewal for our schools and community. The teachers' union was ecstatic with our previous superintendent's departure. The political establishment and the white community loved our choice of the new superintendent. The black community was pleased with having two minority board members for the first time ever, along with the possibility of eventually electing a third. We were riding a crest of goodwill.

What better time to launch another try for a tax increase? That summer, with the enthusiastic support of our school supporters and employees, we began a campaign promoting the need for additional funding. The board, administrators, union leaders, teachers, PTA members, and others all worked hard and cooperatively. In the fall election, the tax increase proposal passed by a margin of 4,164 to 2,569, a 62 percent positive vote landslide, after three consecutive failures at the polls.

Our schools were entering a two-year period of remarkable accomplishment. We were better funded than ever, and innovative reorganization plans were underway. Buildings were to be renovated,

and extensive new educational and extracurricular programs were being created at every level. As my first year on the board came to a close, our community was enjoying a new era of optimism and support for our public schools. I was excited and energized by my small supporting role.

14

FINISHING MY FIRST TERM
WITH A FIZZLE
(1989–1990 AND 1990–1991)

The 1989–1990 school year was a time of planning and progress for our district. This was the year we began the process to combine our city's two high schools into one school with two campuses It was also the year we extensively remodeled the oldest of our three junior high schools.

Our city's original high school, Old Main, had been built in 1929. Although its physical plant was larger and had been beautifully restored in the early 1980s, the older school generally was considered inferior to the newer Northeast High School, constructed in 1972. Shifting demographics meant the newer school had a more affluent student body. The two schools were seen as being attended by "haves" and "have-nots," and the perception was splitting our community.

A new plan solved the problem by creating a single high school, with all ninth and tenth graders attending the East Campus (formerly Northeast High School) and all eleventh and twelfth graders attending the West Campus (formerly Old Main). This decision, which came after much discussion and community involvement, ultimately was applauded throughout the city. The community

could see that all kids, from every area of town, would benefit from specialized instruction at their grade levels and from significantly increased opportunities for extracurricular involvement.

Also, the athletic boosters' vision of future powerhouse teams in all sports certainly didn't hurt our community's acceptance of the idea. In fact, during my time in office, we eventually did see highly successful teams in every sport except, alas, football.

Our school system spent the 1989–1990 year thoroughly planning the transition to one high school, and we launched the new concept in 1990–1991. The planning sessions were led by administrators and involved the active participation of teachers, parents, community leaders, and students. All phases of school activities were covered, including curriculum, class schedules, discipline, student dress codes, academic honors and societies, student government, athletic opportunities, spirit groups, and much more.

The committees put a great deal of time and effort into developing the many details of the different programs. The school board's role was to hear reports and then vote on recommendations presented by the various groups. We rarely rejected a recommendation, although in the approval process I learned more than I ever cared to know about such topics as cheerleaders' uniforms.

The committee that attracted the most attention was the one designated to come up with a new mascot for the combined high school. Parents and other community members were more active here than any other area, with graduates of the two high schools equally represented. The older school's mascot was a fierce and feisty wildcat, and the newer school was represented by a charging steed emblazoned with a lightning bolt. The respective graduates were known as Wildcats and Chargers. The committee met for months.

Finally, they came to the board with a recommendation: The new mascot should be a mean-looking cat who would be menacingly clenching a lighting bolt in an upraised paw while riding a charging stallion. The new mascot was to be known as a Charging Cat. The

school board, happy to have the decision made, rubber stamped the committee's recommendation—and the town erupted in dismay. Most of the community's lifelong residents were older folks who had graduated from the original high school. The Wildcat was part of their identity. Our next school board meeting was standing room only. The crowd probably numbered 200, most of whom I had never seen before or since. I recognized their leader, though, our elderly former mayor. Several people made presentations, but the mayor was designated to deliver their summation.

He spoke of the tradition of the Wildcat symbol and its importance to the community. He said the group wanted the new mascot to be called a Charging Wildcat, not just a Charging Cat. I can't recall all of his exact words, but I remember his final plea verbatim: "Please. I'm begging you. Give us back our Wildcat."

The board, glad once again to get the decision behind us, quickly voted for the revised mascot name. The crowd jumped to their feet in loud, sustained applause and congratulated each other for our decision and their victory. Perhaps they were reliving the delight of Wildcat victories in their youth. Nearly ten years would pass before we again saw such a large crowd with so much spontaneous happiness in our board room. I thought, "How can we transfer this enthusiasm for the mascot to enthusiasm for educating kids?" I never did figure that one out.

The new mascot name stuck this time. The old Wildcats in our town were just too numerous for the young Chargers. The only place where Wildcats and Chargers were represented equally had been on the committee that made the original flawed recommendation.

The 1990–1991 school year saw the exciting opening of the newly combined North Little Rock High school. The new school made a dramatic impact on our community and ultimately on my family. My oldest son and the girl who eventually became his wife met in class for the first time as seniors. If we hadn't combined the two schools, they likely never would have gotten to know each other, because originally they were slated to attend the separate

high schools. I was elected president of the board for the first time that year, which allowed me the pleasant duty of handing diplomas to all the graduates, including my future daughter-in-law.

We had a relatively calm time in the schools and among board members that year. The union leadership would press as much as possible for their membership, and the board and administration would try to accommodate them whenever possible. The board began having to say "no" a little more frequently, though generally the relationship was a cooperative one.

The board's most urgent concern was the continuing litigation that originally began with the lawsuit to consolidate the three county school districts into one large district. The seemingly endless legal battles mostly revolved around court jurisdiction of various school activities and around financial settlements to be paid by the state to the school districts. However, the amount of compensation to be paid *by* the school districts to the intervening attorneys was one of several controversial, distracting side issues.

The courtroom proceedings inspired in me an idea that turned out to be politically impossible. My thought was that the three districts should consider the possibility of cooperatively consolidating down to only two school systems, with the Arkansas River as the dividing line. The cities of Little Rock and North Little Rock are like Minneapolis and Saint Paul in that they are separate governmental entities divided by a river in one contiguous urban area. The river is both a geographic and cultural dividing line because the two groups of residents largely view themselves as distinct from one another.

The two cities' school districts are encircled by the donut-shaped Pulaski County Special School District that encompasses outlying areas on both sides of the river. The concept of reorganizing into two school districts separated by the river wasn't new, but no one had ever seriously attempted to study the idea.

The timing for a study seemed right in early 1991, when the talk of consolidation lingered in the air and federal court intervention

in our schools was a reality. Also, the potential advantages for everyone in the county seemed attractive enough to justify at least giving the two-district concept a thorough look, even if we ultimately decided against it. I thought the merit was particularly obvious for the North Little Rock school district.

Our district boundaries were totally enclosed by those of adjoining communities, making geographic growth impossible. Internal growth was unlikely because little available land existed for new housing development. Our city's population growth had been stagnant since the 1960s when we reached 60,000. The school district's student count had been shrinking as our population aged, and the white percentage of the student body had been declining as the black percentage increased.

The demographic shift was obvious. The black birthrate had outpaced the white birthrate, and young families with children looking for new homes were forced to live outside our borders as few new homes were built in our community. The established white neighborhoods were filled with older folks whose kids had gone to the public schools and then moved away to find new homes, leaving the parents behind with their empty nests.

In 1979, our student population of about 12,000 was 75 percent white and 25 percent black. By the 1990–1991 school year, though, the total had declined to 9,264 with 53 percent white students and 47 percent black. Although no hard confirming numbers existed in 1991, the local private schools didn't seem to be enrolling an inordinately large number of white children. The informal totals, voluntarily furnished by the private institutions and occasionally reported in the newspapers, weren't extraordinarily high.

That's not to say that white flight to private schools was nonexistent, but the primary reason for our decline in white students appeared to be lack of available space for young families in our community. My concern was that the continued demographic population shift eventually might drive the racial makeup of the student body to a tipping point that would cause an escalation of

white flight. Thinking this problem might be solvable by extending our school district boundaries outside the city limits into the northern reaches of the county, I decided that conducting a study to at least consider the concept was clearly wise.

I expected that the idea would get solid support from various leaders in the community. In particular, I thought the teachers' union would back a proposal to conduct the study and would want to actively participate. Their parent group, the Arkansas Education Association, had supported the original lawsuit for consolidation to one district and also had intervened in the case.

However, the word "consolidation" was anathema for many people. After all, we recently had spent a lot of money on legal fees to prevent our school district from being consolidated into one large, countywide district. The superintendent and I discussed the two-district concept at length and decided that before going to the public with the proposed study, we first should try to build support quietly among community leaders and in the other school districts.

The Arkansas Freedom of Information Act prevents members of a school board from meeting together privately, so I was unable to discuss my thoughts extensively with my fellow board members. However, I did meet privately with all fourteen members of the other two school boards. The superintendent kept our board informed through separate conversations with individual members. We both met with various leaders in the community and received cautious but positive receptions from everyone we included in our discussions.

After a few months, we held a private, invitation-only meeting of perhaps 50 people representing a hand-picked cross section of our community. We purposely avoided inviting individuals whom we expected to be immediately negative to the idea of conducting the study, so again, the feedback was encouraging.

With the support of this group, we decided to take the proposal to the public. We held an open forum meeting in an auditorium at the East Campus of the high school. The superintendent briefly

presented the two-district study concept and then asked for people to come to the microphone to express their reaction. And did they ever react! Parents, former students, past school board members, and others told us in no uncertain terms they didn't want to consider anything that might cause us to share decision making for our schools with anyone else.

They were distrustful of any discussion that would envision the city's schools belonging to a larger governmental body. They liked the schools just the way they were, thank you, and didn't want to consider a change that addressed problems that might lie ahead. In fact, they wouldn't even acknowledge that a problem might be looming in the not-too-distant future. The degree of their distrust was fueled by the way we went about developing the proposal. Many of them, of course, had heard beforehand about our private meetings and were offended that we had not come to them with the plan sooner.

Of course, a few people spoke in favor of considering the two-district concept. Their words were reasoned and unemotional, but their voices were drowned out by the vehemence and numbers of the opponents. The community leaders, whose help had been promised early in the process, mostly stayed away from the public forum. Their support evaporated in the heat of the objections they heard before the public session was held.

The teachers' union, too, declined to endorse the idea of studying the two-district concept. When I asked why, one of their leaders gave me the nonsensical answer that they wanted to hold out for consolidation to one district, which at that point clearly wasn't going to happen. I never learned the real answer. Perhaps they had seen the writing on the wall that the study wasn't going to take place, or maybe their decision had something to do with political infighting among the union leadership in the three districts.

School district consolidation also would have meant consolidating the three groups of union leaders into two. In any case, I was bitterly disappointed that they wouldn't lend their support to even

talking about an issue potentially beneficial to the schools, especially after I had worked so hard on their behalf during my three years on the board.

Soon after the public forum, the superintendent told me he was throwing in the towel on this issue, and I agreed that conducting the study was impossible. My immediate reaction was frustration, disappointment, and anger. I had invested considerable time and energy into an idea that had gained us nothing. In truth, the idea was half-baked from the beginning. The two-district study never stood a chance, but I wasn't politically astute enough to understand that at the outset. I made a serious mistake in doing far too much talking privately before going to the public. My approach only wasted time, created distrust and rumors, and offended my fellow school board members who couldn't be a part of the private discussions.

Also, I didn't realize private support could dissipate so quickly in public, nor did I think about offending people or creating distrust. I simply didn't know what I was doing. I was naive. To their credit, the other board members didn't chastise me or act offended, though they couldn't have appreciated my actions. They hadn't elected me board president to take the district on a politically impractical, needless waste of time and energy.

Years later, I apologized to the superintendent for getting him into something so unproductive, and I also apologized to several of the remaining board members. My apologies were delayed because I didn't understand how badly I had mishandled the whole episode until later. I still consider this experience to be my greatest blunder as a school board member.

Time, though, has proven me at least partially correct about the continued decline in our district's enrollment and the changes in its racial makeup. By 2003 when I left the board, the student count was down to 8,796 with a racial distribution that was 58 percent black and 42 percent white. Since then, our enrollment has increased for four consecutive years to 9,842 in 2006. The district's

racial composition has remained fairly stable at 59 percent black and 41 percent white.

In 2005, the Arkansas State Legislature mandated a study to review the possibility of realigning Pulaski County's school district boundaries, including consideration of the two-district concept. The study group found merit for realignment, but the federal court's continued supervision, along with previous legal settlements, made implementation impossible.

I'm not sure switching to a two-district concept would have made any difference in the numbers of students attending our city's schools. The best way to draw students is through outstanding educational and extracurricular programs, which is exactly what our district did with some degree of success, beginning not long after the demise of the two-district study proposal.

Various desegregation programs implemented in the early 1990s allowed Pulaski County students to voluntarily enroll in school districts outside their areas of residence. Our schools routinely attracted several hundred more students each year than we lost to the surrounding districts. Today, I doubt that manipulating school district boundaries for demographic reasons makes any sense at all.

My first term on the school board was drawing to a close as the school year ended. I was concerned that my advocacy of the two-district study might attract an opponent in the upcoming election, but most of my activity promoting the study had been out of public view. Also, the community generally was quite happy with the schools, the administration, and the board. A third factor in my favor was that our change to board representation by geographic zone meant that only people who lived in or near my neighborhood, where my support was the strongest, were eligible to run for my seat.

So the deadline for candidates to file for the upcoming election eventually passed, and I breathed a sigh of relief when no one entered the race against me. As the year ended, I was grateful I'd be returning to the board without having to campaign but disappointed that my work on the two-district concept had come to nothing.

In spite of all our board and the school district had accomplished, I felt the disappointment of my recent failure more strongly than I felt the pleasure of our successes. Looking back, it's clear my priorities were out of order, and not just concerning school issues. Why had I invested so much time and emotional energy into something that, in hindsight, was so politically explosive and impractical as school district consolidation? I didn't know the answer at the time, though now it seems obvious that my misguided efforts were at least partly an unconscious diversion from problems at home, which were worsening as the school year drew to a close. I thought things were bad as the year ended, but I had no idea how much worse they soon would become. For the next few years, school board concerns would be a minor part of my life.

⓯

MY SCHOOL BOARD
DARK AGES (1991–1994)

My second three-year term on the board began with my unopposed reelection in September 1991. A month later, my 22-year marriage came to an end. My wife moved out, leaving the kids with me. School board matters became completely unimportant, quite literally overnight. If my marriage had dissolved a few weeks earlier, I probably wouldn't have placed my name on the ballot, and my school board career would have ended after one term.

This isn't the place to discuss events leading to the end of a long marriage. It's relevant, though, to consider the role my school board–related activities might have played. Did time spent on school issues cause me to neglect my marriage? Or was my desire to be involved in the schools a symptom of a troubled relationship? Was I trying to escape problems at home by going to board meetings?

I couldn't have answered those questions objectively soon after the end of my marriage. Stunned and disoriented, my divorce caused me to question everything about the way I conducted my life. Since then, much time and mental energy have gone into deciding how I want to live out the remainder of my days.

A small part of what I've decided is that finding a meaningful way to serve society is important and beneficial to me. My trouble at home caused me to act irrationally and direct my emotions toward a secondary target at times, and certainly school issues were a welcome distraction from personal problems. However, school board activities neither caused the difficulties in my first marriage nor were they primarily an escape mechanism.

The proof lies in the success and happiness of my second marriage. I remarried two years into my second school board term. My new wife encouraged me to remain on the board, saw me through a tough reelection campaign, and lived with me during my last ten years on the board.

I had my hands full just trying to survive the first year or so immediately following my divorce. My eight-year-old daughter was in third grade, my younger son was 15 and a high school sophomore, and my older son had just left home for his freshman year in college. Parenting became the focus of my life. I had custody of the kids and responsibility for meeting their daily needs.

I took care of their meals and got them to school and to their various activities. I learned to shop for little girls' clothes (one time, I tipped a clerk at The Gap $5.00. My desperation was deep and my gratitude sincere) while I was dealing with an unhappy teenager who was acting out in ways that typically result from a dysfunctional marriage. My older son, out of state and away from home for the first time, needed me, too. He was on my mind but unfortunately received little of my time.

Meanwhile, the demands of my work life increased substantially and simultaneously. Within a week of my wife's departure, the Little Rock newspaper war came to an end when my employer, the *Arkansas Democrat*, bought the assets of our longtime rival, the *Arkansas Gazette*. Suddenly, we were taking over the other newspaper and working furiously to combine the operations of two companies.

About this time, I received a letter informing me that I had been selected for jury duty. Several long paragraphs explained that I could not be excused and detailed serious penalties if I refused. No exceptions would be made for anyone, I was told. The notice gave me a date and time to report. I laughed and scrawled in pencil a note at the bottom that read, "I'm in management with the *Arkansas Democrat*, and we're in the middle of acquiring the *Arkansas Gazette*. I'm on the North Little Rock School Board. My wife just left, and I have the kids. You really don't want me right now." I mailed the letter back to the courts' office and didn't even consider showing up at the scheduled time. I never heard from them again, so they must have agreed they really didn't want me on a jury.

Our superintendent told me years later that during this period when I came to school board meetings, I looked like I didn't want to be there. He was right. My mind was elsewhere.

A year into my second term, our district's relationship with the teachers' union hit a low point. The recession of 1992 caused a serious shortfall in tax collections, so the State of Arkansas notified all school districts of significant reductions in funding for the coming year. In North Little Rock, we had to cut our budget by $1.7 million to avoid deficit spending. We spent the summer of 1992 whittling our planned expenditures in all areas but couldn't find enough to cut without reducing personnel costs.

The plan we developed included (1) reductions in administrative and teaching staff, (2) eliminating the normally routine annual pay-scale increase, (3) holding longevity increases to a modest level, and (4) reducing the length of all certified employees' contracts by two workdays. The net effect of these changes was that most employees would receive approximately the same salary as the previous year because their small longevity increases would be offset by the loss of two days' pay. Even so, some staff members would be taking a small decrease in pay. This unhappy situation applied equally for administrators, teachers, and other employees.

The leaders of the teachers' union, of course, were opposed to the administration's proposal. They pointed out that a relatively small operating fund balance from the previous year was available to pay for teacher raises in the coming year. This option would allow near-normal raises but would mean deficit spending with no plan to meet salary requirements beyond the next year. This choice seemed viable to them because they had no responsibility for the financial solvency of the school district. Moreover, they distrusted the administration's evaluation of the funding shortfall.

The discord inherent in the collective bargaining process had been increasing toward the end of my first term, but the district's financial crisis brought the disharmony to a roar. Neither the union nor the administration felt they could budge from their positions on the proposed budget changes. The union was adamant that pay scales should be increased, and the superintendent insisted that the district shouldn't take the risk of depleting our meager reserves. He said raises weren't impossible, but that agreeing to the union's demands would be fiscally irresponsible.

I wasn't putting much time into decisions at this point in my school board career. However, even if I had studied the situation for hours, I wouldn't have concluded that our budget should have called for higher expenses than revenue. In brief private conversations, the superintendent assured me that we couldn't and shouldn't meet the union's demand.

We discussed the possibility of a teacher strike triggered by denying their request, but he remained committed to his position. He said he was willing to face the possibility of a walkout. This left me with an uneasy feeling even though, along with the rest of the board members, I voted to support the superintendent's position. We unanimously denied the union request.

Shortly after school began in the fall, the union leaders did in fact call for a teachers' strike. Two days into the walkout, the superintendent relented. He changed his recommendation to the board and agreed to meet the union demands. Although he had been res-

olute about facing a possible strike, no forethought had been given to keeping the schools open. Teachers were on picket lines and angry parents were calling. Against the superintendent's better judgment, his only option was to give the union what they wanted.

Our rapid reversal of position stuck in my craw. The board and the administration had badly erred by failing to plan for the union's predictable reaction to our stance. If we were incapable of dealing with a teachers' strike, we shouldn't have been so intransigent in negotiating. I didn't like what had happened, but my only choice was to acknowledge our mismanagement. My disengagement had been a contributing factor.

I attempted to ignore the disagreeable situation but didn't forget our error. The superintendent, as he later proved, didn't forget it either. We had learned a valuable lesson about being prepared to back up decisions. I privately resolved never to make the same mistake again. I wasn't alone.

The local union leaders unwittingly furthered our resolve by gloating over their victory. They were pictured and praised in a feature article in a nationally distributed union newsletter, which of course they made certain was seen throughout our schools.

Bickering over wages and personnel policies reached a new level of intensity during the 1992–1993 school year. Teachers' salaries in Arkansas were then—and still are today—far too low, as are resources for funding education. The reality was that the administration and school board frequently had to tell the union we couldn't meet their requests for pay increases and personnel policy changes. So collective bargaining turned nasty.

Going into the negotiating process, the union would ask for more than they actually expected to realize, and the administration would start with offers lower than they ultimately expected to meet. The unreasonableness of the both sides' positions was readily apparent, but the participants were forced to defend their ground.

Often the bargaining discussions would turn into power struggles and personal attacks. In no way was the process helping us

provide a quality education for our students. In fact, the opposite was true. Without a doubt, our collective bargaining process was detrimental to students. It wasted the time and energy of talented teachers and administrators and created an acrimonious atmosphere in the schools.

The solution to our dilemma seemed clear: The school district absolutely required increased funding if we were to provide normal salary increases to the staff and meet other needs. So, in the fall of 1993, four years after our last successful tax increase campaign, we decided once again to go to the public to ask for a tax increase. Incredibly, the union leaders decided *not* to support our campaign to the public! This made no sense, except in the context that they were angry with us and wanted to see the tax increase proposal fail as an embarrassment to the board and the administration.

Publicly, the union leaders took a neutral stance on the proposal. Privately, they appeared to work against it. On a major thoroughfare on election day, I saw (and took down) an elaborate homemade sign, done artistically in black, asking voters to cast their ballots against the tax increase. I couldn't prove the sign was the work of the union, but it had all the earmarks.

Predictably, after a dispirited campaign, the tax increase proposal did go down to defeat. The vote was 1,373 in favor to 1,496 against. The caustic atmosphere had taken its toll. Our board was demoralized. However, we were reaching the bottom of our downward spiral. This loss at the polls turned out to be the only tax increase proposal defeated during my 15 years on the board.

The union saw the proposal's defeat as another victory for them in their ongoing power struggle with the board and the administration. They publicly told us that we couldn't win an election without their support, and they said the voters had sent a message that our policies and treatment of the union weren't appreciated. We scraped through the 1993–1994 school year, depleting our financial and emotional reserves.

About midway through my second school board term I met the woman who was to become my second wife. Betti and I began dating in early 1993, and we married in January 1994. She had been divorced about the same length of time as I. Her two daughters attended a private elementary school.

The oldest daughter, Kayce, was a sixth grader and would be changing schools to go to a junior high the following fall. I played no role in Betti and Kayce's decision concerning which school to attend, but in the end, after much investigation and discussion, they decided Kayce would enroll in our neighborhood's public junior high. Their decision came down to which school could provide the best educational and extracurricular opportunities.

Kayce entered our junior high school in the fall of 1994, the final year of my second term on the school board. Once she made the move to public school, I decided, with Betti's encouragement, to file for a third term. Had Kayce chosen a private school, I probably wouldn't have put my name on the ballot. So, once again, my school board career continued when my tenure easily might have ended.

Although unaware at the time, I just then was coming into my own as a board member. I was on the verge of grasping what it means to serve successfully and to truly enjoy being on a school board.

16

MY SCHOOL BOARD
RENAISSANCE (1994–1997)

My third term on the board began in the fall of 1994. I had the good fortune once again to be reelected without opposition, due to having done nothing controversial during the past three years and to the public's general satisfaction after the teachers' strike had been settled. The year began with some good fortune for the school district, too. We tried once again to pass a tax increase—this time successfully. Everyone appeared to work harder than before. At least, I certainly did. I was starting to come alive again as a board member.

Going into the year, the school district truly was in desperate shape. Not only had the economic recession of 1992 driven us to deficit financing, but we were scheduled to lose even more money, because short-term funds provided through the desegregation lawsuit settlement were coming to an end. The city leaders helped us promote the message that, without a tax increase, our town's schools could be designated as financially distressed and taken over by the state. The teachers' union pitched in to help this time, too, maybe because we obviously had everyone else's support. The public didn't disappoint us, approving the tax increase by a vote of 2,718 to 2,283.

Two issues stand out in my mind from my third term. First, gang activity and the resulting street violence of the early 1990s were becoming an alarming factor for supporters of our schools. A 1994 HBO television special, *Gang Wars: Bangin' in Little Rock*, highlighted our local problems for a national audience and exacerbated the fear in our community. The publicity caused us to increase our focus on the problem, but no doubt also caused the gangs to be even more attractive to some of our students. Parents were anxious about possible gang problems in the schools, and some of their concerns were justified.

In a cooperative effort with the city a year or so earlier, we already had placed uniformed police "resource" officers on the two campuses of our high school. We were the first school district in the county to make this move. The others soon followed our lead. We tightened our student dress codes to prevent wearing gang colors, and we implemented tighter disciplinary policies for gang activity. In addition, we strengthened our efforts in the schools to find and remove weapons and drugs.

The number of student expulsions increased. The school board routinely held hearings to review and approve all expulsions recommended by the administration. We interviewed the students, their parents, and any advocates the families wanted us to meet. Often we would hear mumbled excuses or statements of contrition from the students and pleas for leniency from their families.

One particular case stands out in my mind, though. A young man of perhaps 16 years had been accused of firing a handgun at another student in the parking lot at our high school's East Campus. When we interviewed him, he readily acknowledged firing the gun but claimed in a low uncooperative tone that the other student was lying. He said he wasn't firing the gun at the other student. He was shooting at someone else.

I was dumbfounded and wasn't sure I had heard him correctly, so I asked him directly: "Are you telling us you did fire the gun in the parking lot, although at another person?" He answered in the

affirmative and again said his accuser was lying about being the intended target. He seemed to have no understanding he was admitting to attempted murder. I didn't say another word to him but sat quietly in disbelief. Of course, the police were working this case, and his expulsion was perfunctory. I hoped the authorities had an efficient system for dealing with him. I had no doubt that, set free, he would soon be in serious trouble again.

Our board reviewed from two to ten expulsion recommendations every year during my 15-year tenure. This case was the only one that left me feeling concerned for my own safety. I certainly was cautious in the darkened parking lot outside the school offices after we had voted to expel the young gunman that evening.

The gang scare reached its peak in 1994 as the police department and the schools reacted by increasing security. Actually, though, crime statistics show that gang activity topped out in 1993. The gang problem already had begun to decline when the public's reaction was the strongest.

The other especially important issue from my third term was a chain of events that led to the most unusual and significant contribution our board made to the schools during my 15 years. In short, we dropped recognition of the teachers' union and terminated collective bargaining. This action immediately meant a contested election for me against a union-backed opponent and eventually led to another call for a teachers' strike. Here's how it all came about.

Arkansas law allows governmental bodies, including school boards, the option of recognizing unions at their sole discretion. I learned this fact from our original superintendent in my first year on the board. He was upset at something the union had said or done and told me we really didn't have to "put up with it." Until then, I had assumed, like most everyone else, that the law always required union recognition whenever a majority of employees voted to form a collective bargaining unit. Not so in Arkansas, though, for public employees.

Our school district participated in collective bargaining with the union only because our board had agreed to do so in the early 1970s when the AEA made its transition from professional association to union. When I heard this fact initially, I didn't think we would ever want to make a change. I recently had been elected with the union's help and was convinced then we could work together cooperatively for the betterment of our schools. However, I confirmed the superintendent's understanding of the law and mentally filed away the knowledge.

Toward the end of my first term, my mind opened to the idea of dropping union recognition. I was beginning to reflect that perhaps, at least in our school district, the never-ending acrimony of the collective bargaining process might be counterproductive to the education of children. Also, the union leaders' failure to support the two-district initiative caused me to question their priorities.

Then, in my second term, I became convinced the district should free itself from the union when their leaders declined to support a tax increase after forcing us into deficit spending. Clearly, the well-being of students was not at the top of their priorities.

At this point, I privately and regularly began encouraging the superintendent to consider dropping recognition of the union. In truth, he didn't require much encouragement to consider the idea, although he did need to be certain the board would back him. Over a period of about three years he quietly talked separately with each of the seven board members and slowly realized a consensus.

During this period, the union leaders gradually alienated the entire board (as well as many parents and teachers) by their antagonistic behavior. They never let up. Satisfying them was impossible. We had a dedicated, highly regarded superintendent and a motivated administrative staff who were continually under union attack.

Toward the end of the 1996–1997 school year, the superintendent informed us individually that he had the board's unanimous support for dropping recognition of the union. All seven board

members had been endorsed by the union in past election bids, and now all seven were ready to sever our ties! Two of us were former teachers, and a third member was a *current* teacher in the adjoining Pulaski County Special School District.

So in early June, a day or so after the end of the school year, we gave the required three hours' notice to the media and the teachers' union that we were calling a special board meeting. In about ten minutes, we voted to terminate our relationship with the union and to create a new eight-member personnel policies committee as required by Arkansas law. The law stipulated that the new committee would consist of five teachers elected by the teachers as a whole and three administrators named by the board. The new committee would recommend and review personnel policies but would have no authority to bargain. Board decisions would not be subject to mediation of any kind.

A newspaper article the next day quoted me saying, "It's a great day for public education in North Little Rock." That was exactly how I felt, although if I had to reexperience that moment, I would temper my enthusiasm and comments for the media. I hadn't liked hearing the union leaders gloat earlier, and now I could be accused of doing the same thing.

School was out, so we had all summer to prepare for a strike attempt we thought likely to come at the beginning of the next school year. I also had the summer to get ready for my upcoming reelection campaign in the fall. I wouldn't have the ease of an uncontested election this time, as the teachers' union surely would find an opponent for me.

Two of us were up for reelection that fall, and the union did find opponents for both of us. Recruiting candidates was difficult, though, because the change we made to board representation by geographic zone meant that our opponents had to live within our zone boundaries. The change also meant that fewer people would turn out for the election. The ballot had no school tax issues that

year, so voters would be coming to the polls only to vote for one school board candidate.

My opponent was a woman who was retired from teaching in the neighboring Little Rock School District. She lived on the opposite side of my board zone, not really in my neighborhood. Also, she was not well known in the community. I campaigned just as hard in this election and used all the same techniques in my earlier contested campaigns in 1986 and 1988. This time, my three daughters and their friends canvassed for me, distributing hundreds of cards as my sons had done before.

Generally, campaigning was easier, though, because I didn't have to cover the entire city as in the past. Also, the limited area allowed a new technique. I used a cross-reference directory of residential listings to make certain my supporters and I contacted everyone we knew in my geographic zone. I asked people to write a brief note on a prestamped, premessaged postcard to be sent to personal acquaintances. We mailed the postcards en masse just before election day.

One person who wrote notes to about 25 of her friends was the incumbent board member from my first campaign, the woman who defeated me in 1986. She had left the board in 1991 but had kept up with happenings in the schools. She had decided the board was on the right course and wanted to help. I was surprised to hear from her and delighted to have her on my side. I genuinely appreciated her unexpected support and forgiveness.

I felt confident going into the election but had learned nine years earlier that my electoral perceptions could be startlingly incorrect. We didn't hold an election night party. I can't remember what I did that evening. I think the night was fairly normal, which meant lots of activity with kids coming and going. My home with the three daughters was much livelier than when my two sons were in their teenage years. My one indelible memory of the evening is seeing some union leaders on a busy street at rush hour with signs

encouraging people to vote against me. One sign read, "Anyone other than Lynn Hamilton."

When the ballots were counted that evening, I had 215 to my opponent's 178. I received about 55 percent of the total; however, my margin of victory was only 37 votes. I won my own precinct by 54 votes. So, I actually *lost* in the other four precincts of my zone. The union's work had been effective everywhere except in my immediate neighborhood. I'm unsure I could have won reelection in a citywide contest.

My fellow incumbent member, who had served only one term on the board, was defeated by her union-backed opponent. She gave him some help, though, by failing to get enough valid voter signatures on her filing form for the ballot. Some of the signatures she obtained weren't registered voters, so she was forced to run a write-in campaign. Her opponent received 66 votes to her 34 in a school board zone scarred by economic deprivation and noted for voter apathy.

⑰

A RESOLUTION TO CONFLICT WITH THE TEACHERS' UNION (1997–2000)

The teachers' union leaders seemed dispirited after my reelection. They had worked hard, expecting to beat me as they had beaten most school board candidates they had targeted over the years. The anticipated strike attempt didn't materialize, at least not in the 1997–1998 school year. Instead, union leadership focused on using the new personnel policies committee to further their cause.

All the committee's teacher positions were filled by union leaders and their supporters, and they routinely spoke at school board meetings, asking that we reinstate recognition of the union and collective bargaining. And the board, of course, routinely denied their requests. The animosity between the board and union leaders increased significantly as the 1997–1998 school year progressed. The union leaders were angry with us, and they weren't shy about showing it. In turn, some of the board members also weren't timid about lecturing the union.

For the most part, I tried to keep quiet during these exchanges. I knew the union leaders couldn't hurt us except by evoking inflammatory statements from board members.

Early in the summer, at the end of the 1997–1998 school year, our board made another controversial decision, just as we had the summer before. Upon the superintendent's recommendation, we voted six to one to thoroughly revise our old personnel policies manual. The one negative vote belonged to the newest board member who had been elected with the union's support. The new set of policies gave the administration greater flexibility in such areas as employee discipline and naming replacements for retiring teachers.

Many of the outdated, restrictive policies had been the result of extended negotiations through collective bargaining over the past two decades. Naturally, the union leaders were incensed by the board's summertime activity. I was nearly certain our action this time would evoke a strike attempt.

The new set of policies went before the union-led personnel policies committee as the 1998–1999 school year began. State law required the board to submit our changes to the committee for their review before implementation, but didn't require us to heed any of the committee's recommendations. This review process set the stage for the strike threat we had been expecting.

The union leaders spoke angrily at length against our action, and the board grew tired of listening. In mid-autumn, we voted to limit any presentation to the board to ten minutes. When the motion was made by one of our most irate board members, I almost spoke up to voice an objection. I've never liked the idea of limiting anyone's input and wasn't particularly bothered by listening to the union leaders' repeated complaints.

Personally, I found their reaction understandable under the circumstances, even though I didn't agree with their position. I opened my mouth to speak against the restriction, and then abruptly changed my mind. I made a split-second intuitive decision that we should temporarily enact the time limit. I wanted to show support for our board members who were most upset by the union leaders' continuing complaints. We could remove the restriction whenever we wanted.

Our new time limit on presentations to the board gave the union leaders an effective public relations club with which to beat us. They told teachers that we didn't want to listen to them, and they offered this new restriction as proof that we were dictatorial. The animosity between the board and the union escalated to a previously unscaled height. The discord finally had reached what was to become its peak.

In late January 1999, the union leaders called a meeting of all teachers to consider a strike vote. They filled a small auditorium on the high school's East Campus with several hundred teachers. A picture in the *North Little Rock Times* weekly newspaper showed a packed room. The media weren't allowed to hear the discussions, but the union leaders related afterward that a standing vote overwhelmingly favored a strike. An actual count of the votes wasn't taken. The teacher walkout was set to take place in one week, the following Tuesday, if the board didn't reinstate collective bargaining and the previous personnel policies.

The superintendent was ready this time, and the board stood firm in its support.

The day after the union leaders called for a strike, the board met in a special Wednesday night session, voting six to one to reaffirm its position and to keep the schools open during the planned walkout. One of my board colleagues brilliantly explained in the meeting that the union leaders didn't have the claimed overwhelming support of teachers.

She had counted the seats in the auditorium where the strike vote was taken and had determined that the favorable vote could have included no more than 300 teachers, or slightly less than half the district's total of about 620. The superintendent publicly stated his intent to operate the schools by hiring substitutes for teachers who failed to show up for class. The showdown between the union and the board was on.

Thursday morning, the superintendent announced that the district would offer substitute teachers $100 a day to fill vacancies, or

twice the normal $50 rate. A front-page photograph in Friday's newspaper pictured applicants applying for the positions. An accompanying article said the district had 610 total substitutes available. By Friday afternoon, the strike attempt was crumbling.

Teachers in tears were calling administrators to say they really didn't want to strike. They had been pressured to vote for the walkout, they said. The union leaders were hearing from teachers too, and didn't know how to react. They hadn't expected the board and administration to stand up to their threat. We had never stood firm in the past.

In truth, the union leaders could take no action to improve their situation, because the majority of the district's teachers trusted the superintendent and the board more than they did the union. Many of the classroom teachers had known the superintendent throughout their careers, and others were friends of the board members. We consistently had treated teachers with genuine personal regard for years.

Months earlier, the board had scheduled a retreat on the following Saturday for long-term planning. (One union leader had expressed particular glee at our using the word "retreat.") Originally, a few board members had wanted to conduct the session at an out-of-town resort. Others of us, though, preferred to hold the meeting at our local administrative office. We feared a remote resort might appear frivolous to the public and that some people perhaps would consider the retreat an unnecessary expense.

The board had settled on the local site, and as things turned out, our timing and location were perfect. Our retreat was converted into an all-day Saturday meeting with the union leaders, nonunion teachers, and other interested individuals. The doors were opened to anyone who wanted to attend.

The atmosphere in our daylong session was calm and cooperative. The union leaders were polite, and the board was conciliatory. No one wanted to see the strike move forward. The union leaders

desired a graceful exit, and the board wanted a return to normalcy without reverting to collective bargaining or the previous restrictive personnel policies. We listened to the union leaders and their supporters' comments throughout the morning, and I took notes on each person's statements.

Early on, I realized that a settlement instrument could be written that would give all parties what they needed to preempt the strike. I made this observation to the group, and they asked me to continue with my notes, suggesting that I create such a document. The discussion extended through a working lunch into the early afternoon. After everyone had their say, I left the group to formulate a proposed settlement while the board covered some of the long-range planning material originally scheduled for the day's meeting.

Drafting the document took a little more than an hour. Afterward, the gathering reviewed the initial work and suggested a few revisions. By late Saturday afternoon, I had incorporated their suggestions, and the final instrument was approved by the group. The settlement stipulated that teachers would observe a one-year moratorium on further strike considerations, and the administration would honor a similar moratorium on future personnel policy changes.

The thrust and tone of the document was that the board would renew its efforts to listen to teachers. A key point was that we would rescind our rule limiting presentations to ten minutes. Inclusion of this concession gave the union leaders a victory on an emotional point, while the board prevailed on the major issues, all of which went unmentioned in the document. The extensive personnel policy revisions of the past summer would remain and collective bargaining would not be reinstated. The ten-minute time limitation was poor policy from the beginning, but its existence allowed the board to make a concession symbolizing our willingness to listen and to be reasonable.

The settlement document was perhaps my best tangible work as a school board member. It is reproduced below.

A Proposal to the Teachers
of the North Little Rock Public Schools
January 30, 1999

In the spirit of mutual respect and compromise, the North Little Rock School Board asks teachers to adopt with us the following proposal.

1. We propose a moratorium on future teacher strike considerations until at least February 1, 2000. The Board will not pursue a court-ordered decree preventing the occurrence of a strike during the moratorium period. This moratorium will allow us to finish the current school year and begin the next one peacefully. It also will allow time to improve communications with each other.

2. The Board will not initiate any further personnel policy changes, excluding salaries, through February 1, 2000, unless a change is necessitated by law or an emergency. However, in accordance with state law, any recommendations by the Personnel Policy Committee (PPC) will be considered. The Board acknowledges the need to ensure a period of stability and comfort in teachers' working conditions and environment.

3. The Board immediately will name new administrators to the PPC. We recognize that past experiences may be affecting the present actions of the administrator members of the committee as well as the teacher members.

4. We encourage and endorse the participation of all teachers in the process to elect members to the PPC.

5. The Board will fund the cost of a facilitator (a trained, professional, impartial individual) to build consensus and communication between the administrator and teacher members of the PPC. The facilitator will be named through mutual agreement of administrator and teacher members of the committee. The PPC will decide the timing of the initial facilitation process and determine the

length and frequency of its future use. The Board will participate in a planning retreat for the process if the PPC so desires.

6. The Board will rescind its recently adopted policy which set time limits on presentations in Board meetings. This new policy created the impression for some teachers and others that the Board was unwilling to listen to their concerns. The Board wishes to dispel that impression.

7. The Board will renew its commitment to listen to teacher concerns. Board members will ensure their accessibility to all teachers throughout the District, but especially to teachers in their geographic zones of representation.

The adoption of this proposal will not only allow time to improve our communications, but it will allow time for the democratic process to work through the upcoming school board election and through election of teacher representatives to the Personnel Policies Committee.

Most importantly, though, the adoption of this proposal and avoidance of a strike will allow us all to prove we are serious when we say our students come first.

We notified the media that a proposed settlement had been reached and that the school board would go into a formal session at 7:00 p.m. to vote on the proposal. The local television stations included the announcement on their 5:00 p.m. news programs. As a result, our Saturday night meeting was packed with the largest, most emotionally involved throng since the mascot-naming fiasco almost ten years earlier.

The crowd was in an excited, high-spirited mood, anticipating an end to the threatened strike. The board president told them we were going to present our proposed settlement and that we would take comments and answer questions from anyone who wanted to speak that evening. We made several hundred copies of the settlement document, but before we distributed them, I read the proposed

agreement to the entire group. The crowd cheered when I came to the first sentence in point six, explaining we would rescind the time limit on presentations to the board. And they laughed when I responded, "I'll remind you about midnight that you liked that provision."

A few comments were made and questions asked, but not many. The meeting ended after about 30 minutes. Afterward, the atmosphere was like a celebration, or perhaps a homecoming event. Union leaders, teachers, administrators, parents, and board members smiled, laughed, cried with relief, even embraced, and offered congratulations all around. I realized a satisfying sense of accomplishment that evening and in the weeks that followed.

That Saturday was a watershed for our school district. A few days later, the union leaders held a well-attended meeting of their membership in the high school's large West Campus auditorium and recommended that teachers accept the settlement proposal. This time, the positive vote was truly overwhelming. For the remainder of my board tenure, our relationship with the union was mostly cordial.

That's not to say disagreements didn't arise, but both sides were less antagonistic than in any period during the previous 12 years. The older, more recalcitrant union leaders took a backseat or retired, and the younger, newer leaders were more understanding of the administration's limited ability to make changes in teacher compensation and working conditions. Union disputes ceased to be major distractions from our primary purpose of educating children.

(18)

ENDING ON A HIGH NOTE
(2000–2003)

In September 2000, I was reelected without opposition to my final three-year term on the school board. Serving one more time was important because my two youngest children would graduate during those years. I announced up front that this would be my last appearance on the ballot.

I thoroughly enjoyed those final three years. The hard work we had done the previous 12 had paid off. The school district was peaceful and productive. We renovated some of our oldest elementary schools, built a beautiful state-of-the-art gymnasium at the high school's West Campus, and implemented an innovative charter school concept at one of our three middle schools.

The building renovations and new construction were made possible by the voters' consent to refinance some bonds and extend the existing level of taxation into the future. Over the years, the public approved (and never rejected) similar proposals four times. This particular tax extension, though, was our largest and most important request for bond refinancing.

In March 2001, President George W. Bush paid us a visit to recognize the improvement in test scores at one of our elementary

schools (a political favor by someone in North Little Rock also may have figured into his consideration). He brought along the secretary of education. This was just six months before the terrorist incidents of September 11. Education issues were the primary focus of the president's administration at that time. In fact, he was making a similar visit to an elementary school when he first heard the news of the attacks on the World Trade Center.

The presidential recognition was a red-letter date in the history of our school district. The event excited our entire community. The seven school board members had front-row seats in the elementary school's small, freshly painted cafeteria. The president and the secretary of education spoke from the same stage where my children had performed years earlier.

My final term was perhaps the best of my time on the board, because I was able to put into use what I'd learned in the past. Being conciliatory in a conflict became easier. I paused often to appreciate the company of the other board members and school district employees. The board chose me to serve as president the first and third of the three years. I delighted in handing diplomas to the graduating seniors those years and in being a part of the dedication ceremony for the new gymnasium.

Our superintendent had talked about retirement for some time but finally made up his mind that the year following my departure would be his last. I didn't participate at all in the selection of the school district's next leader, though I saw firsthand how important it was to our superintendent that he find a quality successor.

The question of who would succeed him came up frequently in our private conversations. Even before I left, he had identified the bright young superintendent of a smaller Arkansas community as his hoped-for successor. The young superintendent was well spoken, well known, and well liked throughout Arkansas education circles. He was a rising star. A year after my departure, the board unanimously selected him to lead our schools. I had nothing to do with his hiring, although I'm sure I would have voted for him too.

He left our district after only one year to accept a better paying job that his father had held previously. No one really blamed him, but neither did anyone think the turn of events good for our schools or community.

So what did our board do next? In a genuinely wise move, the members had the collective good sense to look within our school district. They reached down to the high school's East Campus principal, a low-key, soft-spoken man in his 50s who had spent his entire 33-year career in our district.

The board first named him interim superintendent and then made the appointment permanent beginning in 2006. I have no doubt he'll be a long-term success, even though he had been disregarded in the hiring process just two years earlier. A homegrown superintendent looked much more attractive after a disappointing experience with a seemingly superior outsider.

Sometimes history does repeat itself. The board I joined in 1988 had gone through essentially the same learning experience 18 years earlier. Both times our community could have been spared considerable turmoil if we more quickly had recognized the talent we had at home and the importance of commitment to our local schools. Today, our newest superintendent continues to grow in stature.

In leaving the schools, I also wanted to find a high-quality successor for my spot on the board. Fortunately, I didn't have far to look because one of my neighbors was interested. He and his wife were bringing up a large family in the same community where they had been reared themselves. Their children attended several of our schools. He was well liked, bright, articulate, and honest. In short, he was exactly what all school boards need. Encouraging him to be my replacement was one of my final acts in support of our district, and knowing he would be on the board made my departure easier.

When I departed in the fall of 2003, my tenure was the longest in our school board's recent history. I can't say every minute of those 15 years was enjoyable, but perhaps no one ever appreciated the entire experience more.

APPENDIX:
HISTORICAL NOTE
AND TIMELINE

In September 1957, the Little Rock public schools erupted into the nation's consciousness. The city became the site of the first true test of the U.S. Supreme Court's landmark 1954 decision in *Brown versus the Board of Education of Topeka, Kansas*. The *Brown v Board* ruling struck down the court's long-standing "separate but equal" doctrine and deemed state-mandated racially segregated schools to be fundamentally unconstitutional.

President Eisenhower sent federal troops to Arkansas to enforce the law and gain entrance for nine African American students into Little Rock's Central High School. The nation followed the month-long series of events on television, a then still new medium.

For the past 50 years, racial relationships in both Little Rock and adjoining North Little Rock have been conducted in the shadow of this historic occurrence. Monuments, museums, and anniversary celebrations frequently recall our past. School officials are ever aware of our community's role in the early days of the equal rights struggle.

The timeline on the following pages summarizes the related desegregation happenings in the two cities' school districts. Much of

the list is reprinted from the Little Rock Central High 40th Anniversary website by permission of Craig Rains/Public Relations. The North Little Rock History Commission also assisted in the listing's development.

TIMELINE OF SIGNIFICANT DESEGREGATION EVENTS IN THE LITTLE ROCK AND NORTH LITTLE ROCK PUBLIC SCHOOLS

May 24, 1955   The Little Rock School Board votes unanimously to implement a plan of gradual integration starting at the high school level in 1957.

July 14, 1955   The North Little Rock School Board adopts a plan similar to Little Rock's to begin integration in 1957.

September 2, 1957   Arkansas Governor Orval Faubus calls out the National Guard to surround Little Rock Central High School purportedly to preserve peace and avert violence.

September 4, 1957   Nine black students attempt to enter Central High but are turned away by the National Guard.

September 9, 1957   Six black students try to enter North Little Rock High School but are rebuffed by an unruly white crowd. The students are permanently denied admittance. North Little Rock High School would not be integrated for nine more years.

September 20, 1957   A federal judge rules that the governor had used troops to prevent integration, not preserve law and order as

	claimed. The governor removes the guardsmen, and the Little Rock Police Department takes over security duties outside Central High School.
September 23, 1957	A crowd of approximately 1,000 forms in front of the school as the nine black students enter the building through a side door. The throng becomes agitated upon learning the students are inside, and police fear they will not be able to maintain control. The black students are led surreptitiously out a back exit.
September 24, 1957	President Eisenhower announces he is sending 1,000 members of a U.S. Army Airborne division to Little Rock. He also federalizes the Arkansas National Guard.
September 25, 1957	Under escort of the Army troops, the nine black students enter the school.
May 27, 1958	Ernest Green becomes the first black student to graduate from Central High School.
August, 1958	Governor Faubus calls a special session of the state legislature to pass a law allowing him to close public schools to avoid integration and to lease the closed schools to private school corporations.
September 15, 1958	The governor orders the three Little Rock high schools, two black and one white, closed.
September 27, 1958	In a specially called referendum, voters support the governor, rejecting integration by a vote of 7,561 for and 19,470 against.

September 28, 1958   Little Rock public high schools close for the year, sending the city's 3,698 students to seek alternatives. Many attend school in neighboring communities, but the North Little Rock School District shuts its doors to Little Rock students who don't move into the city limits. Only about 30 Little Rock students relocate.

November 12, 1958   Five of six members of the Little Rock School Board resign in frustration, having been ordered by a federal appeals court to proceed with integration even though the board had no high schools open to integrate.

December 6, 1958   A new school board is elected with its membership evenly divided between those favoring compliance with the court's orders and those favoring resistance.

May 5, 1959   Segregationist members of the school board attempt to fire 44 teachers and administrators suspected of integrationist sympathies. The board's action serves as a wake-up call to the white community.

May 25, 1959   Two newly formed activist organizations narrowly win a recall election and replace the three segregationist board members with moderates.

June 18, 1959   The new Little Rock School Board announces it will reopen the high schools in the fall. A federal court earlier declared the state's school-closing law unconstitutional.

August 12, 1959   Little Rock public high schools open a month early. A small number of black students attend Central High and the one other of the city's two traditionally white high schools. Carrying American flags, about 250 white people march to Central High School to protest. The Little Rock police quickly arrest 21 and call for fire hoses to be turned on the remaining crowd, which disperses.

September, 1964   A small number of African American students enroll in North Little Rock elementary schools.

March 1966   The first African American is elected to the Little Rock School Board.

September 1966   Black students attend North Little Rock High School for the first time.

March 1968   The first African American is elected to the North Little Rock School Board. He is the father of one of the six students denied admittance to the city's high school 11 years earlier.

September 1972   All grades in the Little Rock and North Little Rock Public Schools finally are integrated. Student busing plans are implemented in both districts.

October 1972   In a bid for reelection, North Little Rock's lone black school board member is defeated by a white challenger whose principal campaign platform is opposition to busing students for purposes of desegregation. North Little Rock's board remains all white for the next 16 years.

March 1975	The Little Rock School Board names an African American president for the first time.
November 30, 1982	The predominantly black Little Rock School District sues state officials and the predominantly white neighboring North Little Rock and Pulaski County school districts for consolidation of the three districts as an end to racial segregation in the schools.
April 13, 1984	A federal judge rules that the three districts must consolidate.
November 7, 1985	The 8th U.S. Circuit Court of Appeals overturns consolidation as too extreme a remedy.
October 1988	An African American is elected to the North Little Rock School Board for the second time in the body's history and is named president in the first meeting he attends.
March 1, 1989	Attorneys for the state board of education, the three school districts, and the NAACP sign a proposed settlement to the seven-year-long desegregation lawsuit.
December 11, 1989	A federal judge accepts the financial aspects of the settlement but requires the three school districts to remain under court supervision.
August 1993	The first African American is named permanent superintendent of the Little Rock School District. A black interim superintendent had been appointed five months earlier.

September 1995   The North Little Rock School District is
                      released from court supervision of its
                      student assignment plan. The district
                      returns to a semblance of neighbor-
                      hood schools.

December 2000   A North Little Rock School Board mem-
                      ber is elected as the first African Amer-
                      ican president of the Arkansas School
                      Boards Association.

February 23, 2007   The third federal judge to oversee the
                      desegregation case declares the Little
                      Rock School District unitary. The city's
                      school district is released from the
                      court's supervision, 25 years after filing
                      suit for consolidation as a remedy for
                      segregation and nearly 50 years since
                      the initial integration of Central High
                      School.

April 9, 2007   Attorneys for the African American inter-
                      venors file a notice of appeal to over-
                      turn the finding for unitary status. A
                      ruling by the 8th Circuit Court is
                      awaited as this book goes to press.

ABOUT THE AUTHOR

Lynn Hamilton attended pubic schools in and around Fresno, California. A bachelor's degree in mathematics from the University of Central Arkansas was followed by a brief teaching experience at an inner-city high school in Memphis, Tennessee. He enlisted as an officer in the U.S. Navy when a low military draft number terminated his time as a teacher.

Later, hc earned an MBA from the University of Arkansas at Little Rock and began a long career with the *Arkansas Democrat-Gazette* newspaper, where he remains employed as vice president of operations. He stayed close to public education during 15 years as a member of the North Little Rock, Arkansas, school board. In 1999, the *North Little Rock Times* named him one of the city's "Agents of Change" for his role in averting a threatened teachers' strike.

DATE DUE